Envisioning Landscape

One World Archaeology Series
Sponsored by the World Archaeological Congress
Series Editors: Joan Gero, Mark Leone and Robin Torrence

One World Archaeology volumes contain carefully edited selections of the exemplary papers presented at the World Archaeology Congress (WAC), held every four years, and intercongress meetings. WAC gives place to considerations of power and politics in framing archaeological questions and results. The organization also gives place and privilege to minorities who have often been silenced or regarded as beyond capable of making main line contributions to the field. All royalties from the series are used to help the wider work of the organization. The series is published by Left Coast Press, Inc. beginning with volume 48.

54 *Archaeology and Capitalism*, Yannis Hamilakis and Philip Duke (eds.)
53 *Living Under the Shadow*, John Grattan and Robin Torrence (eds.)
52 *Envisioning Landscape*, Dan Hicks, Laura McAtackney and Graham Fairclough (eds.)
51 *Rethinking Agriculture*, Tim Denham, José Iriarte and Luc Vrydaghs (eds.)
50 *A Fearsome Heritage*, John Schofield and Wayne Cocroft (eds.)
49 *Archaeology to Delight and Instruct*, Heather Burke and Claire Smith (eds.)
48 *African Re-Genesis*, Jay B. Haviser and Kevin C. MacDonald (eds.)

Previous volumes in this series, available from Routledge:

47 Indigenous Archaeologies
46 Archaeologies of the British
45 Natural Disasters and Cultural Change
44 Matériel Culture
43 The Dead and Their Possessions
42 Illicit Antiquities
41 Destruction and Conservation of Cultural Property
40 Madness, Disability & Social Exclusion
39 The Archaeology of Dry Lands
38 The Archaeology of Difference
37 Time and Archaeology
36 The Constructed Past
35 Archaeology and Language IV
34 Archaeology and Language III
33 Cultural Resource Management in Contemporary Society
32 Prehistory of Food
31 Historical Archaeology
30 The Archaeology and Anthropology of Landscape
29 Archaeology and Language II
28 Early Human Behaviour in the Global Context
27 Archaeology and Language I
26 Time, Process and Structured Transformation in Archaeology
25 The Presented Past
24 Social Construction of the Past
23 Sacred Sites, Sacred Places
22 Tropical Archaeobotany
21 Archaeology and the Information Age
20 The Archaeology of Africa
19 Origins of Human Behaviour
18 From the Baltic to the Black Sea
17 The Excluded Past
16 Signifying Animals
15 Hunters of the Recent Past
14 What's New?
13 Foraging and Farming
12 The Politics of the Past
11 Centre and Periphery
10 Archaeological Approaches to Cultural Identity
9 Archaeological Heritage Management in the Modern World
8 Conflict in the Archaeology of Living Traditions
7 Animals into Art
6 The Meaning of Things
5 Who Needs the Past?
4 State and Society
3 Domination and Resistance
2 The Walking Larder
1 What Is an Animal?

Envisioning Landscape:
Situations and Standpoints in Archaeology and Heritage

Edited by
Dan Hicks, Laura McAtackney and Graham Fairclough

Left Coast Press Inc.

Walnut Creek, California

Left Coast Press Inc.

LEFT COAST PRESS, INC.
1630 North Main Street, #400
Walnut Creek, CA 94596
http://www.LCoastPress.com

ISBN 978–1-59874–281–7 hardcover

Library of Congress Cataloging-in-Publication Data:

Envisioning landscape : situations and standpoints in archaeology and heritage / edited by Dan Hicks, Laura McAtackney and Graham Fairclough.
p. cm. — (One world archaeology series; 52)
Papers originally presented at the 5th World Archaeological Congress in Washington, D.C. in June 2003.
Includes bibliographical references and index.
ISBN-13: 978-1-59874-281-7 (hardback : alk. paper)
1. Landscape archaeology. 2. Landscape assessment.
3. Urban archaeology. 4. Archaeology and history.
I. Hicks, Dan, 1972- II. McAtackney, Laura, 1977- III.
Fairclough, G. J. (Graham J.), 1953- IV. World Archaeological Congress
(5th : 2003 : Washington, D.C.)
CC75.E585 2007
930.1—dc22
2007022702

Printed in the United States of America

∞™ The paper used in this publication meets the minimum requirements of American National Standard for Information Sciences—Permanence of Paper for Printed Library Materials, ANSI/NISO Z39.48–1992.

07 08 09 10 11 5 4 3 2 1

Cover design by Joanna Ebenstein

Contents

List of Illustrations 7
Acknowledgements 11
1. Introduction: Landscapes as Standpoints 13
Dan Hicks and Laura McAtackney
2. The Contemporary Politics of Landscape at the Long Kesh/Maze Prison Site, Northern Ireland 30
Laura McAtackney
3. Facing Many Ways: Approaches to the Archaeological Landscapes of the East African Coast 55
Sarah Croucher
4. Landscape Archaeology in Lower Manhattan: The Collect Pond as an Evolving Cultural Landmark in Early New York City 75
Rebecca Yamin and Joseph Schuldenrein
5. Cultural Landscapes, Communities and World Heritage: In Pursuit of the Local in the Tsodilo Hills, Botswana 101
Susan O. Keitumetse, Geoffrey Matlapeng and Leseka Monamo
6. Common Culture: The Archaeology of Landscape Character in Europe 120
Sam Turner and Graham Fairclough
7. Landscape Archaeology and 'Community Areas' in the Archaeology of Central Europe 146
Martin Kuna and Dagmar Dreslerová
8. Historical Archaeologies of Landscape in Atlantic Africa 172
Kenneth G. Kelly and Neil Norman
9. Landscape, Time, Topology: An Archaeological Account of the Southern Argolid Greece 194
Christopher L. Witmore
10. A Landscape of Ruins: Building Historic Annapolis 226
Christopher Matthews and Matthew Palus
11. Colonialism and Landscape: Power, Materiality and Scales of Analysis in Caribbean Historical Archaeology 251
Mark W. Hauser and Dan Hicks
Index 275
About the Contributors 299

List of Illustrations

Figures

Figure 2.1	Nissen hut at the Long Kesh Internment Camp.	32
Figure 2.2	H Block at HMP, the Maze.	33
Figure 2.3	Plywood Celtic Cross, undated. This artefact was confiscated from unnamed prisoners at Long Kesh/Maze when it was discovered that its filling contained hidden bullets. Held at the NIPS Museum, Millisle, Northern Ireland (photograph: Laura McAtackney).	40
Figure 2.4	Banner used for Orange Order–style marches by Loyalist prisoners.	41
Figure 2.5	Republican H Block mural flanked by Irish Republican flags.	44
Figure 3.1	Map showing the East African coastline and some key Swahili archaeological sites.	56
Figure 3.2	Map showing the clove plantations on Unguja and Pemba, and the four regions of the Zanzibar Clove Plantation Survey 2003.	63
Figure 3.3	Site types in the Zanzibar Clove Plantation Survey 2003.	66
Figure 3.4	Plan of Trench A, at Mgoli, Pemba (2004).	67
Figure 3.5	Wattle-and-daub house at Bweni.	68
Figure 4.1	Location map of the Collect Pond and surrounding topographic features during the early Euroamerican periods (1650–1805).	76
Figure 4.2	Oil painting of the Collect Pond attributed to Alexander Robertson, 1798.	78
Figure 4.3	Detail of the 1754 plan of New York.	79
Figure 4.4	Plan view of 18th century landforms and geographic features, Lower Manhattan.	81
Figure 4.5	Semi-schematic historic stratigraphy, Metropolitan Corrections Center tunnel.	86
Figure 4.6	Detail of Metropolitan Corrections Center tunnel trench, 1995.	87
Figure 4.7a	Diachronic model of land use and occupation, Lower Manhattan (AD 1750–present).	91

Figure 4.7b	Diachronic model of land use and occupation, Lower Manhattan (15,000 BP–AD 1650).	92
Figure 5.1	Location of Tsodilo Hills World Heritage Site, Botswana.	103
Figure 5.2	Tsodilo Hills and other geographical features.	104
Figure 5.3	Rock art at Tsodilo Hills World Heritage Site.	105
Figure 6.1	Holbeton, Devon (UK).	123
Figure 6.2	Agricultural landscape near Budingen, Hessen (Germany).	125
Figure 6.3	A typical 'fieldscape' palimpsest in the UK.	126
Figure 6.4	A landscape view of Tinhay Down, west Devon, generated from the Devon HLC database.	130
Figure 6.5	A landscape view of Whimple, east Devon, generated from the Devon HLC database.	133
Figure 6.6	Prehistoric rock art site at Lordenshaws, Northumberland (UK).	134
Figure 6.7	The pastoral landscape of the Limousin at Chéronnac, Haute Vienne (France).	135
Figure 6.8	Devon HLC: publicly accessible web-based GIS version.	138
Figure 7.1	Map of Bohemia in the Western part of the Czech Republic.	158
Figure 7.2	The Loděnice project area – an Iron Age industrial zone.	159
Figure 7.3	Burial mounds in the landscape, the Schwarzenberg deer parks area at Hluboká nad Vltavou, distr. České Budějovice.	164
Figure 7.4	A GIS model of prehistoric tumulus cemeteries in the area of the Schwarzenberg deer parks (detail), distr. České Budějovice.	165
Figure 7.5	A predictive model of 'suitable' versus 'unsuitable' land for prehistoric settlement sites within the Schwarzenberg deer parks area.	166
Figure 8.1	Map of African sites mentioned in chapter.	173
Figure 8.2	Ocean view of Elmina Castle.	179
Figure 8.3	Early 18th century view of the Savi Palace Complex.	183
Figure 9.1	Map of Melos, Greece, with the site locations.	199
Figure 9.2	Map of locations discussed in this chapter.	207
Figure 9.3	Photograph of 'B20' wellhead with plastic bucket.	215
Figure 10.1	The 1743 redraft of Stoddert's 1718 survey of Annapolis, Maryland.	229
Figure 10.2	View of the William Paca Garden from the lowest terrace.	232
Figure 10.3	African-American gardener behind William Paca House, 1860.	239

Figure 10.4 Power lines along East Street in the historic district of Annapolis. 244
Figure 11.1 Map of Caribbean showing the islands discussed in the chapter. 252
Figure 11.2 Ideal model of coffee plantation landscape (*from Laborie 1798*). 255
Figure 11.3 Locations of the Jamaican sites discussed in the chapter. 256
Figure 11.4 Avery Trim (St Lucia National Trust) and Rebecca Craig (Museum of London Archaeological Services) undertaking a drawn landscape survey at Balenbouche Estate, St Lucia, in January 2001. 266

Tables

Table 7.1 The basic concepts of the community area theory. 155
Table 7.2 Burial mounds in the landscape, the Schwarzenberg deer parks area at Hluboká nad Vltavou, in the district České Budějovice. 162
Table 7.3 Burial mounds in the landscape, the Schwarzenberg deer parks area at Hluboká nad Vltavou, in the district České Budějovice. 163

Acknowledgements

We are indebted to our contributors for their excellent contributions and for their efficient responses as we assembled this volume. We thank the editors of the One World Archaeology series – Joan Gero, Mark Leone and Robin Torrence – for inviting this book, and for their support for the project. We are also very grateful to the two referees who commented on an earlier version of the manuscript. We are grateful to the Conference Fund of the Faculty of Arts at the University of Bristol, which funded Dan Hicks' participation in the World Archaeological Congress in Washington, DC, in June 2003, from which this volume has developed. Laura McAtackney's research is supported by the Arts and Humanities Research Council. Thanks also to Mitch Allen and Left Coast Press for their editorial support. And finally, we thank each other.

This volume is dedicated to Mark P. Leone.

CHAPTER 1

Introduction: Landscapes as Standpoints

Dan Hicks and Laura McAtackney

INTRODUCTION

Archaeologists 'listen to landscapes' inspired by the 'late medieval soundscape of church bells' in Polhograjsko hribovje, Slovenia (Smith 2004). 'Sacred landscapes and social memory' are studied through inscriptions at Oaxaca, Mexico (Foundation for the Advancement of Mesoamerican Studies 2005). New 360-degree panoramic photographs of Native American pictographs in the American West allow 'the viewer to see the pictographs up close as well as the entire landscape which surrounds them' (Trujillo 2005). A meeting in South Carolina considers 'vernacular settlements, early industrial places, sacred Indigenous sites, places of memory, sites of conscience and of the recent past, plus once invisible or miniscule sites whose thematic values are reinforced by being linked together in cultural landscapes, heritage areas and cultural corridors' (Araoz 2004). In China, archaeologists gather to explore how heritage routes can represent 'meta-landscapes' (Smith 2005), while others meet in Nevis in the eastern Caribbean to debate 'the historical archaeology of colonial or shared landscapes of the Caribbean' (Townsend 2005). Another symposium explores how 'humans have always interacted with their environment and helped to create and modify the landscapes in which they live' (Regan 2004). A Japanese contribution reflects upon how, for former prisoners of war revisiting the Burma-Thailand Railway, it is a landscape filled with 'lived memories' (Nakao 2005).

This sample of postings to the World Archaeological Congress (WAC) email listserver between July 2004 and July 2005 demonstrates some of the many, contrasting uses of ideas of 'landscape' in contemporary world archaeology. Highly diverse theoretical and methodological approaches to landscape have developed in archaeology over the past 40 years, from the archaeology of 'settlement patterns' (Chang 1972) and the 'spatial archaeology' of David Clarke (1977) to post-processual ideas of space as 'socially constructed and constitutive of social relations' rather than 'a passive backdrop for action' (Robin and

Rothschild 2002: 161), more empirical approaches that developed from British traditions of 'local studies' (Aston 1985), and even archaeologies of 'natural places' (Bradley 2000). Still more diverse are the situations in which ideas of landscape have been developed and used by archaeologists. In the Middle East, some archaeologists have been attracted to landscape studies, drawing upon geophysical survey and aerial photography, because of a desire to combine targeted excavation with 'larger operations on a scale commensurate with massive urban sites' (Postgate 2002: 402). In studies of the British Neolithic others have used phenomenology to seek to grasp the experiential dimensions of monuments in the landscape – the 'multisensory experience of being out in the open' (Ingold 2005: 122; see Tilley 1994). Denis Byrne (2003) has explored the 'nervous landscapes' of racial segregation in New South Wales. Alice Gorman has tested the physical limits of landscape still further by exploring how the 'spacescape' of human exploration of space since the late 1950s ranges from terrestrial space sites such as the Woomera rocket range in South Australia to space junk in Earth orbit and planetary landing sites (Gorman 2005). At the same time, landscape archaeologists have tested disciplinary boundaries (Layton and Ucko 1999: 15), especially with geography and anthropology. 'Landscapes', as Barbara Bender has put it, 'refuse to be disciplined' (Bender 2006: 304).

Perhaps the archaeological notion of landscape is so broad that it is 'vacuous' (Thomas 1993: 20). And perhaps we should find this diversity and ambivalence in definitions of landscape troubling. Kurt Anschuetz, Richard Wilshusen and Cherie Scheick, for instance, have called for 'a landscape paradigm' that would define a single, coherent 'landscape approach' with 'a common terminology and methodology' (Anschuetz, Wilshusen and Scheick 2001: 157). There are, of course, already many commonalities between alternative landscape archaeologies: they employ a range of (mainly non-intrusive) methods, operate at multiple scales of analysis and seek to move beyond a focus upon apparently bounded entities like monuments or 'sites'. But our point of departure in bringing together this collection of essays is that diversity – of method, field location, disciplinary influences and contemporary voices – is a principal characteristic of landscape archaeology (Hicks 2003: 326). The ambivalence of archaeologists' ideas of landscape can, perhaps, be 'useful' (Gosden and Head 1994: 113). In this introduction, discussing three overlapping themes that emerge from the papers collected here – Heritage, Temporality and Situations – we suggest that archaeologists' own conceptions of 'landscape' might represent a significant tool in the important task of building upon the acknowledgement of diversity in contemporary world archaeology in order more adequately to theorise the situated nature of our knowledge of the past – envisaging landscapes as 'standpoints'.

LANDSCAPES AND HERITAGE

Landscape archaeologies are often explicitly political: distinguishing how 'people, differently engaged and differentially empowered, appropriate and contest their landscapes' (Bender 1993a: 17; cf. Bender and Winer 2001). By recognising the political dimensions of landscapes and heritage, archaeologists have used the idea of landscape to capture the complex intersections between the human, archaeological and geographical situations in which they work. Indigenous archaeologies have pioneered these approaches to the permeable nature of conventional distinctions between people, places and the past, particularly in North America (Rubertone 2000) and Australia (Smith 1999), as have the approaches to 'cultural landscapes' in heritage management and public archaeology (Turner and Fairclough this volume). Two chapters presented here directly address this theme.

Laura McAtackney (Chapter 2) studies one of the most politically contested landscapes in the United Kingdom: the Long Kesh/Maze site in Northern Ireland. From the start of its re-use as a prison in 1971 until its decommissioning in 2000, this landscape was used as an incarceration centre for paramilitary prisoners connected with 'the Troubles'. McAtackney's research combines interviews and oral histories with material engagements with the landscape and a range of objects connected with it. She suggests that a landscape approach can avoid unhelpful divisions between the human and material dimensions of the site through the study of the significance, representations (through the media or in political murals) and diverse experiences of the landscape in the past and the present. In all these respects, McAtackney evokes a complex reciprocal and historical process of ebbs and flows in which the landscape has played a part in the political process on a small-scale, intimate and often emotional level as well. Thus, her work contributes to a growing body of work that seeks to move the archaeology of institutional landscapes away from Foucauldian notions of constraint to feminist studies of embodiment (De Cunzo and Ernstein 2006). She demonstrates how through the decisions over the future of the site in the post-conflict state, the Long Kesh/Maze landscape continues to play a complex role in contemporary Northern Irish politics.

The relationships between communities and archaeological landscapes are explored further in the study by Susan Keitumetse, Geoffrey Matlapeng and Leseka Monamo of the Tsodilo Hills UNESCO World Heritage Site in Botswana (Chapter 5). Here, the archaeological landscape and its significance are visible and interpreted according to international standards. Nevertheless, the authors reveal complexities in reconciling local and official connections to the historic environment. Through interviews, the study reveals that the disenfranchisement of the local population from the management of the site means that the value, maintenance and preservation of the World Heritage Site is generally discussed in terms of tourist income rather than any close connection with the landscape. Here, the authors argue, a lack of involvement of local communities has led to an indifference to the historic environment that is counterproductive in attempts to maintain and enhance the site as a community resource.

Both of these studies seek to capture the close connections between people, archaeological heritage and the everyday lived environment. They suggest that heritage studies can use archaeological ideas of landscape as a way of revealing the attachments and political relationships that develop between landscape and communities, and point towards how heritage management might develop methods of recognising how quotidian human life, as well as material things, forms part of the contemporary historic environment.

LANDSCAPES AND TEMPORALITY

The temporal dimensions of landscape raised in these studies of contemporary heritage are explored further in four chapters of this volume. This has been a central theme in landscape archaeology over the past 15 years. Roland Fletcher (1995) has observed how the built environment constrains the long-term development of settlements. Richard Bradley (1993) has explored the 'afterlives' of European prehistoric monuments, while Cornelius Holtorf (1998) has examined their 'biographies' or 'life histories'. Historical archaeologists have explored the relationships between landscapes and memory (Holtorf and Williams 2006). However, the most influential contribution to such studies has been Tim Ingold's (1993) discussion of 'the temporality of landscape'. Ingold suggests that 'temporality' (as opposed to history or chronology) emerges in a 'rhythmic' manner from the 'pattern' of human activities or 'dwelling' in the landscape. In such a view, events can be seen to 'encompass a pattern of retentions from the past and protentions for the future' (Ingold 1993: 157). Ingold introduces the idea of the 'taskscape' to denote the temporal and emergent nature of human dwelling in the landscape, and extends this concept to archaeological practice itself:

> *[T]he practice of archaeology is itself a form of dwelling*. The knowledge born of this practice is thus on a par with that which comes from the practical activity of the native dweller and which the anthropologist, through participation, seeks to learn and understand. For both the archaeologist and the native dweller it tells – or rather *is* – a story. It enfolds the lives and times of predecessors who, over the generations, have moved around in it and played their part in its formation. To perceive the landscape is

> therefore to carry out an act of remembrance, and remembering is not so much a matter of calling up an internal image, stored in the mind, as of engaging perceptually with an environment that is itself pregnant with the past. To be sure, the rules and methods of engagement employed respectively by the native dweller and the archaeologist will differ, as will the stories they tell, nevertheless – in so far as both seek the past in the landscape – they are engaged in projects of fundamentally the same kind. (Ingold 1993: 152, original emphasis)

Such an approach contrasts with the emphasis in the 'interpretive' archaeologies that emerged in the 1980s and 1990s upon historical and contextual dimensions of 'meaning' (Barrett 1999: 24–25; compare Deetz 1990), instead seeing landscapes as emergent and embodied entities that bind together past, present and future.

In this light, the radical archaeological approach to the temporality of landscape presented in Chris Witmore's study of the South Argolid in Greece (Chapter 9) is of particular interest. Witmore draws upon philosopher Michel Serres' account of the 'percolation' of time, which sees the flow of time as 'turbulent' and chaotic, leaving material traces of the past that are 'folded' together in the present (Serres and Latour 1995: 58, 60–61). Through a discussion of the results of the Argolid Exploration Project – an archaeological programme conducted in 1972 and between 1979–1981 – Witmore argues that landscape archaeology has distinctive methods that can capture the simultaneous (rather than straightforwardly successive) nature of the experience of time in the landscape. Thus for Witmore, time is not an external parameter along which landscape change can be ordered and demarcated chronologically, but a quality of landscape that emerges from momentary engagements with it, including archaeological engagements.

A more conventional sequence of landscape change is presented in Rebecca Yamin and Joseph Schuldenrein's study (Chapter 4), which combines excavated evidence (often overlooked in landscape archaeology) with documentary and cartographic sources to present an account of the role of the Collect Pond, on Manhattan Island, in the long-term development of the urban landscape of New York City. But here too the authors demonstrate how the past and the present fold into one another. In this case, the focus is upon the many contrasting, sometimes conflicting, histories of the landscape as a place of industry, recreation, domestic life or death and burial: the presentation of the Five Points neighbourhood in the Scorsese movie *Gangs of New York*, the descriptions of Charles Dickens and the excavation of the African Burial Ground on the edge of the Collect. Here, political acts of remembering people and places elide with the persistent presence and influence of this large archaeological feature in the urban landscape.

The political dimensions of the temporality of landscape are also at the forefront of Christopher Matthews and Matthew Palus' account of the changing landscapes of the city of Annapolis, Maryland, from the 17th century to the present (Chapter 10). Having traced how the design and development of the urban landscape during the 17th and 18th centuries were bound up with claims of power by city elites, they suggest that during the 19th and 20th centuries similar processes were worked out through the negotiation of the city's heritage. History was 'extracted' from the landscape as a commodity, and through the effects of the historic preservation movement Annapolis became a 'landscape of American ruins'. This invention of 'historic Annapolis' provided a material symbol for American history for tourists and locals alike, designed to evoke the 'spirit of the revolution'. Matthews and Palus eloquently demonstrate how apparently straightforward debates among preservationists over the removal of the 'wirescape' of above-ground utilities in the contemporary urban landscape are bound up with these long-term political processes of power and historical contingency, in which the complexity and changing nature of the historic landscape is denied.

These issues of change and preservation are central to Sam Turner and Graham Fairclough's account of the theory and practice of historic landscape characterisation (HLC) in Europe (Chapter 6), which is based on examining 'the historic dimension of the present day landscape'. Building upon the European Landscape Convention's broad definition of the contemporary political and social dimensions of landscape, HLC seeks to build a landscape approach into the practical requirements of heritage management. They suggest that through its use of the idea of 'landscape character', HLC holds the potential to develop more politically engaged and democratic practices in heritage management – acknowledging the lived, everyday and (importantly) changing environments of heritage rather than particular sites that require protection. Thus, the HLC approach has much in common with anthropological approaches to landscapes as 'time materializing' in which 'landscapes, like time, never stand still' (Bender 2002: S103).

These chapters explore the radical potential of landscape approaches to underline the contemporary nature of archaeological practice, by seeing landscapes as folding together people and things, past and present, in simultaneous processes of 'time materialising'. In this way, landscape archaeologies are beginning to make substantive contributions to other social science disciplines that have sought to define the contingencies and auras of particular places. The permeabilities between people, temporality and place are strikingly evoked by environmental sociologist Michael Mayerfeld Bell's description of landscapes as being filled with 'ghosts', or 'the sense of the presence of those who are not physically there' (Bell 1997: 813):

> Ghosts ... help constitute the specificity of historical sites, of the places where we feel we belong and do not belong, of the boundaries of possession by which we assign ownership and nativeness. Ghosts of the living and the dead alike, of both individual and collective spirits, haunt the places of our lives. Places are, in a word, personed – even when there is no one there. (Bell 1997: 813)

In this vein, one recent innovative study by archaeologist Rodney Harrison (2004a) has traced the re-emergence of elements of antiquarian discourse in the ways in which Aboriginal people in New South Wales describe their relationships with archaeological sites. Building on his study of 'shared landscapes' (2004b), Harrison suggests that conventional archaeological discourse too often serves to 'erase the aura of artefacts' and sites – thus reducing the potential for acknowledging multiple perspectives or for community engagement. He calls for alternative approaches to cultural heritage management that allow for the 'special' qualities of places, akin to a sense of authenticity or memory, to be acknowledged. The studies of temporality and landscape presented here contribute to such attempts to weave together humanistic and material conceptions of archaeological landscapes, seeking to throw light upon the character and contemporary power of landscapes of archaeological heritage.

LANDSCAPES AS SITUATIONS

In world archaeology, the acknowledgement of the particular disciplinary traditions, social contexts and possibilities of regional archaeologies is a major contemporary challenge that must moderate the desire for international standards, conventions and unity (Hicks 2005). By considering the regional development and reception of ideas of landscape in archaeology, four chapters in this volume explore how different 'situations' have informed alternative ideas of landscape in world archaeology.

Martin Kuna and Dagmar Dreslerová (Chapter 7) consider landscape archaeology and settlement archaeology in central Europe (cf. Galaty 2005). They eloquently demonstrate how ideas and approaches from Britain and elsewhere were partially adopted in this region, and how a distinctive range of theoretical and applied approaches have been developed here. They outline the changing approaches to 'settlement' that include concepts of *Landesaufnahme* and *Siedlungskammer*, and Evsen Neustupný and Martin Kuna's notion of 'community areas'. Through two case studies – the Loděnice project in central Bohemia, which focuses on an Iron Age industrial site, and a study of Bronze Age tumuli in South Bohemia – they show how such theoretical ideas have been applied through the use of fieldwalking and Geographical Information Systems. Kuna and Dreslerová highlight the importance of finding theoretical and practical approaches that can meet the specific challenges of particular landscapes – in this case regions in which above-ground landscape remains have often been destroyed through intensive arable farming.

Sarah Croucher (Chapter 3) focuses on past and present approaches to archaeological landscapes on the East African Swahili coast. She suggests that a 'vernacular' tradition of landscape archaeology has developed in the region. Where previous studies had focused only on the impressive stone buildings – assuming that urban life was a recent introduction to the region from outside and neglecting rural landscapes – recent landscape surveys such as Mark Horton's work at Shanga in Kenya (1996) have demonstrated the antiquity and Indigenous development of Swahili urban landscapes, tracing a complex sequence of timber, thatch and earth structures from the 8th century AD. Croucher calls for such perspectives to be extended to study the roles of landscape in the construction of identity, and, drawing on her recent fieldwork carried out on the islands of Unguja and Pemba, explores the potential of such a focus through a study of the everyday experience of landscape by plantation owners and slaves on 19th century clove plantations. She demonstrates how by working to move beyond the 'Western gaze' of conventional archaeology in a reflexive manner, archaeological research can play a crucial role in generating histories of social control and social differentiation in the plantation economies of 19th century Zanzibar.

Reflexivity is also a central theme of Ken Kelly and Neil Norman's discussion of historical landscape archaeology of Atlantic Africa (Chapter 8). Their landscape approach focuses on 'the contested and imagined locales that are in a constant state of cultural construction, deconstruction and reconstruction' with an emphasis on 'complex and often multiethnic dimensions'. Drawing upon fieldwork undertaken at the trade entrepôts of Savi and its successor, Ouidah, in coastal Bénin, Kelly and Norman demonstrate the complexities of colonial interactions in West Africa. They consider the manifestations of power for both the Hueda and Dahomey states in their control and restriction of space that European traders could occupy and fortify. Their study reminds us of the variety and complexity of the landscapes of the historical Atlantic world, and of the need for more nuanced archaeologies of colonial encounters and interactions.

Mark Hauser and Dan Hicks (Chapter 11) consider the potential of developing postcolonial approaches to landscape in the historical archaeology of the anglophone Caribbean. Building upon Chris Gosden's (2004) approach to colonialism as a material process, they consider the close historical relationships between colonialism and landscape, which have affected the development of the disciplines of archaeology and anthropology, and they point to interdisciplinary calls to complement purely ideational, interpretive studies with an acknowledgement of the material dimensions of landscape. Considering recent debates over domination, resistance and power in Caribbean plantation archaeology, and building upon Hicks' study of landscape archaeology in the eastern Caribbean (Hicks 2007), they suggest that landscape archaeology's distinctive methodologies hold the potential to be used to develop studies of the complex materialities of colonialism, which might complement previous studies of 'ideas of landscape' – which in colonial contexts have tended to overdetermine the power of the coloniser, and to define the agency of the colonised only in terms of resistance.

LANDSCAPES AS STANDPOINTS

The chapters presented here emphasise landscape archaeology's material engagements with temporality, community heritage and public archaeology, along with the details of particular situations and the complex permeabilities between human and nonhuman dimensions of landscape. Together, they underline what might be termed the hybrid nature of archaeological conceptions of landscape (cf. Whatmore 2002). While this rather empirical focus – upon the material as well as the ideational – is, perhaps, unfashionable, it encourages us to take stock of the diversity of landscape archaeologies in different situations. Moreover, it points to the potential of reorienting the humanistic consensus of Anglo-American 'post-processual', 'interpretive' or 'social' archaeologies of landscape that has represented a powerful voice in mainstream world archaeology during the past decade (see reviews by Ashmore 2004, Thomas 2001). To understand the 'received view', we may turn to Matthew Johnson's recent review of *Ideas of Landscape* (2006). Johnson contrasts an 'empirical school' of landscape archaeology, represented by the work of British archaeologists such as Mick Aston (1985) and Tom Williamson, with the studies of anthropologists and cultural geographers whose 'discussions of the meanings of landscape sit at the forefront of theoretical debate':

> It is easy to peruse the pages of Landscape History, Journal of the Medieval Settlement Research Group, and Landscapes and conclude that landscape archaeology remains firmly in the grip of the most unreflective empiricism in which 'theory' is a dirty word and the only reality worth holding onto is that of muddy boots – a direct, unmediated encounter with the 'real world'. (Johnson 2006: 2)

Johnson does not seek to criticise any of these writers, but the account of landscape studies as having a 'double nature … simultaneously one of the most fashionable and avant-garde areas of scholarly enquiry, and also, paradoxically, one of the most theoretically dormant areas' (Johnson 2006: 1) passes perhaps too quickly over issues of situation, method, practice and materiality. The strong humanistic focus of interpretive archaeology famously engendered a dissatisfaction with the dry empiricism of some landscape archaeology, and its 'anecdotal' discussions of human agency (Thomas 1993: 26). The argument was clearly set out by John Barrett:

> To describe the landscape as a history of things that have been done to the land results in a cataloguing of the material transformations wrought upon the land. This procedure conforms with current archaeological expectations. To understand the landscape as inhabited demands a significant shift in our perceptions, and it is one that will not carry current methodological procedures with it. To inhabit the landscape is to look about, observe, and to make sense of what one sees; it is to interpret. (Barrett 1999: 26)

Like Barrett, Julian Thomas suggests that 'overwhelmingly empiricist' approaches in landscape archaeology have developed accounts that 'often see[m] quite remote from the past human lives that were lived in these places' (Thomas 2001: 165). But despite this dissatisfaction with the methods of landscape archaeology, the focus in interpretive archaeology upon 'reading' landscape (after Daniels and Cosgrove 1993: 1) has, when it comes to field practice, led only to the generation of phenomenological studies which have been criticised by both post-processual and processual archaeologists alike for parochialism, nostalgia, romanticism or poor methodological standards (Hodder 2004; Johnson 2005; Fleming 2005).

Others, however, have suggested that alternative 'romantic' approaches to 'landscape' and more empirical approaches to 'settlement' both form important aspects of spatial archaeology (Sherratt 1996). By underlining the active role of landscapes in social life, archaeology's interpretive turn problematised the 'epistemological metaphors' (insights, perceptions, focuses, views) that give social scientists the sense that knowledge is revealed, or made visible, rather than constructed (Salmond 1982: 66), but it simultaneously turned away from landscapes' material biographies, and the engagements, attachments and entanglements with complex 'landscapes' of many people, places and material things through which archaeological knowledge is constituted. A number of recent studies, informed especially by perspectives from ethnography and science studies, have started to redress this imbalance (Yarrow 2003; Edgeworth 2006). A focus on landscape in its broadest sense – the heterogeneous, constantly shifting networks of places, people, institutions and objects – reveals how archaeology is a relational process, rather than purely descriptive and discovering, or purely creative or interpretive. As Barbara Bender has argued, we must learn to acknowledge mess, complexity and contradiction, 'disorder and untidiness' (Bender 2006: 310), not only in the remains of past landscapes that we study but in our contemporary disciplinary landscapes as well. Doing so can represent a political move, which uses archaeological techniques to expose the positionality and situated nature of our contemporary knowledge of the past.

How, then, might the studies collected here begin to contribute to the recognition of the heterogeneous situations (both human and nonhuman) in which world archaeology is practiced? While archaeologists and anthropologists of landscape have often self-consciously rejected Western traditions of landscape as empiricist, painterly or viewed (see contributions to Bender 1993b), the same studies have clearly demonstrated the strength of landscape approaches to generate alternative archaeologies. We suggest here that landscape archaeology might be used to frame distinctive kinds of reflexive archaeologies, which seek to use the material engagements of archaeology to move beyond the Euro-ethnocentrism that is inherent in conventional ideas of landscape to forge alternative archaeologies that acknowledge the material diversity of landscapes, as well as just the multiple ways in which landscape is conceived or understood (cf. Schmidt and Patterson 1995: 23–24). We suggested at the start of this chapter that diversity represents a principal strength of landscape archaeology, inherent in its methods and practices. The distinctive thing about this pluralism is that it is not simply relativism, but is materially situated. The radical potential of landscape archaeology, then, lies in its ability to generate distinctively archaeological perspectives upon positionality, and the situated nature of all archaeological knowledge. Landscape archaeology has seen a strong critique of the 'jeweller's eye' approach to landscape (Edmonds 1999: 9), in which knowledge appears to be constructed from nowhere, or from everywhere. This is like what Donna Haraway (1991: 189) has termed the 'God trick' of infinite vision. We look down impossibly upon plotted distribution maps or culture historical movements, generated out of attitudes to landscape and technologies of field survey largely developed through European colonialism (Higman 1988: 19–64; cf. Hauser and Hicks this volume).

In recognising how landscapes emerge from human action in particular situations, some landscape archaeologies have come close to two distinctive feminist positions: Sandra Harding's 'standpoint epistemology' (Harding 1993) – which built upon Nancy Hartsock's conception of 'a specifically feminist historical materialism' (Hartsock 1987) and has been more recently combined with postcolonial theory in science studies (Harding 1994, 1998) – and Donna Haraway's notions of 'situated knowledges' and 'the privilege of a partial perspective' (Haraway 1991). We want to suggest that seeing archaeological landscapes as standpoints – that is, as situations in which material conditions, human life, political contexts and research practice are bound up together – represents one way of developing the potential of landscape archaeology to acknowledge diversity in the archaeological past and the disciplinary present.

Standpoint epistemologies remain little explored within archaeology, mainly due to their previous tendencies towards essentialist conceptions of identity (cf. Visweswaran 1997: 609–610). However, as Alison Wylie has observed, they hold significant potential to contribute a sense of contingency and subject-specificity to our understanding of the construction of scientific knowledge (Wylie 2003). In gender archaeology, third-wave approaches have set the agenda here, particularly:

> Feminist approaches ... [that] share a focus on local, empirical data. ... Especially within Americanist gender archaeology, a feminist epistemology seems to have emerged that concentrates on the small-scale, on everyday occurrences and relations between people, on subtle shifts in power and relations of production. ... Drawing on their own perspectives, some feminists are creating an archaeology concerned less with hierarchies and meta-narratives and more with the observation of detail, complexities, and local or personal experience. (Gilchrist 1999: 29–30)

How might such perspectives operate at the multiple scales of analysis that a landscape perspective encourages? The challenge for interpretive perspectives in archaeology around the world is to find ways of expressing more than simply a sense of how archaeological data are 'mediated by interpretive theory' (Wylie 2002: 172), building instead upon the 'mitigated objectivism' that Alison Wylie (2002: 177) suggests characterised both the processual and post-processual archaeologies – the awareness of the particularity and contingency of archaeological knowledge. Understood in this way, perhaps landscape archaeology is a good place from which to think through interpretive archaeology's calls for reflexive approaches (Hodder 1999): providing one way through which archaeologists can acknowledge, as some social anthropologists have, that the 'connections' we make are always 'partial' – in both senses of the word, neither total nor impartial (Strathern 1991). Here, we follow Wylie's suggestion that:

> Political self-consciousness enforces a critical awareness of the contingency of knowledge production that does not (necessarily) entail a politically and epistemically paralysing cynicism about the process of inquiry.... [The challenge is one of] negotiating the tensions created by a commitment to use the tools of systematic empirical enquiry to rigorously question the authority and presuppositions of scientific inquiry, to turn

> science and history against themselves when they serve as tools of oppression, and to reclaim their emancipatory potential. (Wylie 1995: 272)

CONCLUSIONS

From a humid, rainy July evening in southwest England, how can we envisage landscapes in world archaeology? At one scale of analysis world archaeologists are unified in a number of political and theoretical programmes, while at a more fine-grained scale human and material diversity emerges. The lesson that we learn from landscape archaeology is that both scales co-exist, simultaneously (Hicks in press). They emerge as we enact world archaeologies. Like any world system, contemporary world archaeology

> ... is a social system, one that has boundaries, structures, member groups, rules of legitimation, and coherence. Its life is made up of the conflicting forces which hold it together by tension and tear it apart as each group seeks eternally to remold it to its advantage. (Wallerstein 1976: 229)

But in editing this volume we have tried to put across our sense that archaeologies of landscape offer a powerful range of tools that can be appropriated and used to place diversity at the centre of the theory and practice of world archaeology (cf. Restrepo and Escobar 2005, Ribeiro and Escobar 2006). By approaching the boundaries between people, situations, heritage and temporality as permeable, contingent and emergent, an awareness of the hybridity of archaeological landscapes and of the multi-sited practices of landscape archaeology can help us to move beyond essentialised or nationalist notions of regional or local archaeologies, and to celebrate the diversity of contemporary archaeology (cf. Hicks 2003: 324–326; Hicks and Beaudry 2006: 7–9).

This position risks the charge of 'eclecticism', and the accusation that 'contradictions and incompatibilities ... can arise from juxtaposing fragments from different theoretical programs' (Schiffer 2000: 3). But we prefer to see the potential of world archaeology as similar to that of 'unity in diversity' in world anthropology (Krotz 2006). This involves a shift that Nick Shepherd has described as leading from 'One World Archaeology' to 'One World, Many Archaeologies' (Shepherd 2005). In the process, recognising that situations in which archaeology is practiced are never purely social, and that archaeological knowledge is never simply a social construction, is crucial. The landscapes of world archaeology are landscapes of complex and uneven materialities. Where many past and present voices are silenced or erased, our conception of landscapes as standpoints seeks to emphasise that a focus upon 'materiality' can be double-edged: combining the study of *material things* with a sense of *human significance*. Situated archaeologies can confront these silences, highlighting things that 'matter'. They can not just accommodate, but can celebrate, the contingent diversities of contemporary world archaeology. In this respect, archaeologies of landscape can represent not only politically engaged archaeologies, but also archaeologies of hope.

REFERENCES

Anschuetz, K.F., R.H. Wilshusen and C.L. Scheick 2001. An Archaeology of Landscapes: Perspectives and Directions. *Journal of Archaeological Research* 9(2): 157–211.

Araoz, G. 2004. Fw: Abstracts/Call/Llamado/Appel. Email sent to wac@flinders.edu.au, 6 December 2004.

Ashmore, W. 2004. Social Archaeologies of Landscape. In L. Meskell and R.W. Preucel (eds) *A Companion to Social Archaeology*. Oxford: Blackwell, pp. 255–271.

Aston, M. 1985. *Interpreting the Landscape: Landscape Archaeology in Local Studies*. London: Batsford.

Barrett, J.C. 1999. Chronologies of Landscape. In P. Ucko and R. Layton (eds) *The Archaeology and Anthropology of Landscape*. London: Routledge, pp. 21–30.

Bell, M.M. 1997. The Ghosts of Place. *Theory and Society* 26: 813–836.

Bender, B. 1993a. Landscape – Meaning and Action. In B. Bender (ed) *Landscape: Politics and Perspectives*. Oxford: Berg, pp. 1–17.

Bender, B. 1993b. *Landscape: Politics and Perspectives*. Oxford: Berg.

Bender, B. 2002. Time and Landscape. *Current Anthropology* 43 (Supplement): S103–S112.

Bender, B. 2006. Place and Landscape. In C. Tilley, W. Keane, S. Küchler, M. Rowlands and P. Spyer (eds) *Handbook of Material Culture*. London: Sage, pp. 303–314.

Bender, B. and M. Winer (eds) 2001. *Contested Landscapes: Movement, Exile and Place*. Oxford: Berg.

Bradley, R. 1993. *Altering the Earth. The Origins of Monuments in Britain and Continental Europe*. Edinburgh: Society of Antiquaries of London.

Bradley, R. 2000. *An Archaeology of Natural Places*. London: Routledge.

Byrne, D. 2003. Nervous Landscapes: Race and Space in Australia. *Journal of Social Archaeology* 3(2): 169–193.

Chang, K.C. 1972. *Settlement Patterns in Archaeology*. Menlo Park, CA: Addison-Wesley (Modules in Anthropology 24).

Clarke, D.L. (ed) 1977. *Spatial Archaeology*. London: Academic Press.

Daniels, S. and D. Cosgrove 1993. Introduction: Iconography and Landscape. In D. Cosgrove and S. Daniels (eds) *The Iconography of Landscape*. Cambridge: Cambridge University Press, pp. 1–23.

De Cunzo, L. and J.H. Ernstein 2006. Landscapes, Ideology and Experience in Historical Archaeology. In D. Hicks and M.C. Beaudry (eds) *The Cambridge Companion to Historical Archaeology*. Cambridge: Cambridge University Press, pp. 255–270.

Deetz, J. 1990. Landscapes as Cultural Statements. In W.M. Kelso and R. Most (eds) *Earth Patterns: Essays in Landscape Archaeology*. Charlottesville: University of Virginia Press, pp. 1–4.

Edgeworth, M. (ed) 2006. *Ethnographies of Archaeological Practice*. Walnut Creek, CA: AltaMira Press.

Edmonds, M. 1999. *Ancestral Geographies of the Neolithic: Landscapes, Monuments and Memory*. London: Routledge.

Fleming, A. 2005. Megaliths and Post-modernism: The Case of Wales. *Antiquity* 79(306): 921–932.

Fletcher, R. 1995. *The Limits of Settlement Growth: A Theoretical Outline*. Cambridge: Cambridge University Press.

Foundation for the Advancement of Mesoamerican Archaeology 2005. Zapotec Writing, essay by Javier Urcid and Ndaxagua Project Report. Email sent to wac@flinders.edu.au, 23 May 2005.

Galaty, M.L. 2005. European Regional Studies: A Coming of Age? *Journal of Archaeological Research* 13(4): 291–336.

Gilchrist, R. 1999. Archaeological Biographies: Realizing Human Lifecycles, Courses and Histories. *World Archaeology* 31(3): 325–328.

Gorman, A. 2005. The Cultural Landscape of Interplanetary Space. *Journal of Social Archaeology* 5(1): 85–107.

Gosden, C. 2004. *Archaeology and Colonialism: Culture Contact from 5000 BC to the Present*. Cambridge: Cambridge University Press.

Gosden, C. and L. Head 1994. Landscape – A Usefully Ambiguous Concept. *Archaeology in Oceania* 29: 113–116.

Haraway, D.J. 1991. Situated Knowledges: The Science Question in Feminism and the Privilege of a Partial Perspective. In *Simians, Cyborgs, and Women: The Reinvention of Nature*. London: Free Association Books, pp. 183–201.

Harding, S. 1993. Rethinking Standpoint Epistemology: What Is Strong Objectivity? In L. Alcoff and E. Potter (eds) *Feminist Epistemologies*. London: Routledge, pp. 49–82.

Harding, S. 1994. Is Science Multicultural? Challenges, Resources, Opportunities. *Configurations* 2: 301–330.

Harding, S. 1998. Is Science Multicultural? *Postcolonialisms, Feminisms and Epistemologies*. Bloomington and Indianapolis: Indiana University Press.

Harrison, R. 2004a. Objects of Desire: Stone Artefacts, 'Aura' and Archaeological Heritage Management in Australia. Paper presented at Australian Archaeological Association Conference, University of New England, Armidale, New South Wales, 13–15 December 2004. http://www.anu.edu.au/culture/aaa/Harrison.pdf (Consulted 15 July 2005).

Harrison, R. 2004b. *Shared Landscapes: Archaeologies of Attachment and the Pastoral Industry in New South Wales*. Sydney: University of New South Wales Press (Studies in the Cultural Construction of Open Space 3).

Hartsock, N.C.M. 1987. The Feminist Standpoint: Developing the Ground for a Specifically Feminist Historical Materialism. In S. Harding (ed) *Feminism and Methodology*. Bloomington: Indiana University Press, pp. 157–180.

Hicks, D. 2003. Archaeology Unfolding: Diversity and the Loss of Isolation. *Oxford Journal of Archaeology* 22(3): 315–329.

Hicks, D. 2005. 'Places for Thinking' from Annapolis to Bristol: Situations and Symmetries in 'World Historical Archaeologies'. *World Archaeology* 37(3): 373–391.

Hicks, D. 2007. *The Garden of the World: A Historical Archaeology of Sugar Landscapes in the Eastern Caribbean*. Oxford: Archaeopress (Studies in Contemporary and Historical Archaeology 3).

Hicks, D. in press. Commentary: The Multiscalar Debate in Historical Archaeology. *International Journal of Historical Archaeology*.

Hicks, D. and M.C. Beaudry 2006. The Place of Historical Archaeology. In D. Hicks and M.C. Beaudry (eds) *The Cambridge Companion to Historical Archaeology*. Cambridge: Cambridge University Press, pp. 1–9.

Higman, B.W. 1988. *Jamaica Surveyed: Plantation Maps and Plans of the 18th and 19th Centuries*. Kingston: Institute of Jamaican Publishers.

Hodder, I. 1999. *The Archaeological Process*. Oxford: Blackwell.

Hodder, I. 2004. British Prehistory: Some Thoughts Looking In. In I. Hodder *Archaeology Beyond Dialogue*. Salt Lake City: University of Utah Press (Foundations of Archaeological Inquiry), pp. 125–239.

Holtorf, C.J. 1998. The Life-histories of Megaliths in Mecklenburg-Vorpommern (Germany). *World Archaeology* 30(1): 23–38.

Holtorf, C.J. and H. Williams 2006. Landscapes and Memories. In D. Hicks and M.C. Beaudry (eds) *The Cambridge Companion to Historical Archaeology*. Cambridge: Cambridge University Press, pp. 235–254.

Horton, M.C. 1996. *Shanga: The Archaeology of a Muslim Trading Community on the Coast of East Africa*. London: British Institute in Eastern Africa.

Ingold, I. 1993. The Temporality of Landscape. *World Archaeology* 25(2): 152–174.

Ingold, T. 2005. Comments on Christopher Tilley 'The Materiality of Stone: Explorations in Landscape Phenomenology' (with Reply to Comment by Chris Tilley). *Norwegian Archaeological Review* 38(2): 122–129.

Johnson, M.H. 2005. On the Particularism of English Landscape Archaeology. *International Journal of Historical Archaeology* 9(2): 111–122.

Johnson, M.H. 2006. *Ideas of Landscape: An Introduction*. Oxford: Blackwell.

Krotz, E. 2006. Towards Unity in Diversity in World Anthropology. *Critique of Anthropology* 26(2): 233–238.

Layton, R. and P. Ucko 1999. Introduction: Gazing on the Landscape and Encountering the Environment. In P. Ucko and R. Layton (eds) *The Archaeology and Anthropology of Landscape*. London: Routledge, pp. 1–20.

Nakao, T. 2005. From Japan (tomoyo). Email sent to wac@flinders.edu.au, 9 April 2005.

Postgate, N. 2002. The First Civilisations in the Middle East. In B. Cunliffe, W. Davies and C. Renfrew (eds) *Archaeology: The Widening Debate*. Oxford: Oxford University Press and the British Academy, pp. 385–410.

Regan, M. 2004. WAC eNewsletter vol. 4. Email sent to wac@flinders.edu.au, 19 October 2004.

Restrepo, E. and A. Escobar 2005. 'Other Anthropologies and Anthropologies Otherwise': Steps to a World Anthropologies Framework. *Critique of Anthropology* 25(2): 99–129.

Ribeiro, G.L. and A. Escobar (eds) 2006. *World Anthropologies: Disciplinary Transformations in Systems of Power*. Oxford: Berg.

Robin, C. and N.A. Rothschild 2002. Archaeological Ethnographies: Social Dynamics of Outdoor Space. *Journal of Social Archaeology* 2(2): 159–172.

Rubertone, P. 2000. The Historical Archaeology of Native Americans. *Annual Review of Anthropology* 29: 425–446.

Salmond, A. 1982. Theoretical Landscapes: On a Cross-cultural Conception of Knowledge. In D. Parkin (ed) *Semantic Anthropology*. London: Academic Press, pp. 65–88.

Schiffer, M.B. 2000. Social Theory in Archaeology: Building Bridges. In M.B. Schiffer (ed) *Social Theory in Archaeology*. Salt Lake City: University of Utah Press (Foundations of Archaeological Inquiry), pp. 1–13.

Schmidt, P.R. and T.C. Patterson 1995. Introduction: From Constructing to Making Alternative Histories. In P.R. Schmidt and T.C. Patterson (eds) *Making Alternative Histories: The Practice of Archaeology and History in Non-Western Settings*. Santa Fe: School of American Research Press, pp. 1–24.

Serres, M. and B. Latour 1995. *Conversations on Science, Culture, and Time* (trans. R. Lapidus). Ann Arbor: University of Michigan Press.

Sherratt, A.G. 1996. 'Settlement Patterns' or 'Landscape Studies'? Reconciling Reason and Romance. *Archaeological Dialogues* 3: 140–159.

Shepherd, N. 2005. From 'One World Archaeology' to One World, Many Archaeologies. *Archaeologies* 1(1): 1–6.

Smith, C. 1999. Ancestors, Place and People: Social Landscapes in Aboriginal Australia. In P.J. Ucko and R. Layton (eds) *The Archaeology and Anthropology of Landscape*. London: Routledge (One World Archaeology 30), pp. 189–205.

Smith, C. 2004. No Subject. Email sent to wac@flinders.edu.au, 25 November 2004.

Smith, C. 2005. Call for Papers – China GA. Email sent to wac@flinders.edu.au, 2 February 2005.

Strathern, M. 1991. *Partial Connections*. Savage, MD: Rowman and Littlefield.

Tilley, C. 1994. *A Phenomenology of Landscape*. Oxford: Berg.

Thomas, J. 1993. The Politics of Vision and the Archaeologies of Landscape. In B. Bender (ed) *Landscape: Politics and Perspectives*. Oxford: Berg, pp. 1–17.

Thomas, J. 2001. Archaeologies of Place and Landscape. In I. Hodder (ed) *Archaeological Theory Today*. London: Polity, pp. 165–186.

Townsend, A. 2005. SPMA 2005 Conference. Email sent to wac@flinders.edu.au, 20 May 2005.

Trigger, B. 1967. Settlement and Archaeology: Its Goals and Promise. *American Antiquity* 32: 149–160.

Trujillo, J. 2005. [arte_y_rpestre] Panoramic photos reveal ancient native culture. Email sent to wac@flinders.edu.au, 25 November 2004.

Visweswaran, K. 1997. Histories of Feminist Ethnography. *Annual Review of Anthropology* 26: 591–621.

Wallerstein, I. 1976. *The Modern World-System: Capitalist Agriculture and the Origins of the European World-Economy in the Sixteenth Century*. New York: Academic Press.

Whatmore, S. 2002. *Hybrid Geographies*. London: Sage.

Wylie, A. 1995. Alternative Histories: Epistemic Disunity and Political Integrity. In P.R. Schmidt and T.C. Patterson (eds) *Making Alternative Histories: The Practice of Archaeology and History in Non-Western Settings*. Santa Fe: School of American Research Press, pp. 255–272.

Wylie, A. 2002. 'Heavily Decomposing Red Herrings': Middle Ground in the Anti-/Postprocessualism Wars. In *Thinking from Things: Essays in the Philosophy of Archaeology*. Berkeley: University of California Press, pp. 171–178.

Wylie, A. 2003. Why Standpoint Matters. In R. Figueroa and S. Harding (eds) *Science and Other Cultures: Issues in Philosophies of Science and Technology*. London and New York: Routledge, pp. 26–48.

Yarrow, T. 2003. Artefactual Persons: The Relational Capacities of Persons and Things in the Practice of Excavation. *Norwegian Archaeological Review* 36(1): 65–73.

CHAPTER 2

The Contemporary Politics of Landscape at the Long Kesh/Maze Prison Site, Northern Ireland

Laura McAtackney

INTRODUCTION

The Long Kesh/Maze prison site is probably the most significant contemporary political landscape in Northern Ireland. It is best known for its role during the recent conflict in Northern Ireland, colloquially known as 'the Troubles', as the prison that incarcerated terrorists deemed most dangerous. During the Hunger Strikes of 1980–1981, in which 10 Republican prisoners starved to death as part of a protest against attempts to criminalise them, the prison became a powerful media image around the world. Meanwhile, within Northern Ireland it represented one of the most prominent examples of the militarisation of society that continually escalated during the course of the Troubles. Long Kesh/Maze was a prison landscape for 30 years (1971–2000), and during this time has only been experienced as a high-security zone. It was, and remains, a place apart. As probably the most prominent icon of the Troubles, Long Kesh/Maze has been largely experienced without any actual physical interaction – that is, as an imagined landscape. It has been central in the political landscape of Northern Ireland without actually being part of the state apparatus. Whilst highly visible, it is simultaneously inaccessible.

This chapter examines the Long Kesh/Maze as an archaeological landscape, but also considers how, through its role in the recent and contemporary political landscapes of Northern Ireland, the boundaries of its 'landscape' have been constantly extended and broken down. In doing so, the physical experience of the landscape in the past and the present will be examined alongside the many different ways in which it has been imagined and understood. I aim to present an account of the ways in which the landscape was experienced in the past together with a sense of how contemporary politics are determining how it will be experienced in the future. I shall combine first-hand prisoner testimonies and archaeological and documentary evidence to study the complex histories of the landscape. In this way, unlike many recent archaeological studies of prisons that have focused on architectural planning and Foucauldian ideas of the enforcing of discipline, I shall seek to consider the diverse and unforeseen ways in which the prison landscape was experienced, understood and imagined.

Examining the Long Kesh/Maze prison as a landscape requires a diachronic approach, and inevitably highlights multiple perspectives, myriad contexts and the subjective nature of experiencing such a site. Rather than using the idea of landscape to evoke static, unchanging entities, at this late 20th century site landscape archaeology does not provide a window through which a single past can be observed and described, but instead reveals how landscapes are diverse and often contested entities constituted by the events of the past, the interventions of the present and the potential of the future. Studying Long Kesh/Maze reminds us of how contemporary politics shapes our experience of historic landscapes, and (through the practice of heritage management) also defines, and can attempt to control, what can be remembered of the past in the future.

THE BIOGRAPHY OF THE LONG KESH/MAZE LANDSCAPE

Long Kesh/Maze has a short, yet incident-filled, history. It was first utilised as the Long Kesh Internment Camp from August 1971 as a short-term overspill facility to be used during a period of civil unrest. This short-term measure became *de facto* permanent as the Troubles escalated and the site continued to house predominantly paramilitary prisoners until its closure in September 2000. During these three decades, the prison developed greatly in terms of extent, structure, philosophy and profile.

The 360-acre site, including an adjacent army base, developed from the infrastructure of a World War II Royal Air Force (RAF) base to the encampments of the Long Kesh Internment Camp, at which internees were held within 'cages' (so-called because the wire fencing that surrounded each section resembles a cage) within nissen huts that had initially been constructed by the RAF (Figure 2.1). The prisoners who were held at the site knew it as 'the Kesh' or 'the Cages', whereas the authorities labelled it 'Maze Compounds'. Each cage held the personnel of a single paramilitary organisation, as prisoners were placed on their arrival with those with which they self-identified, thereby perpetuating the sectarian divisions of society at large. The nissen huts continued to be used until it became apparent to the authorities not

Figure 2.1 Nissen hut at the Long Kesh Internment Camp. *Courtesy of John O'Neill.*

only that was the civil unrest worsening, thereby creating more inmates to fill the site, but also that in the wider community the conditions of their imprisonment meant that those incarcerated were gaining a form of kudos. As officials recognised at an early stage of its utilisation, the site was wholly inappropriate for its new function: 'The short point appears to be that the camp was built on a design more appropriate for a prisoner of war camp than an internment camp' (Burns 1971). Therefore, the construction of a new prison was proposed by the Gardiner Report (1975), which expressed the need to combat the paramilitaries through criminalising their prisoners. To do so, a new regime, and therefore a new structure to facilitate implementation, was needed.

Thereafter, from the mid-1970s, the cages co-existed with a new addition to the prison landscape, the concrete H Blocks of Her Majesty's Prison (HMP) Maze. The infamous H Blocks of HMP Maze (officially known as 'the Maze' or the 'Maze Cellular', and by prisoners as 'the Blocks') were constructed in haste between 1975 and 1976 (Figure 2.2). This addition to the prison landscape saw the introduction of newly built structures and policies that served to criminalise the political prisoners: most visibly through the provision of individual cells rather than communal areas and of concrete foundations rather than tiles over dirt floors, and by the wearing of prison uniforms rather

Figure 2.2 H Block at HMP, the Maze.

than civilian clothes. The number of visits and food parcels were reduced to that of ordinary prisoners. Most importantly, prisoners of different paramilitary organisations were integrated and freedom of association (the right of prisoners to associate with whom they wanted, when they wanted) was stopped. The reaction to such policies was widespread protest. The first prisoner to be sent to the H Blocks, Republican prisoner Kieran Nugent, refused to wear the prison uniform and therefore entered the prison donning only a blanket. This act instigated a series of escalating protests against prison authority policies by Republican prisoners that began with the 'blanket protests', in which the prisoners wore only blankets, and escalated to the 'dirty protests', where prisoners wearing only blankets were confined to their cells all day and smeared the walls with their own excrement. These protests culminated in the Hunger Strikes.

The Hunger Strikes of 1980–1981 were probably the most significant event in the biography of the prison landscape. The death of the first Hunger Striker, 27-year-old Bobby Sands, on 5 May 1981, brought with it the culmination of media and public interest in the site. So seminal was this event that it has been estimated that the rioting that ensued after his death claimed the lives of 20 civilians and security personnel throughout Northern Ireland. Two days later, somewhere between 70,000 and 100,000 people attended his funeral. The effect of Bobby Sands' death on recruitment of volunteers to the Irish Republican Army

(IRA) was immense, and in part ensured that the armed conflict continued into the 1990s.

Despite the Hunger Strikes not immediately achieving their stated aims (the return to the political status they enjoyed during the period of internment), the prisoners increasingly exerted control over the prison: by 1982 segregation returned to the wings, and by the late 1980s the prisoners had regained freedom of association. A frequently quoted example of the extent of the power of prisoners is the visit of the then Secretary of State for Northern Ireland, Dr Mo Mowlam MP, to the prison in 1998. At the time there was a growing fear that the Loyalist paramilitary ceasefires were on the verge of collapse. The fact that Mo Mowlam publicly visited the Loyalist leaders within the H Blocks to negotiate their continued ceasefire provides evidence of official acknowledgement of where the real power of the paramilitaries lay.

The use of the name 'Long Kesh/Maze' in this chapter also relates to the complex and contested political landscape of the prison. Multiple, highly-charged names exist for the same landscape, with different connotations for the site and its political status. Today ex-prisoners and their communities tend to use names connected to 'Long Kesh', and its association with internment and the political status of prisoners, whereas officials and the wider public tend to use names connected to the H Block period of criminalisation, namely 'the Maze'.

The prison was finally closed in September 2000 as a result of the conditions of the Belfast (Good Friday) Agreement, which was signed by all the major political parties in Northern Ireland in 1998. Many commentators would state that 'the Troubles' ended in 1998 (e.g. Amstutz 2005: 170), and that the ongoing political difficulties and protracted peace process are merely the manifestations of a still-divided society attempting to move to a post-conflict state. However, as part of this attempt to take the 'first steps to normalisation' (*Irish News* 2005), the site of the Long Kesh/Maze has become increasingly important. One of the major manifestations of the peace process has been the decreasing prominence of the security infrastructure associated with the strife-ridden streets of Northern Ireland. Many police stations, army bases, watchtowers and security posts have disappeared almost overnight with little discussion of their future or recording of their remains prior to destruction. To a large extent this has been welcomed by the general public. However, it has been noted by some commentators that 'some of Ulster's vanishing fortresses are now regarded with something approaching nostalgia' (Clarke 2005: 4). Long Kesh/Maze has been central in this current conflict over the future life of the now-defunct, but still emotive, elements of the physical and political landscapes of Northern Ireland, and as such it is appropriate to examine the extent of their physical remains.

THE LONG KESH/MAZE AS AN ARCHAEOLOGICAL LANDSCAPE

The Long Kesh/Maze prison landscape comprises the remains of the two very different prison structures: the nissen huts and cages of the Long Kesh Internment Camp (1971–mid 1980s), and the H Blocks of HMP, the Maze (1975–2000). The built fabric of both regimes has survived thanks to the policy of the Northern Ireland Prison Service (NIPS) simply to move on to the next available space when a new element was required. Therefore, defunct parts of the site were left to deteriorate, and despite such a relatively short passage of time since falling into disuse the processes of change are highly visible in the contemporary landscape. The physical remains of the different layers of the prison's history provide different challenges to archaeologists in their roles as interpreters.

The nissen huts, hangars and runways of the RAF base were utilised to accommodate a large number of internees, and later convicted prisoners, for what was thought to be a short-term measure. The nissen huts were surrounded by wire fencing and utilised as cages, the traces of which are still apparent. Problems arise in examining this area archaeologically for two reasons: utilisation and survival. As the nissen huts had been used by both the RAF and NIPS, the same buildings have been used for very different purposes without many initial structural changes. However, this changed with circumstance and the once open, internal spaces of the nissen huts were eventually divided into individual rooms, and porches were added to the front doors to add extra protection from the weather (M. Meehan 2006). Furthermore, as these areas have not been used, nor maintained, since the mid 1980s they are frequently in a poor state of repair, with some only now existing as block foundations.

The H Blocks were swiftly built from plans designed by a Northern Irish government architect after consultation with some, but not all, security sources (D McLaughlan 2007). It was later acknowledged by the British government's Hennessey Report, published in response to the mass escape of Republican prisoners in 1983, that the plans were not specifically designed for this unique type of prisoner:

> The prison is unique … in its population, which is totally dissimilar to the usual criminal recidivist population to be found in the nearest equivalent establishment in England and Wales. It consists almost entirely of prisoners convicted of offences connected with terrorist activities, united in their determination to be treated as political prisoners, resisting prison discipline, even if it means starving themselves to death, and retaining their para-military structure and allegiances even when inside. (Hennessey Report 1984: 10.2)

This style of prison architecture was little more appropriate for dealing with the paramilitary prisoners than that of the nissen huts which it replaced. It is apparent from prisoner testimonies, and from close examination of the buildings, that many elements were substantially changed during the period of use. In particular, the internal layout of some of the communal spaces has changed and security measures were added intermittently as a reaction to unforeseen prisoner interactions with the site. Although the concrete buildings present fewer difficulties for the archaeologist in tracing their structure, their swift erection and presumed short-term nature meant that proper foundations were not provided. Many are today in a poor state of repair, and a high water table on the site has resulted in often waterlogged conditions.

However, probably the greatest difficulty in attempting to examine the site as an archaeological landscape, as well as interpret the individual elements of the site, concerns its contemporary setting. As the site has been closed for just six years at the time of writing, the physical manifestations of the Troubles are still very emotive, with the Long Kesh/Maze site probably being the most controversial Troubles-related artefact of all. As such, access to the physical site, and the official documentation relating to it and to those most intimately connected to it during its period as a prison, is very difficult to gain. Sensitivities about interest in Long Kesh/Maze are such that few have gained even limited access. However, there is much that can be done to access the site as a contemporary landscape, and the politics of this involvement are a major part of the story.

The physical remains and associated artefacts of the site still remain largely *in situ*, but access is not always possible and unlimited access is near impossible. Therefore, a more lateral approach to the study of Long Kesh/Maze as a prison landscape is needed. Less emphasis on recording the actual standing remains, and more care in examining the more subjective elements of the site, including the conducting of archaeologically focused oral testimonies and external representations, are two routes to getting at the experiences of Long Kesh/Maze as a physical and imagined landscape. Of course, this work must be informed by theoretical approaches to landscape, and to emotive sites in particular, and this will be considered in the next section.

STUDYING THE LONG KESH/MAZE LANDSCAPE

Archaeology holds significant potential for the study of contested landscapes such as the Long Kesh/Maze, but this potential remains relatively untapped and has not as yet been fully theorised. As Rebecca Yamin and Karen Metheny have observed, in the United States the institutional frameworks and funding of historical archaeology have meant that landscape archaeology has generally been limited to gardens, and especially gardens that had belonged to well-known historical figures (Yamin and Metheny 1996: xiv). Notable exceptions, however, are the studies of the 'Archaeology in Annapolis' group, who have for a number of years combined landscape archaeology with the explicit study of the politics of the recent past and the present (Leone 2005). In Britain, the post-processual emphasis on the interpretative nature of archaeological study has been explored by a number of scholars in relation to landscape: most notably in the phenomenological approaches of Chris Tilley (1994). While focused primarily on prehistoric landscapes, such work has importantly emphasised the contemporary nature of archaeological interpretation, and the importance of studying how the landscape was experienced in the past, rather than simply seeing landscape archaeology purely as descriptive or methodological. What, then, can the approaches of landscape archaeology bring to the study of the very recent and contemporary past at the Long Kesh/Maze site?

The majority of the academic and popular studies of Long Kesh/Maze see it as a microcosm of the Northern Ireland conflict, because it contained individuals and witnessed events that have had a major impact on the course of the Troubles. In such work, the site is used to illustrate a political history that is already known, often focusing on prominent individuals and events, especially the Hunger Strikes, for example through journalistic works such as David Beresford's *Ten Men Dead* (1994) and Liam Clarke's *Broadening the Battlefield: The H-Blocks and the Rise of Sinn Féin* (1987). The Long Kesh/Maze landscape is also prominent in popular biographies of well-known figures who spent substantial periods incarcerated there, such as *The Billy Boy: The Life and Death of the LVF Leader Billy Wright* (Coogan 2001) and Laurence McKeown's more academic treatment of Republican experiences, *Out of Time: Irish Republican Prisoners Long Kesh 1972–2000* (2001).

In contrast with these people- and event-focused approaches, since the site has been closed a number of works have tried to study the landscape in its own right. One of the most prominent examples is the photographic work of Donovan Wylie, who spent almost 100 days repetitively photographing the prison, 'to understand the psychology of the architecture and its ability to disorient and diminish' (Wylie 2004: jacket notes). There are also a number of first-hand accounts of life within the Long Kesh/Maze site, including Gerry Adams' *Cage Eleven* (1990) – a description of his time as an internee on Long Kesh/Maze, which includes some limited discussion on how the inmates interacted with their surroundings. However, there has been little attempt to examine how the prison landscape was lived – instead the human and material dimensions of landscape have been examined as separate, unconnected entities. This is most visible in the photographic work of Wylie, which focuses solely upon the material remains of the prison.

Rather than seeing the landscape as a mere container for major figures and events in the recent conflict, or studying the site bereft of human actors as an empty artefact, landscape archaeology offers one way of combining humanistic and material engagements. Where more 'people-focused' studies fail to consider the complex effects of the built environment in the past and the present, an excessive focus on the material remains of the site can equally fail to engage with the context or how people actually interacted with the site when it was functioning. Landscape approaches, including those of archaeology, can study people as part of the historic environment, and can start to fill the gap between these respective approaches.

In this light, a number of new studies have emerged from the University of Ulster, especially through the work of Brian Graham and his students, focusing on ideas of place and landscape. In particular, a recent collection of papers, *Senses of Place: Senses of Time* (Ashworth and Graham 2005), focused on the senses and experiences of place and situated many of the chapters within a Northern Ireland context, including papers on collective memory, mapping meanings in cultural landscapes, identity and war, conflict commemoration amongst Protestants and the constitution of place identity in Northern Ireland. This current rich vein of geographically grounded work is of obvious relevance to any exploration of Long Kesh/Maze as a landscape. For example, Byronie Reid has recently explored the significance of place and identity in Northern Ireland with regard to the desire of the two communities to claim territory 'centred on absolute ownership and mutual exclusion' (Reid 2004: 103). In the case of the Long Kesh/Máze site, similar processes are part of the explanation of why there is little published material on the Loyalist experience of the site in comparison with the abundance of Republican material – the Republican movement has effectively claimed 'ownership' of the site. Reid has also more recently examined some of the interface, and thereby contested, areas of the Belfast cityscape through the work of a number of female performance artists, which has helped explore alternative means of experiencing a fundamentally male terrain (Reid 2005). In this respect the field of landscape studies, within archaeology as well as other disciplines, has shown an increasing concern with the significance of how landscapes are mutable, evocative and frequently contested. Landscapes play a significant and reciprocal role in the creation of identity and in political relations, and as such, no landscape can be considered inert and unaffected by the past, the present and the potential of the future.

As such the role of landscape archaeology is not simply to define the complex and constantly changing sequence of the Long Kesh/Maze prison landscape, but simultaneously to consider the relationships between the built environment, designed to control, and its everyday experience and use by those who lived within it (whether prison officers, Republican or Loyalist prisoners, or visitors). This point is emphasised by considering Neil Jarman's (1993) examination of how the landscape of Belfast in Northern Ireland was affected by the Troubles. Jarman usefully contrasts two different views of landscape: 'the view from outside' and the view from the ground in Northern Ireland, the latter of which acknowledges the greater complexities of the situation than the more simplified media-based constructions of the former (1993: 109). He observes how the landscape changes included military infrastructure that ranged from more obvious forms such as army bases and police stations to more subtle manifestations such as estate planning and the use of derelict buildings as shields in interface areas. It is in these details, not seen in the soft focus of distanced or mediatised perspectives, that the contradictions and complexities of political landscapes such as the Long Kesh/Maze site can be seen. As the site is only recently defunct there is a mass of information, through oral testimonies, official and unofficial documentation, artefacts and building remains that can be examined to attempt to get to some of these stories.

THE SIGNIFICANCE OF THE LONG KESH/MAZE LANDSCAPE

Due to its connection with the Troubles, the site of the Long Kesh/Maze prison holds a huge fascination and significance for the people of Northern Ireland as one of the principal icons of the Troubles. Its primary role was as a holding centre for Republican and Loyalist prisoners, with over 10,000 primarily paramilitary prisoners passing through its gates from its inception in 1971 until closure in 2000 (Purbrick 2004: 91). However, while the Troubles shaped the development of the Long Kesh/Maze landscape, the site has had a major impact on political, social and cultural aspects of life in Northern Ireland.

Politically, it cannot be doubted that the concentration of prisoners within such an institution provided an impetus for the politicisation of previously solely military organisations. This is most prominent within the Republican movements, who became involved in organised politics with the election of Bobby Sands as MP for Fermanagh and South Tyrone whilst he was on Hunger Strike in 1981. This show of mass popular support was an obvious impetus to the policy of 'ballot box and armalite', which has allowed Sinn Féin, the political wing of the Irish Republican Army (IRA), to grow to be the second largest party in Northern Ireland, with a 23.4% share of the vote in the elections in 2005.[1] Indeed, since the Belfast (Good Friday) Agreement of 1998, Sinn Féin has grown in popularity in stark contrast to their military wing, the IRA, who were officially 'stood down' in August 2005.

Socially, the Troubles have had a huge impact on Northern Ireland by increasing the segregation of working class communities as well as increasing the military infrastructure in hostile and interface areas. Here, the human impact of the Long Kesh/Maze landscape is witnessed by the sheer numbers of people who were interned or imprisoned, a phenomenon that particularly affected working-class communities. As Irish Republican politician, and Sinn Féin president, Gerry Adams mentions in the foreword to his biography:

> Almost twenty years have passed since Long Kesh was opened and through the years it has been a constant element in the lives of all the members of my family. On any one of the many days since then at least one of us has been in there . . . . Our female family members . . . have spent almost twenty years visiting prisons. Yet for all that, ours is a perfectly normal family, and we are by no means unique. (G. Adams 1990: 11–12)

Perhaps surprisingly, there are also cultural dimensions to the prison landscape's effect on Northern Ireland. One example is the large number of cultural artefacts made by prisoners for use both inside and outside of the prison, including wood carvings, leather work, painted handkerchiefs and paintings (Figures 2.3 and 2.4). Although the prison

Figure 2.3 Plywood Celtic Cross, undated. This artefact was confiscated from unnamed prisoners at Long Kesh/Maze when it was discovered that its filling contained hidden bullets. Held at the NIPS Museum, Millisle, Northern Ireland (photograph: Laura McAtackney).

Figure 2.4 Banner used for Orange Order-style marches by Loyalist prisoners.

authorities would not allow any items with a representation of the prison to leave its confines, many of these items had celtic, paramilitary or personal representations, and many contraband items were smuggled out. These artefacts are often still displayed in the homes of relatives and friends of those who were detained in the Long Kesh/ Maze. Outside the household environment, a number of more visible cultural developments occurred as a result of the Hunger Strikes of 1980–1981. First was the adoption of the wall mural painting tradition by Republicans as a means to provide public, mass support for the Hunger Strikes. Perhaps appropriately, Bobby Sands, the first Hunger Striker to die, is still a popular figure for many murals, especially on the occasion of anniversaries. In addition, mass protest marches were increasingly reutilised, after being abandoned due to the violence connected to the civil rights movement, and Bloody Sunday (30 January 1972) in particular, at this time. Therefore, the significance of the site can be seen in the ways in which the landscape of the Long Kesh/Maze has interacted with the wider cultural landscapes of Northern Ireland.

REPRESENTING THE LONG KESH/MAZE LANDSCAPE

To understand Long Kesh/Maze as a contemporary political landscape, one must examine the different ways in which it has been represented, imagined and enacted by communities. As a high-security zone, a classification that continues to exist despite the site being empty since 2000, any access by the wider community is highly controlled. Its visibility since the late 1970s/early 1980s derives from two sources: its continual depiction in the mass media, and its use as an image in wall murals. Through these means, the Long Kesh/Maze had, and still has, as much significance as an imagined landscape as it does as a physical entity.

The media representations of the site began to appear as a result of the growing interest in the prisoner protests, as both internees and inmates of the H Blocks, in the 1970s. The aerial image of the H Blocks became the predominant identification with the site for people who had never been resident at the site. Despite the significance of the Maze Compound/Long Kesh Internment Camp element of the site, its less obvious visual impact ensured that it was overshadowed by its larger and more visually startling H Block neighbours. The representation of the site through aerial images contrasted with the everyday experience of the site by those incarcerated there, who only experienced small parts of the landscape: prisoners had intimate relationships with their immediate surroundings but they were often swiftly and securely transferred with little ability to interact. As one prisoner recalled,

> I find it strange now because as you come into the circle from outside the H Block you were always rushed through the circle as quick as you can and now you can stand and you have access to any room you want. (D. Adams 2006)

Such experiences of the site were not accidental: the authorities were constantly in fear of breakouts – a fear which was confirmed as founded after the escape of 38 inmates in 1983.

The media images of the site came to dominate Republican representation of Long Kesh/Maze in wall murals (the local term used for the large-scale, public paintings often placed on the gable-end wall of terraced houses in working-class estates in Northern Ireland). Whilst they can be viewed as being 'collectively emblematic of the segregated and polarised working class areas of Belfast and elsewhere in Northern Ireland' (McCormick and Jarman 2005: 49), they are an important means by which to gauge opinions and reactions of the wider working-class communities in Northern Ireland to primarily political, social and military events. As has already been noted, the Republican wall mural tradition has a short pedigree that began as a means of protesting during the Hunger Strikes. It is therefore apt that representations of the Long Kesh/Maze are amongst the most prominent, and startling, examples of this tradition and contrast strongly with the majority of representations of prisons in the Loyalist tradition. As the more established wall mural tradition, it has been noted that Loyalists have been slower in incorporating recent politics and social comment than Republicans (McCormick and Jarman 2005: 51). Therefore, the representations of the site by the two different traditions can be used to succinctly illustrate the differences in the experiences of the two sets of prisoners, and their communities, with the landscape.

Neil Jarman's view that 'mural painting has developed into one of the most dynamic media for symbolic expression in the north of Ireland' (Jarman 1998: 2) is significant in examining the Long Kesh/Maze as an imagined landscape beyond the confines of the prison. Republican representations of the site are most prominent in their number, longevity and, most importantly, location. Jarman has highlighted the fact that the locations of these murals are significant factors to be considered in the wider context and possible replication in the media (Jarman 1998: 3). Therefore, the more visually prominent the site, the more significant and representative of wider society the murals are likely to be. It should be noted that the locations of many of the murals representing the Long Kesh/Maze in Republican areas are longstanding and in prominent locations. Within traditional Republican heartlands of West Belfast, these images tended to be concentrated on the busiest section of the road, at the junction between the Falls Road and the Whiterock Road, and were both stylised in form but innovative in treatment. Of particular note is an unusual mural (in that it is not located on a wall) on a free-standing wooden structure (Figure 2.5). It is placed on a prominent location on a cleared area of ground on the Falls Road and the 'H' figure is filled with images of the dead Hunger Strikers of 1981, and is flanked by Republican flags. This is further personalised with a quotation from the writings of the first Hunger Striker to die in 1981, Bobby Sands. This is a very popular representation of Long Kesh/Maze for Republicans

Figure 2.5 Republican H Block mural flanked by Irish Republican flags.

through the frequency of the use of the aerial view of the H Blocks and the focus on the human impact of the Hunger Strikes, especially using images of the 10 young men who died on hunger strike.

This is in contrast to the murals depicted in Loyalist areas, which are smaller in both number and types of representation. When the prison is depicted it is often as a backdrop to more general paramilitary images, namely of hooded gunmen surrounded by paramilitary insignias. The representations of the prison are also more generic, with the aerial view of the H Blocks largely absent from these representations, and the most common image being that of a watchtower with a surrounding wall and barbed wire. In this respect, these less specific images are not so much representative of the unique conditions of the Long Kesh/ Maze; rather they are reminiscent of prisoner of war camps. This connection with war is most easily understood with regard to another common mural theme of Loyalist paramilitaries: World War I, and in particular the slaughter of Ulster soldiers at the Battle of the Somme (1916). The connection made between Loyalist paramilitaries and the Somme is most evident in the murals found within the nissen huts, which feature them prominently. This therefore indicates that the relationship with the site was very different not only for Loyalist and Republican paramilitaries but also for the communities from which they were drawn from and which commemorated them.

An important aspect of the experience and interaction of the wider public with the prison landscape of Long Kesh/Maze is its imagined quality. This connection to an imagined landscape was necessitated by a lack of easy access and has been mediated both externally and internally to working-class communities in Northern Ireland in particular. The representation of the prison landscape that centred on the aerial view of the H Blocks and ignored the previous and simultaneous operation of the Long Kesh Internment Camp, was perpetuated by the media. Simultaneously, internal to many working-class communities these images of the prison were further mediated by the mural painters, many of which are commissioned by local paramilitary and political organisations. The manifestations of these imagined landscapes differ considerably depending on whether the community is Loyalist or Republican. However, this does not explore how the site was actually envisaged or experienced in its immediate landscape context and therefore an examination of the site at a more intimate level must be undertaken.

LANDSCAPE, POLITICS AND LOCAL CONTEXT

As well as connecting with the wider cultural landscapes of Northern Ireland, the Long Kesh/Maze site has a more immediate cultural landscape that is telling of the politics of its setting. The Long Kesh/Maze is set in a greenbelt zone, with very few surrounding buildings and associated infrastructure. Such a location was obviously desirable as the majority of the prisoners were constantly trying, and indeed saw it as their duty, to escape. During the initial period of occupation as a prison, when the prisoners were held in nissen huts with shallow foundations and only tiles to prevent them from burrowing out, there were constant attempts to construct tunnels. As I was told by a former inmate at the site:

> You see the escape attempts, there was that many tunnels built in these grounds, in these compounds that even prison officers fell through them as they were walking around on patrol.... See if you walk around Cage Four there must be about 20 tunnels. (M. Meehan 2006)

In choosing such a site the authorities were attempting to ensure that any escapees had as little cover or potential help from sympathisers as possible. That is not to say, however, that Long Kesh/Maze was completely isolated. As stated earlier, the prison also had an accompanying army base, known as Silver City, although there is now very little trace of this because the army eradicated any remnants of the site once it fell into disuse. Although the two elements were not officially connected, the stationing of the British Army base directly adjacent to the prison was evidently intended to coerce the prisoners into compliance under the constant threat of force. During riot conditions, the prison relied on personnel from the adjacent army base to help quell dangerous situations before reinforcements arrived. Furthermore, ex-prisoners talked about the constant interactions between the two sites that were of a more subtle nature and ensured that the prisoners were constantly aware of the army base presence:

> ... you see in the huts, y'know the corrugated tin? They used to throw bread up at two or three o'clock in the morning and you see when first light came the pigeons and seagulls would come down onto the top of the tins squacking and squacking, that was them.... All them wee stupid, stupid things, y'know, to annoy ya. (M. Meehan 2006)

The surroundings were also of significance in reflecting the contemporary, political nature of the choice of site for such a high-profile prison. Although the site is to some extent isolated, it was also highly visible, in that it can be easily seen from the M1 motorway, particularly when travelling to Belfast. The M1 motorway is the most prominent arterial road in Northern Ireland, connecting Belfast, the capital of Northern Ireland, with Dublin, the capital of the Republic of Ireland. Cars travelling to Belfast from the south and west passed the lights of Long Kesh/Maze along their way. The prominence of the site is evident in Gerry Adams' biography, *Cage Eleven*, which is preceded by a poem that concentrates on the visual impact of Long Kesh/Maze, 'On the Way to Belfast' (1990: vii):

> From afar
> The lights of Long Kesh
> Make a false dawn
> In the night sky.

By locating the prison at this site, with its menacing walls, watchtowers and wires standing unabashed and fortress-like for all to see, this placement could be read as an act of defiance by the authorities. By placing the prison in such a prominent position, the authorities were defying the paramilitaries and their sympathisers by showing that they could maintain control of the prisoners in a highly visible location.

It is clear that the location of Long Kesh/Maze prison shaped how the site was perceived by its inmates, by the authorities and by the general public. The site can be considered as isolated due to the lack of associated infrastructure and its placement in a greenbelt area. To this extent the authorities would have a greater ability to apprehend those prisoners who escaped, especially those who attempted to tunnel their way out, as there was little surrounding coverage to hide them. However, the site was also simultaneously very visible in its location close to the main arterial road in Northern Ireland. This enabled three very different experiences of site: as an isolated prison site for those held inside it, as an area with little coverage to allow easy detection of escapees but also as a statement, readily received by the public, of the ultimate control by the authorities. These different experiences of the wider context of the prison landscape are also mirrored in the experiences of the site at a more intimate level by the prisoners who were held there throughout its 30-year occupation.

DIFFERENT EXPERIENCES OF LANDSCAPE

As already stated, the Long Kesh/Maze prison site encompasses a collection of diverse building types used across different, and sometimes overlapping, periods that come together as a cultural landscape. As such this landscape was experienced differently by people connected to the site depending on a number of factors, including their reason for being on the site, their role within their affiliated group, the length of time for which they were intimately associated with the site, etc. One of the major factors in the experience of the site is whether the person was imprisoned, or at the site through employment or visiting a prisoner. Prisoners' experiences could be vastly different depending on whether they self-identified as a Republican, Loyalist or ODC ('Ordinary Decent Criminal') prisoner. If not a prisoner, then their experience depended upon whether they were a prison officer or a visitor, or if they fulfilled an auxiliary role such as cleaner, workman, etc. To this end, in this section I shall examine the experiences of Republican ex-prisoners of different periods through interviews that I conducted on-site in January 2006. These were with Martin Meehan, who had been interned on Long Kesh Internment Camp on a number of occasions throughout the 1970s and was placed in the H Blocks about 1980–1981, and was released in February 1994 (M. Meehan 2006); Dominic Adams, who came into the H Blocks in 1986 and was released in 1991 (D. Adams 2006); and Patrick Wilson who was imprisoned at Long Kesh in 1987 and was released at Christmas 1990 (P. Wilson 2006).

Within these groups the period of contact is also of significance, as the prisoners had totally different experiences depending on whether they were there during the internment period, the protest period connected to the criminalisation process or the later period of the ascendancy of the prisoner. For example, by the late 1970s the increasing protests against criminalisation led to prison officers being regarded as legitimate targets by both sides, and a number were targeted, and even killed, whilst off-duty. Therefore, this affected the relationships with, and was most obviously manifest in the staff's treatment of the prisoners. Perception of this change in treatment was frequently mentioned during the oral testimony of one ex-prisoner, who had been at the site from the early 1970s for many periods until the 1990s, stating:

> The one thing that struck me was some of the prison officers when they were on the cages were just like ordinary fellas, there was no big deal, no hassle but then they were transferred down to the H Blocks, they were Jekyll and Hyde, you know they were aggressive, arrogant. (M. Meehan 2006)

However, the relationship between prison officers and prisoners had changed completely by the mid- to late 1980s, with the growing ascendancy of the prisoners. Dominic Adams, a prisoner from 1986 to 1991, noted:

> It's funny how times change, cos you see around about '89, '90 when Republicans were more or less running the show, you see when the screws still had to check they would come down on his tiptoes and then lift the flap dead easy cos they didn't want to waken ya, you know. (D. Adams 2006)

The on-site oral testimonies of ex-prisoners indicated how their experiences of the landscape depended on the political situation both within the prison and in the wider context of Northern Ireland. The three interviewees had vastly different experiences of the site. The oldest and most prominent interviewee, Martin Meehan – imprisoned from the inception of the detention centre intermittently until 1994 – recalled much harsher realities of life in the prison, whilst the other two interviewees, who were not imprisoned until the late 1980s and for shorter periods, reported having had an easier time:

> I always consider myself I went into the prison at an opportune time for want of a better term because the blanket was over, the segregation campaign was over, the people who where there before me had won all the conditions that I was walking into. (D. Adams 2006)

While Meehan had a greater understanding of the interconnection of the site, the others were largely ignorant of the layout and usage of the compound element, as they had only ever known the H Blocks whilst they were prisoners. When asked about the differences between the two manifestations of the prison, Meehan was treated as more knowledgeable of the prison's history, and it was notable that a great amount of respect was afforded to him as someone who had witnessed, and suffered, more than they had.

Therefore, the experiences of the prisoners were influenced by the limitations upon the interaction that they had with some parts of the prison landscape. To ensure that strict security was not broken, and perhaps (unsuccessfully) to deter escape attempts, the prisoners were given very little time to get acquainted with elements outside of their actual living quarters. One particularly poignant moment occurred when the interviewees were brought to the hospital as part of the guided tour whilst conducting oral testimonies. This had a noticeable impact on all the interviewees as this was the building that the hunger strikers had died in, and one of those present had never had access to the building before:

> I mean standing here and thinking of the people who came in here and died in here whatever ... a lot of the conditions that we were experiencing at the time are the result of the people who died here that came before then. It is a poignant moment for me, I have never been here. (P. Wilson 2006)

Therefore, it can be said that even within the same paramilitary organisations there was no single experience of the site. This was true for many reasons, including the time period and area of incarceration, a prisoner's esteem within the Republican movement and his treatment by the authorities, as well as other more individual and personal factors. Further research needs to be done with the ex-prisoner community of Loyalist paramilitaries and prison officers in order more fully to discover their experiences of site. It has been shown that Loyalist and Republican prisoners had very different experiences of the site, reflecting their differing relationships with the authorities and their connection to the site now that it is closed (McAtackney 2007). It should be noted that it is difficult to gain access to Loyalist ex-prisoner groups to conduct this research, as at the time of writing they are far less forthcoming than their Republican counterparts. However, it is clear that any attempt to re-create how the site was experienced needs to take into account these multiple perspectives in a subjective approach, as there was not, nor could be, any one, definitive experience of this prison landscape.

POLITICS AND LANDSCAPES OF THE FUTURE

The experience of the landscape of Long Kesh/Maze both through its representation as an image and an idea, and the particular memories of those most intimately connected to it, would be an important place to start when considering the future prospects of the landscape. However, the contentious nature of this landscape, due to its close identification with the Troubles, has created two polar reactions amongst the public: that to *retain* and that to *remove*. Victims, their families and others less connected to the site, such as the middle classes and their political representatives, have all suggested that the site should be converted into something unconnected with its past. An opposing campaign has been fronted mainly by the Republican movement and their prisoner groups, resulting from disconcertion with the fast-paced removal of security elements, and has centred on the future of the Long Kesh/Maze prison site. Since its closure in 2000 the site has evolved into a battleground between the advocates of retention and destruction. This debate has grown more prominent since the all-party Maze Consultation Panel (MCP) published its final report, *A New Future for Long Kesh/Maze*, in February 2005 (MCP 2005). This document has been highly significant in its attempts to appease both sides in the retention/destruction debate, with most of the site being replaced but a prominent element retained as part of an International Centre for Conflict Transformation. The proposal is also significant for its attempts at heritage creation and imposing a new identity on the site as 'a symbol of confidence and hope for the people of Northern Ireland' (MCP 2005: 31).

An updated 'final' report by the same panel was issued with comparatively little publicity in May 2006, Maze/Long Kesh: Masterplan and Implementation Strategy. Executive Summary (Masterplan Consortium 2006). The differences between the two reports are very telling, especially in the changes in emphasis. The new document responded to criti- cisms that the initial reports included heavy-handed approaches to controlling public interaction with the site and the creation of heritage (McAtackney 2005: 10–16). The second document played down the significance and uniqueness of the site, with infrequent mention of its prominent past life. Instead, more concerns with generic issues, such as the potential for the site to become an economic and ecological prototype for post-conflict Northern Ireland, were expressed (Masterplan Consortium 2006). The report also emphasised the transformational qualities of the proposals, seeking to look forward rather than concentrating on the past history of the site. One phrase in particular is frequently repeated: the idea of the landscape 'provid[ing] a physical expression of the ongoing transformation from conflict to peace' (Masterplan Consortium 2006: 2, 5). Those prison buildings that are to be retained are presented in positive terms: as being able to 'provide a range of activities dealing with conflict transformation processes with a special focus on the history of the peace process' (Masterplan Consortium 2006: 9). This change in emphasis between the two reports – the latter being more generic and vague, with little acknowledgement of the overwhelmingly darker or contested aspects of the site – indicates how contemporary politics are attempting to direct the experience of future generations within this landscape.

There is clearly no easy solution to how the prison landscape is to be experienced in the future. The Maze Consultation Panel has obviously attempted to create a proposal that appeases all sides and ensures official control of access to and exploitation of the site. In April 2007, the walls of the site were demolished at the start of a 'fully sustainable' demolition process which, according to the Office of First Minister and Deputy First Minister, will involve 'every brick be[ing] crushed and turned into concrete' (Belfast Telegraph 2 April 2007). At the time of writing, it is unclear how much of the landscape will remain in the future. However, in these plans, which will possibly make way for a new sports stadium, it is clear that there is an inability to view the extensive prison landscape as an opportunity to learn from past experiences, to use the material resources of the past, though they may be complex and dark.

CONCLUSIONS

Examination of the contemporary politics of landscape is of particular significance to Northern Ireland at this time. Due to the implementation of the Belfast (Good Friday) Agreement in 1998, the demilitarisation process has ensured that many prominent previous security sites have become areas of contestation. The lack of consensus, both public and political, about what should happen to the security infrastructure has increasingly focused on Long Kesh/Maze prison, with it becoming a complex political issue that has resulted in much cross-party consultation and much controversy over alternative proposals for its future. Long Kesh/Maze is the obvious landscape to be highlighted in these political negotiations. As a site in which many of the most dangerous and notorious paramilitaries active during the Troubles were detained, it is a highly emotive site for ex-prisoners, their communities, their victims and, in fact, the whole population of Northern Ireland. Materially, for the moment at least, it is simultaneously highly visible, and yet remains a high-security zone without any public right of access, and therefore interests and tantalises the public more than any other security site in the province.

Due to the complex, contested and contemporary nature of this site it is clear that no objective examination and interpretation is possible, nor should it be attempted. As the ex-prisoners' oral testimonies from the site indicate, even within the same organisation the experiences of this landscape can be very different depending on a large number of variables. One must accept that in such an emotive and contentious situation the immediate and associated landscape will be experienced, envisaged and remembered in diverse ways. Long Kesh/Maze is not simply an inert area where recent historical events happened; there is constant interaction between the past, present and future, between the landscape and the people who inhabited it.

The landscape of Long Kesh/Maze will continue to be a contentious site whilst the ongoing wranglings of the peace process continue in Northern Ireland. To this extent the site has been as much a victim of recent political uncertainties as of the deeply felt sensibilities about it. It is clear that the decision-makers do not wish to engage with the more controversial elements of this site, and as such their proposals have become increasingly generic and anodyne. Indeed, for all the potential that the site holds regarding its ability to encompass, explain and demystify the complicated conflict of the past four decades, there is an understandable fear of confronting and learning from those darker and more unsavoury elements of the recent past. However, it can be suggested that Long Kesh/Maze is important in that it highlights that political attempts to control the creation, direction and focus of public remembrance of the recent past are not accepted uncontested. The debates about the future of this site show that on a broader level the will of political parties and the state are not always accepted without question by the populace. Furthermore, it is clear that not all official attempts to control public interactions with place are successful. As Ashworth and Graham have stated, official heritage cannot surpass nor supplant unofficial heritage, and although it can direct experiences of site, it cannot ultimately dictate what we bring, or what we take, from any interaction with place (2005: 5).The changing landscape of Long Kesh/Maze is a telling example of how the contemporary political climate can attempt to affect and control how a site is experienced, interacted with and potentially remembered. However, this manipulation is still confined by the materiality of the site. Only time will tell how the public will react to and negotiate the remains, and the 'afterlife', of this political landscape of the contemporary past.

NOTE

1. See http://www.ark.ac.uk/elections/ (consulted 24 July 2006)

REFERENCES

Adams, D. 2006. Interview by Laura McAtackney with Dominic Adams, conducted 11th January 2006 at Long Kesh/Maze prison site with members of An Coiste na-Iarchimí (Republican Umbrella Ex-Prisoner Organisation).

Adams, G. 1990. *Cage Eleven*. Dingle: Brandon Press.

Amstutz, M. 2005. *The Healing of Nations: The Promise and Limits of Political Forgiveness*. Oxford: Rowman & Littlefield.

Ashworth, G.J. and B. Graham 2005. *Senses of Place: Senses of Time*. Aldershot: Ashgate.

Beresford, D. 1994. *Ten Men Dead*. London: Harper Collins.

Burns, I.M. 1971. I.M. Burns to P. Leyshon Esq., 23 September 1971. (The National Archives, UK: CJ 4/449 TNA, Deployment of prison officers from Great Britain at Long Kesh [Maze prison]).

Clarke, L. 1987. *Broadening the Battlefield: The H Blocks and the Rise of Sinn Féin*. Dublin: Gill & Macmillan.

Clarke, L. 2005. Introduction to Castles of the North: A Photo Essay by Jonathan Olley. *Cold Type Magazine*. http://www.coldtype.net/castles/Castles.LR.pdf (reprint from the [London] *Sunday Times Magazine*, 9 April 2000).

Coogan, T.P. 2001. *The Billy Boy: The Life and Death of the LVF Leader Billy Wright*. Edinburgh: Mainstream.

Galliher, J.F. and J.L. DeGregory 1985. *Violence in Northern Ireland: Understanding Protestant Perspective*. Dublin: Gill & Macmillan.

Gardiner, Lord. 1975. Report of a Committee to Consider, in the Context of Civil Liberties and Human Rights, Measure to Deal with Terrorism in Northern Ireland. London: HMSO.

Hennessy, Lord. 1984. Report of an Inquiry by HM Chief Inspector of Prisons into the Security Arrangements at HM Prison, Maze Relative to the Escape on Sunday 25th September 1983, including Relevant Recommendations for the Improvement of Security at HM Prison, Maze (Hennessy Report). London: HMSO.

Irish News 2005. First Steps to Normalisation (28 August 2005).

Jarman, N. 1993. Intersecting Belfast. In B. Bender (ed) *Landscape: Perspective and Politics*. London: Berg.

Jarman, N. 1998. Painting Landscapes: The Place of Murals in the Symbolic Construction of Urban Space. In A.D. Buckley (ed) *Symbols in Northern Ireland*. Belfast: Institute of Irish Studies, pp. 81–98.

Leone, M.P. 2005. *The Archaeology of Liberty in an American Capital: Excavations in Annapolis*. Berkeley: University of California Press.

Masterplan Consortium 2006. Maze/Long Kesh: Masterplan and Implementation Strategy. Executive Summary: Final Report. Belfast: HMSO.

Maze Consultation Panel. 2005. A New Future for Long Kesh/Maze. London: HMSO.

McAtackney, L. 2005. Long Kesh/Maze: An Archaeological Opportunity. *British Archaeology* 84: 10–16.

McAtackney, L. 2007. The Negotiation of Identity at Shared Sites: Long Kesh/Maze Prison Site, Northern Ireland. Electronic Proceedings of Cultural Landscapes in the 21st Century: The 10th International Seminar: An Inter-Congress of the World Archaeological Congress (University of Newcastle upon Tyne, April 2005). http://www.ncl.ac.uk/unescolandscapes/files/MCATACKNEYLaura.pdf (Consulted 10 April 2007).

McCormick, J. and N. Jarman 2005. Death of a Mural. *Journal of Material Culture Studies* 10: 49–71.

McKeown, L. 2001. *Out of Time: Irish Republican Prisoners, 1971–2000*. Dublin: Beyond the Pale Publication.

McLaughlan, D. 2007. Interview by Laura McAtackney with Duncan McLaughlan, ex-Governor of Long Kesh/Maze prison, on 6 August 2007.

Meehan, M. 2006. Interview by Laura McAtackney with Michael Meehan, conducted 11th January 2006 at Long Kesh/Maze prison site with members of An Coiste na-Iarchimí (Republican Umbrella Ex-Prisoner Organisation).

Purbrick, L. 2004.The Architecture of Containment. In D.Wylie (ed) *The Maze*. London: Granta, pp. 91–110.

Reid, B. 2004. Labouring Towards the Space to Belong: Place and Identity in Northern Ireland. *Irish Geography* 37(1): 103–113.

Reid, B. 2005. 'A profound edge': Performative Negotiations of Belfast. *Cultural Geographies* 12: 485–506.

Tilley, C. 1994. *A Phenomenology of Landscape: Places, Paths and Monuments*. Oxford: Berg.

Tilley, C. 2006. Introduction: Identity, Place, Landscape & Heritage. *Journal of Material Culture Studies* 11: 2–32.

Wilson, P. 2006. Interview by Laura McAtackney with Patrick Wilson, conducted 11th January 2006 at Long Kesh/Maze prison site with members of An Coiste na-Iarchimí (Republican Umbrella Ex-Prisoner Organisation).

Wylie, D. 2001. *The Maze*. London: Granta.

Yamin, R. and K.B. Metheny 1996. Preface: Reading the Historical Landscape. In R. Yamin and K.B. Metheny (eds) *Landscape Archaeology*. Knoxville: University of Tennessee Press, pp. xiii–xx.

CHAPTER 3

Facing Many Ways: Approaches to the Archaeological Landscapes of the East African Coast

Sarah Croucher

INTRODUCTION

This chapter explores past and present approaches to the archaeological landscape of the East African coast, and the ways in which conceptions of this landscape have been tied to notions of the past identities of coastal residents. The area in question is often termed the 'Swahili coast', being home to the Swahili, a cultural group located along a narrow strip of the East African coastline stretching from Mogadishu in Somalia in the north to Sofala in Mozambique in the south. Many offshore islands, including the Comoros and the north of Madagascar, are also included within the geographical extent of the Swahili region. Zanzibar – which comprises the two islands of Unguja and Pemba – also falls within this grouping (Figure 3.1).

While interpretations of the landscape have been central to the way in which archaeological sites along the coast have been understood, there is no tradition of 'landscape archaeology' as such in East African archaeology. No coherent definition of landscape archaeology exists within this regional tradition. Nevertheless, the distinctive archaeological studies of the spatial use of houses, towns and the wider rural landscape that have developed in the region represent, I suggest, a distinctive, 'vernacular' landscape archaeology that focuses especially upon the interpretation of the past identities of coastal residents.

My aims in this chapter are twofold. Firstly, I shall review how conceptions of landscape have been involved in interpretations of the identities of past cultures along the coast. Secondly, I shall provide a detailed case study from my own research into the landscapes of clove plantations on 19th century Zanzibar, showing how the complexity of plantation resident identities can be interpreted through a landscape approach. I hope not only to explain how the theoretical approach that I have adopted to later historical landscapes on Zanzibar builds upon

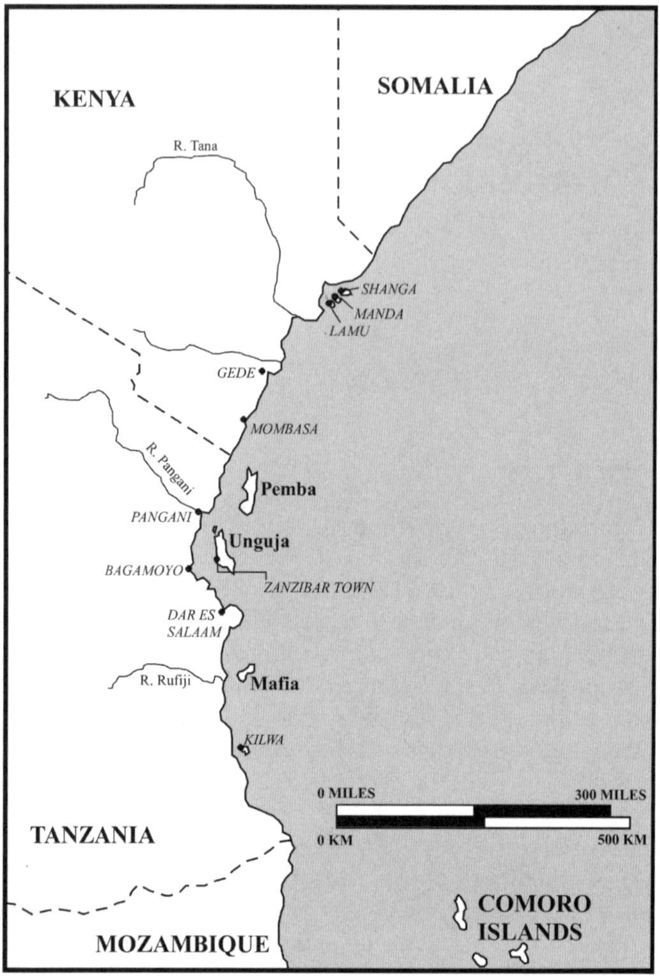

Figure 3.1 Map showing the East African coastline and some key Swahili archaeological sites.

the region's previous archaeologies of landscape, but also to provide a case study that might inform studies focused on earlier periods in the region, and which might contribute to the further development of landscape archaeologies along the East African coast.

PREVIOUS APPROACHES TO SWAHILI LANDSCAPES

Since the early 20th century, most archaeological interest in the East African region has been drawn to the stone-built towns (often called stonetowns) which lie along the coast and date from the 8th to 18th centuries AD. The finely built stone houses and mosques at such sites have been seen as characteristic of Swahili urban landscapes; combined with shared ceramic traditions (Chami 1998) and a shared linguistic history (Spear 2000) they form the historical Swahili landscape traditionally studied by archaeologists. In the first half of the 20th century, these sites were interpreted as evidence of a mercantile colonial culture brought to the coast from the Middle East. At this time, archaeological fieldwork centred only on visible stone-built remains; no survey was carried out in order to find smaller, non-urban sites. Thus 'the Swahili landscape' was conceptualised as comprising only these stonetowns, with everything beyond the extent of the stone architecture being non-Swahili. This coastal landscape was seen as foreign to Africa: a slim strip of civilisation looking outwards to the Indian Ocean. The towns were understood as being different from a perceived African landscape, that is, as being foreign implants from Arab colonists. Fleisher (2003: 37) identifies this attitude in the late-19th century writings of the explorer Sir Richard Burton as he encountered the ruins of Kilwa, one of the most impressive ruined stonetowns of the coast:

> The disjuncture between the past and present, in Victorian eyes, was evidence of discontinuity; surely, they argued, the local inhabitants, living in earth-and-thatch houses, could not have been responsible for such grandeur. Richard Burton's description of Kilwa follows this logic: his description of the 'remarkable' ruins there provides a marked contrast to the contemporary inhabitants, many of whom were dying of cholera. His surprise at such grand architecture and evidence of 'civilisation' is explained by invoking external influences – from Persia. (Fleisher 2003: 37)

This view dominated the perception of stonetowns well into the second half of the 20th century, when systematic archaeological investigations began on major sites such as Kilwa (Chittick 1974), Manda (Chittick 1984) and Gede (Kirkman 1954). These excavations focused upon the stone-built structures of the towns – the mosques and large houses. They presented such urban landscapes as the product of Arab colonial settlers who intermingled with local African populations to produce the Swahili civilisation (Chittick 1984: 220). The very stone fabric of these towns was seen to embody the Islamic values of an urban culture, setting them apart from the continent of Africa within which they were physically located. Within this phase of research the urban centres of the Swahili were seen as tiny nodes of civilisation within a vast uncivilised continent.

Mainland Tanganyika gained independence from British colonial rule in 1961, followed by Kenya and Zanzibar in 1963. In this new, postcolonial setting, Swahili scholarship began to look for the Indigenous origins of Swahili society (see the detailed discussion of the development of Swahili archaeology provided by Jeff Fleisher [2003: 35–61]). Such research was not confined to field archaeology but also included research into documentary, oral and linguistic histories of the Swahili (e.g. Allen 1974, 1993; Nurse 1983; Nurse and Spear 1985). Mark Horton's excavations at the site of Shanga in Kenya dramatically changed understandings of Swahili urban landscapes (Horton 1987, 1994, 1996). The results of Horton's fieldwork clearly demonstrated the continuous development of the town from early structures of timber, thatch and earth in the 8th century AD to later coral-rag-and-lime houses appearing suddenly in the early 14th century AD (Horton 1996: 235–242). Horton suggested that this meant that the stone townscape, which had been so frequently viewed as a foreign landscape, was a 'comparatively recent development' (Horton 1996: 242), developing directly from the earliest wattle-and-daub structures.

Such investigations firmly demonstrated the African origins of the Swahili towns, particularly when coupled with ceramic data (e.g. Chami 1998). Following Horton's investigations of Shanga, further studies have attempted to extend understandings of how such towns functioned within the wider Swahili landscape. In a recent review of the field, Adria LaViolette and Jeff Fleisher (2005: 341) have defined three principal models which archaeologists have used to explain the development of urban centres on the East African coast. These models have shaped the ways in which the Swahili landscape as a whole is understood. The first is that towns developed in order to administer a complex settlement hierarchy of functional centres and to produce goods for long-distance trade (as suggested by Kusimba 1999). The second is that towns were relatively independent mercantile cities that functioned as production and market centres: the kind of model favoured by Horton (1996: 412; Horton and Middleton 2000: 72). A third model suggested that towns formed due to competition between coastal elites who then combined the community ritual of Islam with the power of exotic goods to create centres of power and authority (Wright 1993).

In all of these models, the urban landscape is seen as a functional centre, controlling territory, production and markets or ritual life (LaViolette and Fleisher 2005: 341), within a wider, largely imagined, rural village landscape around the town. But while these interpretations allow for a range of understandings of the urban landscapes of the Swahili, they tend to view urban and rural landscapes as separate entities (LaViolette and Fleisher 2005: 333). This tendency is most explicit in John Middleton's model of clearly separate categories of 'stonetowns' and 'country-towns' (Middleton 1992: 58). Reappraisal by Horton and Middleton (2000: 13) suggested that 'patricians' towns' (i.e. stonetowns) and 'commoners' towns' were merely opposite ends of a continuum of Swahili settlement. In this view, each unique Swahili town could be placed at some point along this scale. However, here the towns of the coast were still viewed as separate from the African landscape around them: areas of civilisation in a wider African context which is (as defined by the Swahili) uncivilised. Thus, the patricians' town as the seat of an *umma*, a community of Muslim believers, was centred on religious buildings (Horton and Middleton 2000: 124). Such an understanding paralleled the perspectives of Henry Wright (1993), who argued that the emergence of towns on the coast was due to such ritual functions rather than mercantile concerns. Whilst all of these models highlight different causal factors in the development of towns on the coast, in their conceptions of the actuality of the Swahili landscape they all still tend to present Swahili towns as 'foreign', and as separate from the wider African landscape. In particular, Islam is emphasised in all the models as crucial to the emergence of urban landscapes (LaViolette and Fleisher 2005: 324), supporting perceptions of the towns as non-African spaces.

Understandings of Swahili landscapes have been reappraised recently, driven largely by two significant changes in fieldwork practices. Firstly, systematic landscape surveys in coastal areas have been carried out with the aim of recognising non-stone-built towns, villages and hamlets (e.g. Fleisher 2003; Fleisher and LaViolette 1999; LaViolette et al. 1999; Wynne-Jones 2005a). Secondly, on Pemba excavation strategies at the Swahili towns of Pujini (LaViolette 1996) and Chwaka (LaViolette 2000) have been designed to investigate how large these towns were when wattle-and-daub sections are included, and the relationships between wattle-and-daub to stone sections of the same settlement. Both these changes in field practice have begun to examine the archaeology of populations that can be termed the 'hidden majority' of the Swahili (Fleisher and LaViolette 1999: 88), since in the past the residents of non-stone buildings on the East African coast would have outnumbered those of stone buildings. Such investigations of a wider range of archaeological remains and sites, rather than simply the highly visible remains of coastal stonetowns, can serve to emphasise how urban centres developed in the context of broader human and material landscapes.

These new approaches represent a radical development in the landscape archaeology of the region. Our understanding of the archaeology of the Swahili landscape has undergone significant reappraisal. These studies contextualise, rather than negate, the impact of Islam on the Swahili, by exploring landscapes as part of wider, complex networks of interactions with the broader African landscape around them. Where projects such as Horton's fieldwork at Shanga (Horton 1996) went some way towards recognising Swahili towns as an Indigenous African development, contemporary Swahili archaeology is moving ever further in this direction. As the archaeologist's conceptions of the Swahili landscape ceases to end at the limits of stone-built sections of towns, a wider population becomes visible. These people were also probably not the merchants who have so dominated interpretations of Swahili populations prior to recent years. Instead these newly recognised areas of towns, as well as the non-urban settlements, may have been home to settled agrarian and craft-producing communities (Fleisher 2003: 344–345).

The new fieldwork has identified and recorded a wide range of new sites in the Swahili landscape. Thirty-three sites dating from the 8th to 18th centuries AD have been identified through survey in northern Pemba (Fleisher 2003: 132), and 66 have been identified in the Kilwa region (Wynne-Jones 2005b). However, the results of this fieldwork are disappointingly still very often interpreted in frameworks that conceive of the Swahili landscape in excessively functional terms, concentrating purely on the functions and economics of sites rather than considering how the landscape was comprehended by those who were resident on these sites in the past. For example, when considering the migration of countryside populations into urban centres, Fleisher simply asks 'what functions were emerging stonetowns providing for the surrounding countryside?' (2003: 415). While important steps have been taken, as sketched above, towards recognising 'still barely visible villagers and urban-dwelling commoners' (LaViolette and Fleisher 2005: 344), in Swahili archaeology the complex, human dimensions of relationships between people and their changing landscapes remain little explored.

In this context, the development of broader archaeological appreciations of the social landscapes of the Swahili is increasingly important. An important contribution here is Stephanie Wynne-Jones' (2005a, 2005b) investigation of the development of urbanism in the Kilwa region of the southern Tanzanian coast (Figure 3.1). Rather than simply analysing the different functions of various settlements in the region, Wynne-Jones also seeks to understand how residents of the area might have experienced these differing forms. The town of Kilwa Kisiwani might appear at first to have represented a relatively fixed point in the landscape, with the short-lived village sites identified through archaeological fieldwork contrasting with the permanence of the stonetown. However, Wynne-Jones considers how this landscape would have appeared to its residents:

> Would this relative permanence have been evident at the timescale of lived experience? A wattle-and-daub settlement that appears transient at an archaeological timescale can in fact exist throughout a lifetime: it is

> thus, in effect, permanent. During the course of survey in the Kilwa region we visited the homes of numerous local residents who had lived their entire lives in a particular wattle-and-daub house. Normally, these had been extensively repaired, renovated and extended, but remained the same essential structure. (Wynne-Jones 2005b: 68)

Wynne-Jones' argument suggests that in order to understand such urban sites and their regional surroundings attention must be paid to 'multiple scales, regions and timeframes' (Wynne-Jones 2005b: 69): examining landscapes at a range of geographical and temporal registers that include those of lived experience.

In developing more nuanced approaches, it is important to remember that Western notions of 'landscape' are intimately tied in with the historical processes of European colonialism. Some representations of landscape form part of a 'Western gaze' (Bender 1999: 31). Through the processes of mapping out archaeological sites on the East African coastal landscape, our attempts to understand past Swahili populations are always grounded in a particular Western perspective, producing ordered subjects within this past who conform to our functional rules of settlement. As Julian Thomas has argued, 'The modern west has developed a particular and distinctive way of looking which is deployed against place as it is deployed against other phenomena' (Thomas 1993: 20).

In this discourse, a fundamental dichotomy between natural and cultural landscapes is presented. In conventional archaeological studies of Swahili settlements, such a division is apparent in the distinction between urban and rural sites. The urban culture of the Swahili has often been viewed in sharp opposition to the natural, rural setting of Africa around it. Critiques of such sharp divisions by scholars such as Wynne-Jones (2005a), who recognise that these divisions may not have been apparent within the lived world of the Swahili, are thus important developments in postcolonial archaeological practice. The recent trends in Swahili archaeology that have sought to make visible all elements of past coastal communities discussed above have remained driven largely by a processual approach: questions have been asked regarding the functions of settlements, and the economic relationships between different sites (Fleisher 2003: 104). Such research goals continue to be driven by intellectual concerns formulated outside of Africa that define which elements of the Swahili past are worthy of investigation. Using archaeology to integrate non-Western, local views of landscape into studies of the Swahili remains an unsolved problem, and is an issue I shall return to later in this chapter. Before doing so, I shall turn to my own archaeological research on 19th century clove plantations in order to explore an alternative approach to investigating identities through the landscape of the East African coast.

ZANZIBAR IN THE 19TH CENTURY

The periods under investigation by archaeologists working on the East African coast commonly terminate at or around the beginning of the 16th century AD (e.g. Fleisher 2003: 24). This is the period marked by the first significant European colonial intrusions to the region. The Portuguese were the first colonial rulers of the coast, beginning with the arrival of Vasco da Gama in 1499 (Gray 1962: 30). From the 16th and 17th centuries small fortified trading posts were established by the Portuguese at coastal locations such as Fort Jesus in Mombasa (Gray 1962: 41). Even at its peak, however, the Portuguese presence in East Africa never totalled more than 1,000 actual Portuguese men (Lofchie 1965: 28), and documentary sources do not mention the presence of Portuguese women at all. From the mid-17th century, local town-based dynasties invited external help from Oman to aid in the eviction of the Portuguese. The Portuguese had ruled Oman in a similar way, with some fortified trading settlements established on the coast, principally at Muscat, the capital of Oman. But their rule ended in 1650 when the Portuguese were expelled from Muscat (Sheriff 1987: 19), and in 1652 they were similarly ousted from Mombasa by a joint effort between Mombasan and Omani forces (Bennett 1978: 11).

Throughout the late 17th and early 18th centuries, Omani forces gradually began to play the role of colonial rulers on the coast (Sheriff 1987: 17). The port town of Zanzibar was the centre of this activity, and eventually became the seat of Omani governance (Sheriff 1995a: 12). Rapid urban expansion occurred from the late 18th century, as East Africa became incorporated into increasingly global capitalist markets (Glassman 1995: 36; Sheriff 1987: 245). Ivory and slaves from East Africa were in demand in Europe, America and Asia, and these demands were met by an increase in caravan trade on the East African mainland. Many of these caravans converged on the international trading entrepôt of Zanzibar Town, and the rapid expansion of trade produced wealth for Arab colonial elites, who then began to invest in agrarian production. From c. 1812 cloves were planted on Unguja (Sheriff 1987: 42). Clove trees were a new crop for the islands; their produce commanded a high price on international markets. Plantations quickly became a boom industry, covering over the majority of the landscape of Pemba island and much of Unguja (Cooper 1977: 48, Figure 3.2). By the 1830s a French naval officer, Captain Loarer, described a clove 'mania' on the islands (Sheriff 1987: 51). The research outlined here begins at the period of establishment of these plantations.

Clove growing requires specific conditions: a combination of deep, fertile soils, a tropical climate and regular rainfall. Not all areas on the islands were suitable for these plantations, and therefore they were

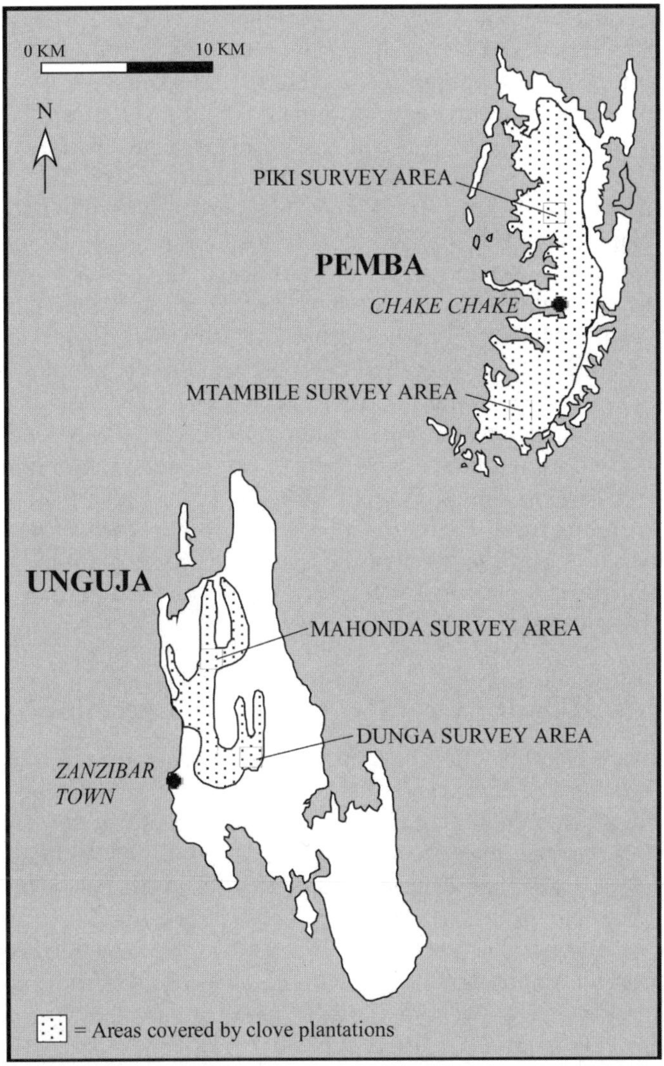

Figure 3.2 Map showing the clove plantations on Unguja and Pemba, and the locations of the four regions of the Zanzibar Clove Plantation Survey 2003. *After Sheriff 1987: 52.*

confined to the northeastern half of Unguja and the hilly ridges of Pemba (Cooper 1977: 48). Indigenous Swahili village settlements continued to be inhabited on both Unguja and Pemba throughout the 19th century (Fleisher 2003: 131), but they were located in the valleys of Pemba, or in the relatively unfertile lands to the east and south of Zanzibar Town (Cooper 1977: 71; Sheriff 1987: 52). Whilst little is known archaeologically of these settlements, at those identified to date there appears to be no significant change in settlement location and form from earlier Swahili sites (Fleisher 2003: 258).

Plantation agriculture was dominated by newly arrived Omani Arabs (Cooper 1977: 70). These ranged from large landowners with up to 10,000 clove trees worked by up to 2,000 slaves, to smaller holdings requiring as few as one or two slaves (Cooper 1977: 68–69). Labour on clove plantations was carried out almost entirely by slaves; they planted and weeded the trees, and harvested the cloves (Middleton and Campbell 1965: 33). These slaves were brought along caravan routes from East-Central or East Africa, and by the mid-19th century the slave trade in this region had touched many of the peoples in eastern and southern Tanzania, northern Mozambique, Malawi and even as far as Zambia (Alpers 2000: 88; Sheriff 1989: 144). By the end of the 1850s, one estimate placed two-thirds of the islands' total population of about 300,000 as slaves (Bennett 1978: 28). From a historical point of view, therefore, the 19th century presents a situation where new Arab colonists settled certain areas of the islands in order to pursue a novel form of agrarian production for the region: one that relied on the unfree labour of slaves.

ARCHAEOLOGIES OF 19TH CENTURY ZANZIBAR

This section examines how archaeological evidence from Unguja and Pemba can be used to elucidate the ways in which clove plantation residents constructed their identities during the 19th century. This approach aims to use archaeology to examine how the study of landscapes can reveal aspects of social life and human identity in the past. The data I present here derives from two seasons of fieldwork: the Zanzibar Clove Plantation Survey 2003 (hereafter ZCPS03, Croucher 2004, forthcoming) and excavations at the site of Mgoli on Pemba in 2004 (Croucher 2006). The survey investigated four different areas of the islands in known clove-growing regions, centred around the villages of Dunga and Mahonda on Unguja, and Mtambile and Piki on Pemba (Figure 3.2), recording a total of 64 sites across these areas. Owing in part to the budget and time constraints of the project, local residents were asked about the locations of past plantation sites as part of the process of site identification. This resulted in a reflection of contemporary Zanzibaris' knowledge of past settlement sites, and their current constructions of the historical plantation landscape. However, this process was followed by the detailed investigations of a single site through survey and excavation, resulting in a more complete archaeological recording of the landscape of a single plantation site.

Sites were classified according to three broad categorisations of 'plantation owner', 'slave' or 'village' sites, and according to whether they contained stone-built architectural elements or not (Figure 3.3). My starting point for examining the landscape of clove plantations was the sites that are today described by contemporary Zanzibari residents as the homes of past plantation owners. These formed 34% of sites overall on the survey. When divided into those with stone-built and non-stone-built elements, stone plantation owners' homes formed only 33% of the total sites associated with plantation owners, whilst 67% had no stone-built elements. These sites were the starting point for examining the intersections between the physical transformation of the Zanzibari landscape through the settlement of clove plantations, and the creation of plantation residents' social identities and conceptions of the landscape.

Stone houses were constructed from the local stone of coral rag, the availability of which varies across the islands. Soils on Unguja are generally thinner than those on Pemba (Milne 1936), and it is on Unguja that a higher proportion of stone settlements were recorded. Despite these differences, significant variations in the type of materials used to build plantation owners' homes, and the proportions of the houses, were identified (Croucher 2006: 221–224). Such variations do not indicate a straightforward index of status differentiation between plantation owners, in which those with larger holdings had stonehouses, whilst those with fewer trees and slaves had houses of wattle-and-daub. Rather, the building of houses was just one element of a broader set of practices concerned with the creation of plantation landscapes, through which a range of plantation identities were also created – from owners to slaves.

Excavations were undertaken at a single plantation site, that of Mgoli on Pemba, located in the eastern edge of the Piki survey region. Complementing the broader characterisation of plantation sites that was provided through survey, excavation gave more detailed evidence of both the structures and artefacts of a plantation site. Interviewees in the Piki area reported that the site had been settled by an Arab named Abdalla bin Jair, who had lived there from approximately the mid-19th century with his wives, slaves, concubines and children (Croucher 2006: 189). A stone-built house dominated the site, and four trenches were opened to investigate different areas around it. One of these, Trench A (Figure 3.4), was situated in front of the stonehouse to investigate a particular aspect of its architecture, the *baraza*. These are important architectural features of East African coastal homes, and consist of stone benches (or earthen ones in the case of wattle-and-daub houses) that run along the front of most of the houses (Donley 1982: 67; Fleisher and LaViolette 1999: 93). It has

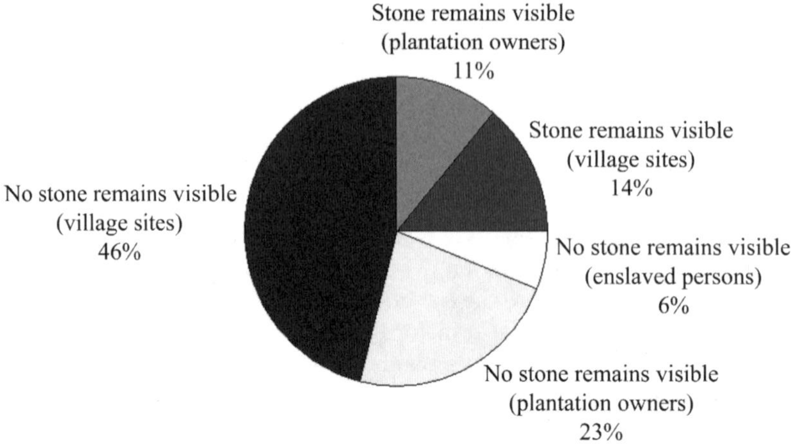

Figure 3.3 Site types in the Zanzibar Clove Plantation Survey 2003.

been argued that *baraza* represented an important space for male social interaction in the upper-class social networks of coastal society, a space where 'men would congregate to engage in the sophisticated pleasures of this literate urban culture' (Glassman 1995: 3). In typical interpretations of this architectural feature in relation to the construction of Swahili domestic space, it is seen to form part of a 'gradient' (Ghaidan 1975: 32) of privacy, whereby the use of space in houses was governed by strict rules relating to Islam (Donley-Reid 1990: 116). As such, the accepted interpretation of this space is that it formed a part of Muslim men's identities at Swahili settlements.

At Mgoli, Trench A was indeed successful in locating the *baraza*, which had been partially cut into by a robber trench, presumably to remove some of the stone used in its construction (Figure 3.4). This feature was not found in isolation, however, abutting it was a packed-rubble flooring surface, which was interpreted as a clove-drying floor (Croucher 2006: 261). No mention of such features is made in the historical literature relating to clove plantations, but evidence of another clove-drying floor was also documented in fieldwork. During the ZCPS03, observations were recorded of a standing building at Bweni, close to Mtambile. The site is currently occupied, and so was not recorded as an archaeological site. Here, a substantial wattle-and-daub building, said to have been a plantation owner's home for 150 years (Croucher 2006: 232), was linked to another smaller building by a large compacted-rubble floor surface, used for drying cloves (see Figure 3.5).

During the picking season, slaves were forced to work harder than usual (Cooper 1977: 157). Following picking, the fresh cloves must be laid out on a clean, dry surface, such as a clove-drying floor and/or

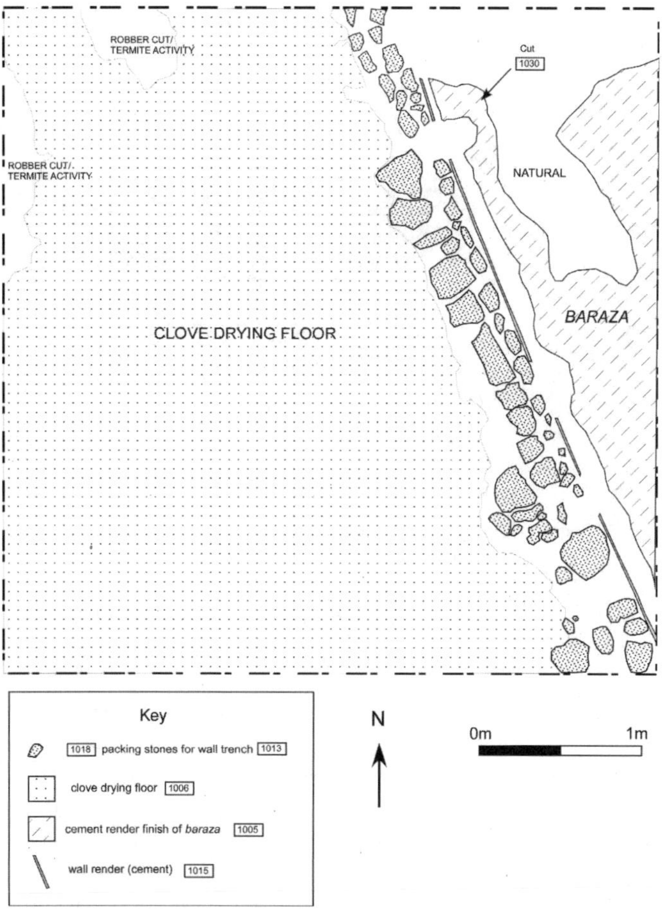

Figure 3.4 Plan of Trench A, at Mgoli, Pemba (2004).

mats, so that they desiccate to the dried form, which is then exported. The drying process thus forms an important step in the overall labour of clove plantations. The occurrence of a clove-drying floor at both Mgoli and Bweni, and the ready identification of this structure by Pembans during the course of the Mgoli excavations, could lead to the conclusion that these were not an unusual part of the plantation landscape. If clove-drying floors were not specifically built, then the area in front of a plantation owner's home was probably the space in which slaves would have laid out cloves to dry, judging by practices today and those shown in archival photographs dating to the early 20th century.

Rather than simply relating to a functional aspect of clove plantation management, the location of a clove-drying floor beside the *baraza*

Figure 3.5 Wattle-and-daub house at Bweni; *baraza* is visible around the house, and the clove drying floor can be seen in the foreground of the picture. *Photograph by A. Brennan.*

can be interpreted as a site at which slaves were observed directly in their working practices, and their production could be monitored. As such, it could be argued to form part of the 'spatialities of control' of Zanzibari plantation owners, an attempt to utilise surveillance to control space and actions (Delle 1998: 156). The production of slaves as working subjects may have formed a constituent part of clove plantation owners' identities, within a world that was increasingly incorporating capitalist economic pressures (Sheriff 1987: 33). Situating this beside the *baraza* meant that not only the identities of male, Muslim, plantation owners, but also those of slaves, were worked out here: indeed, worked out in contrast to each other within the same space. Upper-class men could sit and talk to each other whilst observing their slaves at work. Here, archaeology can provide a glimpse of one aspect of the clove plantation landscape as it was lived.

DISCUSSION

Historical geographer Garth Myers (1995, 1996, 1997) has set out a scale of architectural forms for the study of late-19th- and early-20th-century Zanzibar, defined by the permanence and amount of stone utilised in the houses' construction, which represents an index of the power (*uwezo*) of their inhabitants (Myers 1997: 253). At the bottom of this scale lie impermanent *vibanda* thatch constructions, whilst at the top are permanent stonehouses (Myers 1997: 253–255). Between these lie a number of gradations in architectural form, moving through wattle-and-daub structures to those incorporating lime mortar and increasing amounts of coral rag stone. This kind of scalar approach that links status identities – albeit conceptualised as *uwezo* – to the physical form of houses can be compared to the way in which settlement type has been linked to status identities in Swahili archaeology. Here, differences in the *forms* of architecture between different settlements (i.e. stone or wattle-and-daub) have been argued to rank these settlements (Horton and Middleton 2000: 127).

These historical and archaeological interpretations of East African coastal settlements would suggest that differences in whether the homes of plantation owners were made of stone or wattle-and-daub reflected differences in the social status of plantation owners. This may well have been the case, but status differentiation between plantation owners was only one part of their identities, and cannot be simply read off from the archaeological material. Instead, a focus upon the creation of plantation landscapes highlights how houses stood alongside areas in which labour relations were negotiated, and where daily practices that constituted the norms of plantation behaviour for different residents were undertaken. Slaves were unlikely to be able to sit on the *baraza* alongside their owners as equals, but instead were clearly located in a space for labour. Identifying landscape features such as clove-drying floors – which were recorded at both Mgoli (a stone-built plantation owner's house) and Bweni (a wattle-and-daub plantation) – is an important part of examining the range of social lives of the plantations and the interactions between plantation owners and slaves, rather than simply the status of the house to which they were attached.

Slaves' conceptions of the plantation landscape may have been markedly different from that of their owners. Unfortunately, in contrast to sites relating to plantation owners, those relating to slaves were far harder to locate on the ground, and formed only 6% of the total sites recorded on the ZCPS03. Of the four sites recorded, one related to a Muslim graveyard, and three to possible areas of slave settlement (Croucher forthcoming). The difficulty in locating slave sites probably relates to the ephemeral nature of the structures in which they lived, likely the *vibanda* at the bottom of Myers' (1997: 253) architectural scale. The intangibility of slave archaeology is not confined to Zanzibar. Slave housing in Jamaica, for instance, was also built from impermanent materials, and has proved hard to locate archaeologically (Delle 1998: 144).

Whilst locating further sites of this kind is an important future goal of fieldwork in the area, the current low number of known sites specifically tied to slaves does not necessarily bar interpreting slave landscapes on plantations. The idealised notions of the use of social space by plantation owners and the actuality of spatial use at individual locations may have been quite different (Moore 1986: 105). As mentioned above, one slave site that was identified was a graveyard with coral rag grave markers in a Muslim style. Plantation owners engaged in shared Muslim practices, such as attending mosques. Their conception of the social landscape of Zanzibar thus included their religious landscape, centred on the elite-male-only institution of the mosque (Sheriff 1995b: 5). But slaves were also creating their landscapes through shared Muslim practices. The graveyard recorded shows that they participated in Islamic public rituals (burial in this case), resulting in specific Muslim slave spaces. Whilst sharing one faith, plantation residents did not necessarily have a single, harmonising Muslim identity. Islam was instead linked to many different identities in 19th century Zanzibar, and the adoption of Muslim practices by slaves may have been a vehicle for expressing their autonomy in an oppressive society (Glassman 1995: 85). Plantations were sites of labour for slaves, as they were for plantation owners. The views of plantation owners in relation to the control of slaves' labour, described above, may have been quite different from the slaves' conceptions. Slaves may instead have created landscapes of resistance, transgressing the view of controlled landscapes held by plantation owners. While archaeological evidence relating to these practices may hopefully be recorded in the future, the material presented here still serves to underline how the integrated study of the many elements of a plantation landscape is important part of the development of archaeologies that can acknowledge the diversity of social identities in the past.

CONCLUSIONS

The understandings of landscape by different 19th century clove plantation residents, specifically those of plantation owners and slaves discussed here, were tied to their conceptions of identity. Conceptions of both landscape and identity were not simply a reflection of the physical sites of plantations, but emerged from daily practices undertaken in different parts of the plantation landscape. Thus the labour relations of plantations, and the religious practices of different plantation residents, worked to create changing combinations of elements of plantation identities. Through fieldwork such as that at the clove plantations outlined above, landscape archaeology reveals that there was no single, fixed view of the Swahili landscape. As Barbara Bender, drawing upon the flux, change and potential evoked in Walter Benjamin's descriptions of the urban form of 19th century

Paris, has suggested that 'landscape is never inert': 'people engage with it, re-work it, appropriate it and contest it' (Bender 1993: 3). Similarly, Julian Thomas suggests that if we view landscape as something in which human beings are engaged, it 'can be imagined as an uncompleted process rather than a bounded and static *thing*' (Thomas 2001: 181, original emphasis).

Such approaches stand in contrast to the way in which archaeologists working on the East African coast have engaged with landscapes. Swahili settlements have been characterised mostly according to the functions, size, form and materials of built structures. This produces a certain static conception of landscape, which has then been used to interpret the identity of past coastal residents, their urban or rural complexion, or their mercantile or productive nature. The engagements that past Swahili populations had with their changing landscapes – and the social practices through which these landscapes, and in turn identities, were created – are absent. However, foundations for new ways of interpreting Swahili landscapes are being laid by current approaches to fieldwork, through their emphasis on recognising a complete range of sites, from large stonetowns to short-lived hamlets. A range of surveyed and excavated sites is now increasingly available from which to interpret the contrasting practices of different coastal residents, and to relate these to the way in which landscapes may have been understood. Swahili communities were formed from a range of people, from farmers living in villages to elite traders in stonetowns. But to interrogate these identities the landscape must not be interpreted in a purely functional manner. What did landscapes *mean* to past Swahili populations? How were these meanings involved in the daily practices that helped constitute identities, and how did changes in landscapes and identities interrelate? I have not answered these questions here, but it is clear that East African coastal archaeology is currently beginning an exciting reformulation of the way in which past Swahili landscapes and past Swahili identities are understood.

This chapter has sought to problematise, in the context of Swahili archaeology, the way in which archaeologists' conceptions of landscapes have privileged the Western gaze. An attempt to engage with the ways in which past East African coastal landscapes were understood may be one way in which a distinctive East African type of landscape archaeology can emerge, building on previous approaches developed in the region. Such an approach has the potential to show the non-Western nature of these landscapes, and can try to privilege instead Swahili conceptions of their landscape. This will not be an easy task, but it is one that is increasingly achievable as more and more East African archaeologists undertake work on the coast (e.g. Chami 1998; Kusimba 1999).

ACKNOWLEDGEMENTS

Fieldwork within this paper was undertaken with grants from The British Institute in Eastern Africa, the Emslie Horniman Fund of the Royal Anthropological Institute, the Arts and Humanities Research Council and the Zochonis Special Enterprise Awards of the University of Manchester. My grateful thanks go to the Department of Archives, Museums and Antiquities, Zanzibar for their support of this research, particularly to Abdurahman Juma, whose support and advice was invaluable. Many thanks to all those, too numerous to mention here, who assisted on the fieldwork and were interviewed as part of my research. In particular to Ange Brennan who assisted on the survey element of my fieldwork, and who took the photograph used for Figure 3.5. Many thanks also to Karina Croucher, Steven Price, Edward Tennant and Stephanie Wynne-Jones for their comments and suggestions on earlier drafts of this paper.

REFERENCES

Allen, J. de V. 1974. Swahili Culture Reconsidered: Some Historical Implications of the Material Culture of the Northern Kenya Coast in the Eighteenth and Nineteenth Centuries. *Azania* 9: 105–138.

Allen, J. de V. 1993. *Swahili Origins: Swahili Culture and the Shungwaya Phenomenon*. Oxford: James Currey.

Alpers, E.A. 2000. Recollecting Africa: Diasporic Memory in the Indian Ocean World. *African Studies Review* 43(1): 83–99.

Bender, B. 1993. Introduction: Landscape – Meaning and Action. In B. Bender (ed) *Landscape: Politics and Perspectives*. Oxford: Berg, pp. 1–17.

Bender, B. 1999. Subverting the Western Gaze: Mapping alternative worlds. In P.J. Ucko and R. Layton (eds) *The Archaeology and Anthropology of Lancdscape: Shaping Your Landscape*. London: Routledge, pp. 1–20.

Bennett, N.R. 1978. *A History of the Arab State of Zanzibar*. London: Methuen.

Chami, F. 1998. A Review of Swahili Archaeology. *African Archaeological Review* 15(3): 199–218.

Chittick, N. 1974. *Kilwa: An Islamic Trading City on the East African Coast* (2 vols). Nairobi: British Institute in Eastern Africa.

Chittick, N. 1984. *Manda: Excavations at an Island Port on the Kenyan Coast*. Nairobi: British Institute in Eastern Africa.

Cooper, F. 1977. *Plantation Slavery on the East Coast of Africa*. New Haven: Yale University Press.

Croucher, S. 2004. Zanzibar Clove Plantation Survey 2003: Some Preliminary Findings. *Nyame Akuma* 62: 65–69.

Croucher, S. 2006. *Plantations on Zanzibar: An Archaeological Approach to Complex Identities*. Unpublished Ph.D. thesis, University of Manchester.

Croucher, S. forthcoming. Cultural Identity and Perceptions of Slavery in the Clove Plantations of Zanzibar. In L. McAtackney, M. Palus and A. Piccini (eds) *Contemporary and Historical Archaeology in Theory*. Oxford: Archaeopress (Studies in Contemporary and Historical Archaeology).

Delle, J.A. 1998. *The Archaeology of Social Space: Analyzing Coffee Plantations in Jamaica's Blue Mountains*. London: Plenum Press.

Donley, L.W. 1982. House Power: Swahili Space and Symbolic Markers. In I. Hodder (ed) *Symbolic and Structural Archaeology*. Cambridge: Cambridge University Press, pp. 63–73.

Donley-Reid L. 1990. A Structuring Structure: The Swahili House. In S. Kent (ed) *Domestic Architecture and the Use of Space*. Cambridge: Cambridge University Press, pp. 114–126.

Fleisher, J.B. 2003. *Viewing Stonetowns from the Countryside: An Archaeological Approach to Swahili Regions, AD 800–1500*. Unpublished Ph.D. thesis, Department of Anthropology, University of Virginia.

Fleisher, J. and A. LaViolette 1999. Elusive Wattle-and-Daub: Finding the Hidden Majority in the Archaeology of the Swahili. *Azania* 34: 87–108.

Ghaidan, U. 1975. *Lamu: A Study of the Swahili Town*. Nairobi: East African Literature Bureau.

Glassman, J. 1995. *Feasts and Riot: Revelry, Rebellion, and Popular Consciousness on the Swahili Coast, 1856–1888*. Oxford: James Currey.

Gray, J. 1962. *History of Zanzibar: From the Middle Ages to 1856*. Oxford: Oxford University Press.

Horton, M.C. 1987. Early Muslim Trading Settlements on the East African Coast: New Evidence from Shanga. *Antiquaries Journal* 67(2): 290–323.

Horton, M.C. 1994. Swahili Architecture, Space and Social Structure. In M. Parker Pearson and C. Richards (eds) *Architecture and Order: Approaches to Social Space*. London: Routledge, pp. 147–169.

Horton, M.C. 1996. *Shanga: The Archaeology of a Muslim Trading Community on the Coast of East Africa*. London: British Institute in Eastern Africa.

Horton, M.C. and J. Middleton 2000. *The Swahili: The Social Landscape of a Mercantile Society*. Oxford: Blackwell.

Kirkman, J. 1954. *The Arab City of Gedi: Excavations at the Great Mosque*. Oxford: Oxford University Press.

Kusimba, C.M. 1999. *The Rise and Fall of Swahili States*. Walnut Creek: AltaMira.

LaViolette, A. 1996. Report on Excavations at the Swahili site of Pujini, Pemba Island, Tanzania. *Nyame Akuma* 46: 72–83.

LaViolette, A. 2000. Swahili Archaeology on Pemba Island, Tanzania: Pujini, Bandari ya Faraji, and Chwaka, 1997–1998. *Nyame Akuma* 53: 50–63.

LaViolette, A., W.B. Fawcett, N.J. Karoma and P.R. Schmidt 1999. Survey and Excavations between Dar es Salaam and Bagamoyo: University of Dar es Salaam Field School, 1988 (Part 2). *Nyame Akuma* 52: 74–78.

LaViolette, A. and J. Fleisher 2005. The Archaeology of Sub-Saharan Urbanism: Cities and their Countrysides. In A. B. Stahl (ed) *African Archaeology*. Oxford: Blackwell, pp. 327–352.

Lofchie, M.F. 1965. *Zanzibar: Background to Revolution*. Oxford: Oxford University Press.

Middleton, J. 1992. *The World of the Swahili*. New Haven: Yale University Press.

Middleton, J. and J. Campbell 1965. *Zanzibar: Its History and Its Politics*. Oxford: Oxford University Press.

Milne, G. 1936. *A Provisional Soil Map of East Africa*. Amani: East African Agricultural Research Station.

Moore, H.L. 1986. *Space, Text and Gender: An Anthropological Study of the Marakwet of Kenya*. Cambridge: Cambridge University Press.

Myers, G.A. 1995. The Early History of the 'Other Side' of Zanzibar Town. In A. Sheriff (ed) *The History and Conservation of Zanzibar Stone Town*. Oxford: James Currey, pp. 30–45.

Myers, G.A. 1996. Naming and Placing the Other: Power and the Urban Landscape in Zanzibar. *Tijdschrift voor Economische en Sociale Geografie* 87(3): 237–246.

Myers, G.A. 1997. Sticks and Stones: Colonialism and Zanzibari Housing. *Africa* 67(2): 252–272.

Nurse, D. 1983. A Linguistic Reconsideration of Swahili Origins. *Azania* 18: 127–150.
Nurse, D. and T. Spear 1985. *The Swahili: Reconstructing the History and Language of an African Society*. Philadelphia: University of Pennsylvania Press.
Sheriff, A. 1987. *Slaves, Spices and Ivory in Zanzibar: Integration of an East African Commercial Empire into the World Economy, 1770–1873*. Oxford: James Currey.
Sheriff, A. 1989. Localisation and Social Composition of the East African Slave Trade, 1858–1873. In W.G. Clarence-Smith (ed) *The Economics of the Indian Ocean Slave Trade in the Nineteenth Century*. London: Frank Cass, pp. 131–145.
Sheriff, A. 1995a. An Outline History of Zanzibar Stone Town. In A. Sheriff (ed) *The History and Conservation of Zanzibar Stone Town*. Oxford: James Currey, pp. 8–27.
Sheriff, A. 1995b. Introduction. In A. Sheriff (ed) *The History and Conservation of Zanzibar Stone Town*. Oxford: James Currey, pp. 1–7.
Spear, T. 2000. Early Swahili History Reconsidered. *International Journal of African Historical Studies* 33(2): 257–290.
Thomas, J.S. 1993. The Politics of Vision and the Archaeologies of Landscape. In B. Barbara (ed) *Landscape: Politics and Perspectives*. Oxford: Berg, pp. 19–48.
Thomas, J.S. 2001. Comments on Part I: Intersecting Landscapes. In B. Bender and M. Winer (eds) *Contested Landscapes: Movement, Exile and Place*. Oxford: Berg, pp. 181–189.
Wright, H.T. 1993. Trade and Politics on the Eastern Littoral of Africa, AD 800–1300. In T. Shaw, P. Sinclair, B. Andah and A. Okpoko (eds) *The Archaeology of Africa: Food, Metals and Towns*. London: Routledge, pp. 658–672.
Wynne-Jones, S. 2005a. *Urbanisation at Kilwa, Tanzania, AD 800–1400*. Unpublished Ph.D. thesis, University of Cambridge.
Wynne-Jones, S. 2005b. Scale and Temporality in an Urban Settlement System: Fieldwork in Kilwa Region, Southern Tanzania. *Nyame Akuma* 64: 66–71.

CHAPTER 4

Landscape Archaeology in Lower Manhattan: The Collect Pond as an Evolving Cultural Landmark in Early New York City

Rebecca Yamin and Joseph Schuldenrein

INTRODUCTION

New York, like most major cities, stands on altered ground. Its shorelines along the East and Hudson Rivers have been shifted, and its flattened streetscape is not its original terrain (Cantwell and Wall 2001: 224–241). Since earliest post-glacial times (after 12,000 BP), environmental and climatic changes and human interventions have transformed the city's landscape. Post-Columbian Euroamerican settlement has generated the most radical modifications to the post-glacial environment, especially as the city developed as an administrative and commercial centre from the 18th century.

This chapter explores this complex process of landscape change through a single feature in New York's historical urban form: the 'Collect Pond', also referred to as the 'Fresh Water' (Figure 4.1). This five-acre body of fresh water is a residual feature of the ancient glacial landscape, the site of which today lies on the edge of a recently designated historic landmark: the African Burial Ground. This monumental cemetery for the early slaves of New York appears to have been deliberately placed in the swampy, putrid and most undesirable terrain of lower Manhattan Island, of which the Collect was the central feature at the time of the interments in the 18th century. But the Collect played diverse roles in New York's history before and after that time. In the mid-17th century, it was outside the limits of the Dutch settled city, but as New Amsterdam, and ultimately New York, grew northward, the Collect acquired diverse functions and new meanings. In the 18th century it sustained recreational facilities, commercial operations and community infrastructures, in addition to the burial grounds along its margins. Then in the first

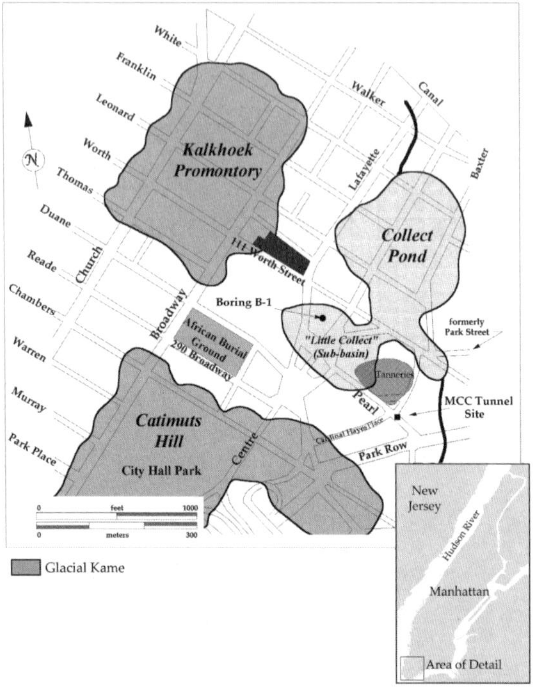

Figure 4.1 Location map depicting the probable position of the Collect Pond and surrounding topographic features during the early Euroamerican periods (1650–1805). The glacial kames south and west of the Collect were named by the Dutch and English settlers. Archaeological sites and landmarks are shown on street grid.

decade of the 19th century it was permanently buried, and its vestiges have remained concealed deep beneath the urban terrain.

American historical landscape archaeology conventionally separates the technical aspects of reconstructing landscapes from their interpretation (Yamin and Metheny 1996: xv), but in synthesising the sequence of physical transformations to the Collect as a landscape feature this chapter aims to combine description with interpretation. We trace the changing role of the Collect in the history of Lower Manhattan over time, considering the interaction between the pond's persisting but changing form and perceived significance. In this way, we explore how landscape archaeology both recovers past meanings and constructs new ones. Many of the changing meanings of the site have been lost in the confusion of myth and memory. The sense of place, which plays such an important role in historical landscape studies, can only be convincingly realised when it is attached to a tangible reality (Hayden 1995), but the Collect's tangible remains are hidden beneath complex layers of urban infrastructure. Nevertheless, they are surprisingly reflected in the contemporary configuration of Foley Square that occupies the site today (Figure 4.1; Schuldenrein 2004). We aim both to peel back the layers of concrete and to deconstruct the memories by combining three bodies of evidence for landscape change: primary and secondary descriptions of the Collect's terrain component and of the pond's significance to colonial and early historic New Yorkers; historical map depictions that indicate how the shape and size of the Collect changed over time; and the results of archaeological and geomorphological research undertaken during the monitoring of the construction of a tunnel near to the eastern outlet of the pond.

THE COLLECT IN MEMORY

The Collect is remembered, and was viewed historically, in many different ways. In the middle of the 17th century it was a 'cultural frontier' (Neville 1994: 23). Dutch farmers had settled north of the pond, but in 1643 and especially in 1655, violent conflicts with local Indians led to a series of city ordinances that required the farmers to move into villages or settlements for their own safety (Neville 1994: 23). To replace them, and to create a kind of buffer against attack, slaves were offered land and temporary manumission if they would settle in the zone just beyond the Fresh Water (Neville 1994: 24). Neville speculates that this began a process of marginalisation that characterised the area until the 19th century.

In spite of the danger, there were several taverns in the vicinity of the Collect in the 17th and 18th centuries. Jasper Danckaerts and his companion visited one of the early taverns in 1679. They described it as a 'delightful place' but deemed it too rowdy, so retreated 'through the salt meadow and over water, upon the trunk of a tree' (James and Jameson 1913: 47). The Bull's Head Tavern, established just north of the pond in 1750, was a centre for all sorts of boisterous entertainment including racing, prizefighting, rat baiting and cockfighting in addition to drinking and eating. The tavern became the Bowery Theatre in 1826, which remained open until 1910 (Neville 1994: 45).

As the city spread northwards under British rule after the 1660s, the noxious industries such as tanning that had been located at its outskirts also migrated. The first tanneries in the area of the Collect date to the mid-1690s (Neville 1994: 35), and tanning continued in the district up to the turn of the 19th century. Other water-dependent industries took advantage of the pond and adjacent streams as well. A hand-drawn map dating to about 1800 and probably made by a black slaughterhouse worker (Milne 2000: 23) shows a cluster of facilities immediately east of the pond, including a tanning shed, a rope walk and the public slaughterhouse. While these industries and others (e.g. breweries, potteries, distilleries) fouled the environment and eventually led to the filling of the pond, another image dating to the same period shows an idyllic landscape (Figure 4.2). Alexander Robertson's painting (1798) is the only known picture of the Collect Pond, and looks across the water from the vantage point of the northern shore to the settled city in the distance. Church steeples are conspicuous features in the distant landscape, but the foreground is framed by natural promontories. The depicted terrain – a pristine environment used by fashionable folk for recreation, a place for Sunday outings, skating in the winter and picnicking and fishing in the summer – would have been unrecognisable to residents half a century later (Yamin 2001: 7).

The 18th century Collect landscape was occupied by two cemeteries: an African ('Negros') burial ground (partially excavated in 1991) located in the ravine to the west of the Collect Pond, and a Jewish burial ground not far to the southeast (Figure 4.3). Both groups were socially marginalised and the placement of their burying grounds outside the city proper was consistent with their treatment as outsiders. When the city needed land for housing the growing working class at the onset of the 19th century, the African Burial Ground was closed. Recent projects have studied both the burials that were removed

Figure 4.2 Oil painting of the Collect Pond attributed to Alexander Robertson, 1798.

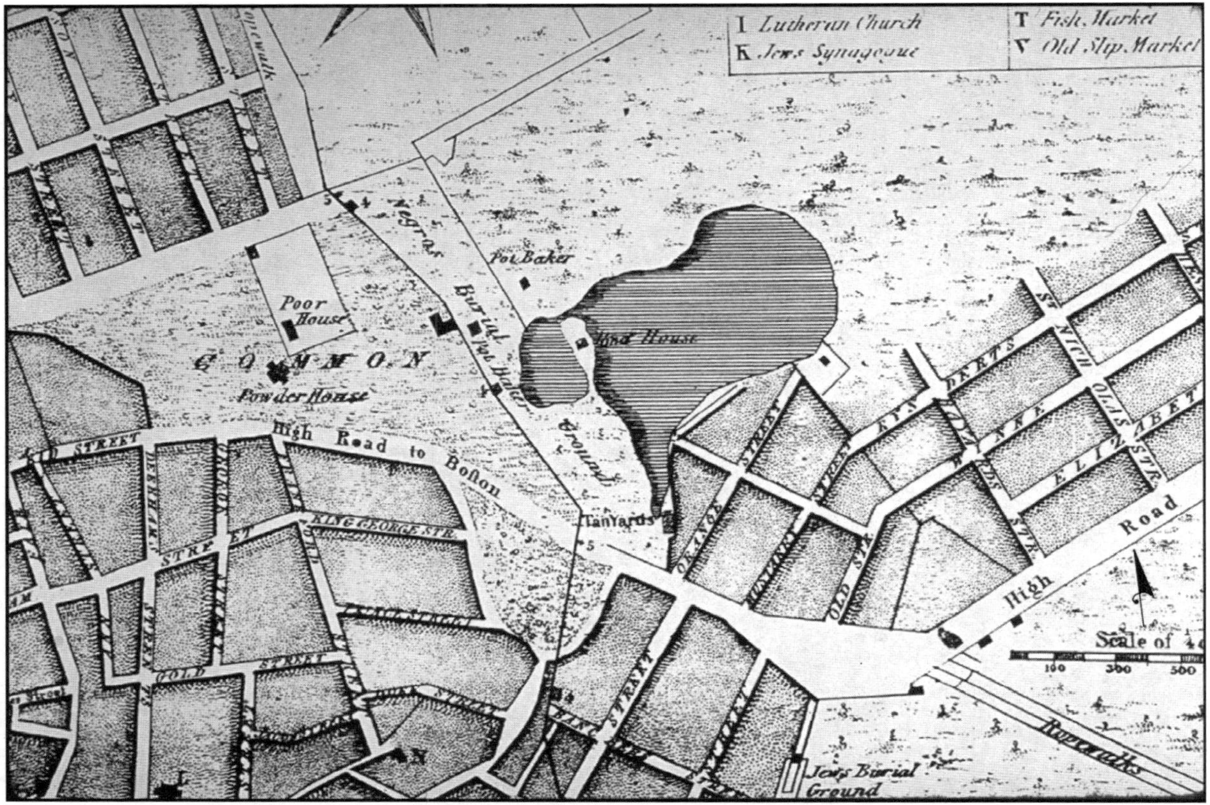

Figure 4.3 Detail of the 1754 plan of New York (Maerschalck 1754). Note locations of 'Negros' and Jews burial grounds (centre and bottom of figure).

during construction of a modern office building on a portion of the burial ground (Blakey and Rankin-Hill 2004; Perry et al. 2006) and the features and artefacts relating to the people who lived on the land after the burial ground was closed (Cheek 2006).

Around the time that the African Burial Ground was closed, a shortage of housing and complaints from the citizenry to the Common Council about the condition of the Collect Pond ('a very sink and common sewer' [Milne 2000: 24]) led to the infilling of the pond and its eventual conversion into building lots. Filling began in 1803 and continued for about eight years (Neville 1994: 72). The resulting neighbourhood, known as the Five Points, was condemned by some as 'America's first and foremost slum' (Ingle et al. 1990: 47), but for many, especially newly emancipated African Americans and German and Irish immigrants, it was a place to get a foothold in America (Asbury 1928; Foster 1990; Yamin 2000, 2001).

The Collect had featured prominently in the lore of early Manhattan and was a conspicuous feature on early city plans and maps such as the 1754 Maerschalck Map (Figure 4.3), which showed the pond as a giant footprint in the middle of the landscape, the 1728 Lyne Plan (Rothschild 1990: 29), the 1767 Ratzer Map (Milne 2000: 17) and the 1797 Taylor-Roberts Plan (Rothschild 1990: 30). But by 1840 it had disappeared from the streetscape (cf. Cantwell and Wall 2001: 191). Today, the space once occupied by the pond is dominated by the courthouses at Foley Square, more recent office buildings on the adjacent blocks and also the first large building to be built on the infilled site – the Tombs, a penitentiary and debtors' prison built in 1834 and provided with inverted arches and heavy timbering to make its foundations secure (Geismar 1993: 58).

The Collect was remembered by New Yorkers in its many manifestations but its exact physical location was forgotten. A composite map made in 1938 shows as many conflicting outlines for the pond as there were maps (Neville 1994, Figures 4.1, 4.4). When several construction projects were planned in the 1990s the New York City Landmarks Preservation Commission (LPC) seized the opportunity to look for physical evidence of the Collect. It was as if the many stories of the pond could not be believed if the pond could not be tied to a specific place with material evidence for its existence. Construction workers' tales of building the Tombs in the 1830s and the subway under Center Street (originally Collect) Street in 1917 were not enough.

THE METROPOLITAN CORRECTIONS CENTER PROJECT

The historical map evidence for the location and shape of the Collect in the urban landscape required confirmation through geological and

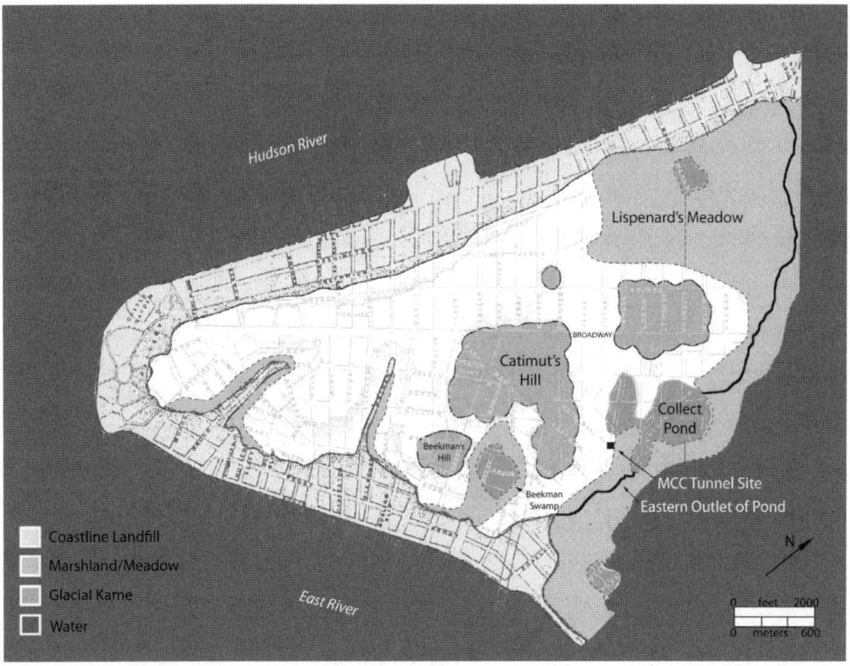

Figure 4.4 Plan view of 18th century landforms and geographic features, Lower Manhattan. The landfilling around the margins of the island continued well into the 19th century.

geomorphic data. Several questions emerged. Was it possible to identify the glacial and post-glacial surfaces and/or origins of the remnant basin-shaped depression that survived as the historic Collect? What was the depth of that basin and how did it change through time? Would it be possible to align the various historic map projections with particular land-use patterns during the Dutch, English and American periods? How were transformations to the immediate physical landscape reflected in the shifting shorelines of the Collect? And what did the changing form and function of the Collect say about the succession of landscapes and attendant lifeways of those who lived and worked along its margins?

Two major archaeological field projects addressed these questions by combining extensive historical research with archaeological and geomorphological analyses: a study associated with the refurbishment of Foley Square led by Joan Geismar (1993), and a study associated with the construction of the Metropolitan Corrections Center (MCC) Tunnel under Pearl Street by John Milner Associates, Inc. (Yamin, Schuldenrein and Schleidt-Peñalva 1995, Figure 4.1), which was specifically designed to reconstruct the history of the buried Collect landscape over 10,000 years.

In the summer of 1994, John Milner Associates monitored the excavation of a 15-metre-long tunnel in Lower Manhattan that connected the MCC with the new federal courthouse on the other side of Pearl Street (Figure 4.1). The courthouse stands in what was once part of the Five Points, a neighbourhood made famous by Charles Dickens (1842) when he visited the United States in 1841, and by Herbert Asbury in 1928 in his pulp history classic *The Gangs of New York* (recently brought back to public view by the 2002 Martin Scorsese movie of the same name). Scorsese and Asbury both portrayed the neighbourhood as a down-and-out slum peopled by petty criminals and gangs with distinctive allegiances (and costumes) who settled their differences violently in the streets. The results of the archaeological excavation at the site of the courthouse provided a more complex view of life at Five Points (Yamin 2000), objectives of the archaeological fieldwork during the excavation of the tunnel were to reconstruct the changing landscape in this part of the Collect basin.

Documentary and Cartographic Evidence

The changing landscape contours, terrain features and attendant land-use histories in the vicinity of the Collect are discernible by tracking map plots from the 17th to the 20th century. The 1639 Vingboom Map, one of the earliest European plots of Manhattan, depicts a wedge-shaped crevasse cross-cutting Manhattan Island, approximately west to east, in the vicinity of today's Pearl Street (Figure 4.1). This trough was the precursor to the later Collect basin. It may represent a remnant glacial outwash tributary to the Hudson River (Newman et al. 1969). Certainly the alignment of historically referenced kames (sandy hills) (Figure 4.4) and eskers (narrow, steep-sided, sandy gravel ridges) – all elements of glacial valley topography – is consistent with late Pleistocene drainage and post-glacial channel history (Fullerton 1992; Schuberth 1968). By late Holocene times the Collect was probably the local catchment for runoff when it only intermittently functioned as an outlet to the Hudson estuary and the antecedent channels of the East River.

The map projections are bolstered by detailed accounts from local geographers. In the mid-17th century the original Dutch colonists referred to the *Kolck* or *Kalchhook* on Pearl Street as the largest of a series of bog-like ponds, surrounded by hills, that spanned the lower portions of the island (Cozzens 1843; Gratacap 1909; Brown 1913). The spring-fed pond was part of a basin that extended over at least five acres, with waters reaching depths in excess of 15 metres (Andrews 1893). The Maerschalk Map (1754) shows a 'tea-water' pump at the location of one of the springs. The pond served as a major source of fresh water for the Manhattan Company, a private municipal waterworks that drew on the basin from the 1790s until about 1840, even after it was thoroughly polluted (Geismar 1993: 36).

The pond was flanked by two marshes, Wolfert's Meadows to the east and Lispenard's Meadows to the west (Figure 4.4). The natural outlet for the Collect was a brook named Old Wreck Brook, which traversed Wolfert's Meadows to the East River. Lispenard's Meadows, the largest of Manhattan's marshes (70 acres), formed the catchment for a number of minor tributaries, and extended continuously westward from the Collect to the Hudson (North) River. The marshy terrain constituted a natural frontier for the southern portion of Manhattan. As its central feature, the Collect Pond proper was subject to numerous hydrographic modifications, chiefly involving large-scale drainage of the natural marshlands and the conversion of the marshland terrain to agricultural fields and pastureland (Cozzens 1843; Andrews 1893). Issachar Cozzens (1843) hypothesised intrusion of saline waters because of encroachment from the Hudson, thus underscoring the magnitude of a task aimed at limiting estuarine sedimentation, which would have constituted a natural impediment to even modest crop yield (see also Gratacap 1909). Even minor rises in the water table would have necessitated extensive landscape-engineering projects, many of which may not have been documented. The stagnant waters of Lispenard's Meadows, and by extension the Collect Pond, evidently created a health hazard (Gratacap 1909). A drain was carved through the swamp, resulting in the partial depletion of the Collect Pond. In 1734, farmers and other landowners along the pond filled in 9–10 metres of the pond's drain in order to prevent further loss of water (Gratacap 1909). The drainage-engineering ventures initiated by the Dutch eventually took hold across Lower Manhattan. By the 1840s the entire perimeter of the island had been built up to similar heights in an attempt to expand the shoreline (and increase the area of buildable land), enhance drainage and minimise the effects of marine degradation (Figure 4.4).

All the historic maps discussed above placed the central axis of the pond along what are now Center, Pearl and Baxter Streets (Figure 4.1), but they neither agree on the precise location of the Collect Pond, nor depict the same shape and size for the impoundment. Thus, one of the earliest colonial maps – the Lyne Plan (1728) – shows that the Collect began as two separate sub-basins within the low-lying terrain of Lispenard's Meadows, when the local landscapes were largely agricultural. The sub-basins were separated by a sill (locally referred to as an island) on which a powder magazine (for the storage of gunpowder) was built in 1728 (labelled 'Powder House' on Figure 4.3), and which became the natural extension of Pearl Street (Gratacap 1909: 48). In the mid-18th century accelerated runoff, perhaps generated by moister climates during the latter stages of the Little Ice Age (Burroughs 1992), established perennial basins or impoundments on both sides of the sill. The 1754 Maerschalck Map (Figure 4.3) presents the Eastern Outlet as a funnel draining the eastern divide of Manhattan Island. Its evolution was apparently linked to the general lowering of water tables by the drainage of Lispenard's Meadows in 1733 and the localised re-channelling of the Eastern Outlet in conjunction with tanning activities. The 1767 Ratzer Plan of the City of New York depicts marshlands, or seasonal impoundments, where the Collect immediately flanked the powder magazine; the deepest waters had migrated to the northern margins of the upper sub-basin. A still later map (Taylor and Roberts' *New and Accurate Plan of the City of New York*, 1797) indicates a depleted lower basin, clearly susceptible to annual or seasonal groundwater fluctuations. Shortly prior to infilling in 1805 the southern sub-basin had receded to a minor cavity. Re-engineering and land-use planning in the latter 19th century precludes accurate reconstructions of the Collect Pond's precise dimensions. The variability between shapes and extents of the depicted basins in the various historic maps suggests that over the course of two centuries the Collect Pond's shorelines receded and transgressed across tens of metres. Moreover, since the margins of the Collect were relatively shallow, its surface area may have varied greatly in size as a result of even minor changes in precipitation and runoff.

The volume of documentation as well as the detail and reliability of observer accounts and historic maps improves after the Dutch period. A distillation of these sources suggests that by the turn of the 18th century, the Collect Pond had become the central processing station for a burgeoning tanning industry, which had continued to migrate progressively northward from the tip of Manhattan Island as it sought to escape from the contaminated swamps and marshes it left in its wake. Specifically, from the 1690s onward the Pearl Street tanneries utilised the Collect Pond and its ancillary effluents (i.e. the Eastern Outlet) for hide processing, possibly draining the springs and influents that supplied fresh water to the catchment. This diversion of water away from the pond, as well as the fouling of its environs, led eventually to the filling of the pond, but first the city had to assume ownership of the land, a process that took over a decade to complete.

Filling of the basins began in earnest in 1803. As noted in a municipal ordinance, 'It is a desirable thing that this low ground [Collect] should be filled up for both the health of the town and also that the ground be applied to some beneficial purpose. It is recommended that contracts be made with suitable persons for digging out this hell and filling up the meadows with earth to a suitable height. …' The probable source of the fill was earth from the foundation of City Hall, which began construction in that year (Cozzens 1843), and also from the surrounding hills that were levelled at the same time. By 1805, the pond was entirely covered over in the wake of municipal debates as to whether a canal should be left running through the basin. By this time, the Collect had become a foul-smelling, stagnant pool in which rubbish and especially animal carcasses and hides were discarded.

Excavated Evidence for Landscape Change

The excavation of the MCC Tunnel explored a complex archaeological sequence over a depth in excess of 5.2 metres, extending through the municipal infill layers and the stratified evidence of the early American, British and Dutch occupations. Beneath the Euroamerican strata a prehistoric buried soil marked the stratigraphic interface with the natural subsoil. The excavation also cut into the glacial sediments antedating the Collect Pond.

Figure 4.5 presents the composite 5.2-metre column, placing emphasis on the glacial through early historic (18th century) sequences, while Figure 4.6 is an expanded view of the complex succession of later historic horizons, underscoring the intricate sedimentation associated with tanning activities and isolating specific filling episodes within Units I and II at the eastern edge of the tunnel trench. The sequence for Units I and II is more typical of the general pattern for historic sedimentation across the greater landform. The sedimentation recorded therefore included material laid down by natural (geogenic) and also anthropogenic archaeological horizons, reflecting human intervention in landscape change. Harris' (1993) matrix classifications were applied to distinguish diagnostic cultural features and horizons with as much resolution as possible. Carbon-14 dates were taken from both geogenic and anthropogenic sediments and features. Radiometric determinations were appropriately calibrated (Stuiver et al. 1998).

The key stratigraphic changes (Yamin et al. 1995: 32–38) are summarised below (Figure 4.5). They began with agriculturally based drainage diversions (17th century; Dutch), proceeded with the mobilisation of stream and spring waters for the tanning process (18th century; British and early American) and ended with a series of infillings of the Collect (19th century; industrial American) and the ultimate build-up of the 'made land' underlying contemporary Manhattan.

The latest geogenic surface furnished a radiocarbon date of 470± 50 BP (Beta-77026; AD 1405–1495) on organic sediments capping a moderately weathered soil, formally a Cambic paleosol ('Ab-E-Bw'

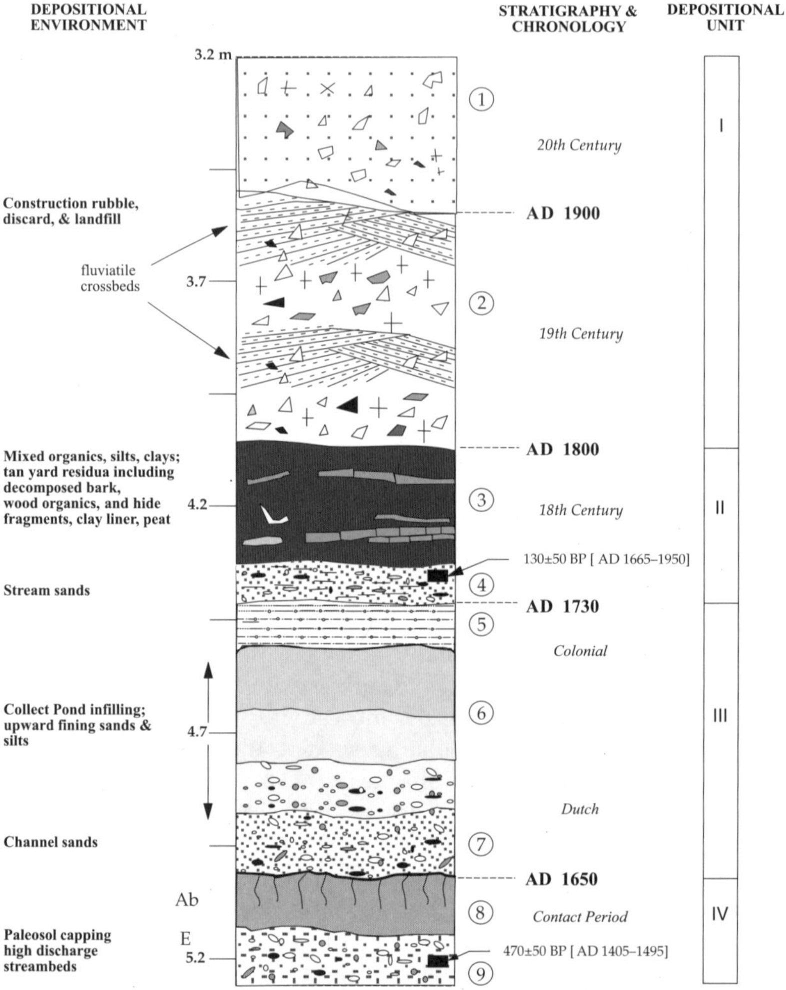

Figure 4.5 Semi-schematic historic stratigraphy, Metropolitan Corrections Center tunnel. Formal Depositional Units (Roman numerals; right) are referenced in text. Principal sources and patterns of sedimentation are identified on the left.

solum) (Holliday 2004). The age of the soil is contemporary with the transition from the Late Woodland (terminal aboriginal) to earliest Euroamerican period in North America. Research elsewhere in the northeastern United States has demonstrated that this was the period of transition to moist and cool climates – the Little Ice Age (Burroughs 1992; Negendank 2004) – that favoured dense settlement on the flanks of modern streams. Late prehistoric sites are typically found in similar soil horizons within the uppermost metre of major stream terraces

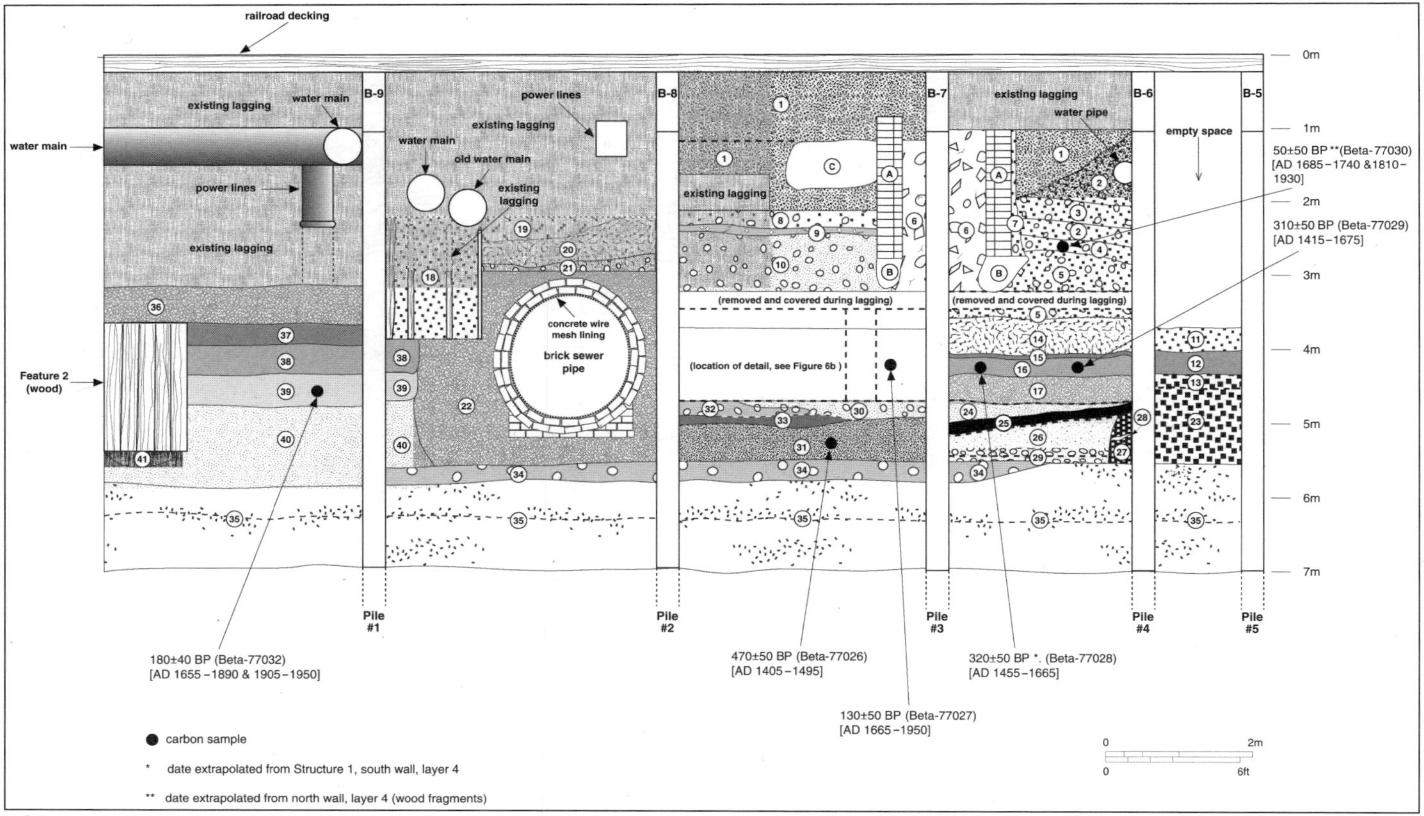

Figure 4.6 Detail of landfill, overburden and historic stratigraphy of eastern wall, Metropolitan Corrections Center tunnel trench, 1995. Microstrata are designated by Harris matrix units (Arabic numerals) housed within to Depositional Units I and II of Figure 4.5.

(Schuldenrein 2003). The sandy texture of the solum, as well as its weathering profile, suggest that the dominant vegetation at the time was coniferous forest. Late Woodland artefacts were not recovered at the MCC Tunnel excavations, but early historical accounts describe shell middens in the general vicinity (Gratacap 1909). The finer-grained estuarine clays of the stabilised shoreline are not present at Pearl Street and might argue that the location was not a near-shore environment during the late Holocene.

The excavations revealed a thick sediment, which was interpreted as evidence of a freshwater stream that extended, between c. AD 1650 and 1730 (Unit III), westward to Lispenard's Meadows and eastward through the Collect Pond and Eastern Outlet (Figure 4.4). It was sealed by the overlying silts of the Collect Pond, which appears to have formed after the beginning of drainage regulation in the 18th century. This stream apparently incised the early Dutch agricultural fields and then systematically aggraded its banks probably midway through the Dutch occupation (c. 1630–1660). Within the sequence, it was clear that the gently upward-fining deposits above the sands signified a transition to quiescent stream flow and intermittent or seasonal impoundment of waters. This, coupled with a rising water table, marked the initial stages of Collect Pond formation. Historical accounts confirm this sequence, pointing to Collect Pond expansion sometime around the turn of the 18th century either by design (i.e. drainage modifications) or because of climatic changes. Runoff accumulated in the basin from the surrounding hills. A sill parallel to the drainage axis (northwest to southeast) created two sub-basins (the 'Little Collect' and the main impoundment to the north; Figures 4.1 and 4.3), which were only breached during periods of high effective moisture; analogous hydrographic transitions for near shore basins have been documented for the 1640s and 1670s in New England (Baron 1988). The main source of fresh water was a series of springs that irrigated Lispenard's Meadows and fed the Collect Pond. The accumulation of soft sediment (fine sands, silts and clays) near the top of Unit III (Figure 4.5) is a product of spring-water circulation and limited sediment reworking. More prominent are isolated peat beds and humic mats that took root along the margins and floor of the Collect Pond and were nourished under aquatic conditions linked to land use, as discussed below.

The deposits excavated as Unit II (AD 1730–1800) related to the Collect Pond and the tanning industry. Unit II, lying around 4.5 metres below the current ground surface, preserved the most complex stratigraphic contexts (Figure 4.6). Within the limited excavation, the exposed sediments were typically 0.35–0.5 metres thick, and related to a number of discrete phases of ponding. These sealed the remains of a tanning vat, which was cross-sectioned. The reconstructed location of the tannery complex places the MCC Tunnel midway between the margins of the Collect and the Eastern Outlet (Figure 4.1). Interpretations of the micro-stratigraphy are reinforced by the hydromorphic properties of the sediment (i.e. oxidation and reduction stains) that are consistent with more protracted submergence of hides during the tanning regimen. Heavier sediments, clays and dense silts somewhat closer to the Collect Basin interior implicated a deeper water body. Hide processing involved soaking and other activities requiring standing water. However, since this location was at the perimeter of the Collect it is not possible to determine the precise margins of the shoreline at the time that the vat was in use. The stratigraphy, and, in particular, the rhythmites, offer preliminary indications that the shore margins of the Collect varied considerably from year to year, possibly accounting for differences in the plotted shapes and extents of the Collect Pond on the historical maps (see Figure 4.3).

The most recent phase of activity identified by the excavation (stratum I) related to the infilling of the Collect, and later processes of grading and levelling (AD 1800-present), as Pearl Street was transformed from a tanning centre to an urban residential neighbourhood (Figure 4.5). Two discrete intervals of land filling can be differentiated chronologically, largely on the strength of the cultural debris incorporated in the fills. The fabric of the lower fill (sub-unit 2) is indicative of open pasture and perhaps even agricultural land (i.e. sod, humic mats). Episodic, high-energy flooding events with signature cross-bedding structures occurred over the course of the infilling. This sequence of events conforms to the initial 'turnover' of the non-urban landscape. The upper fill (sub-unit 1) consisted of redeposited lower fill mixed with sands and gravels clearly hauled in from distant locations. By this time the local neighbourhoods had been sufficiently ensconced to inhibit unlimited local access to landfill. This marks the beginning of Pearl Street's transformation from a local hub of the tanning industry to an administrative centre for a major metropolis.

THE LANDSCAPE ARCHAEOLOGY OF LOWER MANHATTAN

The various sources of geologic, cartographic, informant and historical data abstracted for this study can be merged with field observations to develop a more comprehensive model of landscape archaeology for Lower Manhattan. A diachronic account of landscape evolution, land use, occupation and site preservation is presented in Figures 4.7a and 4.7b. The sections depicted follow a south-southeast to north-northwest transect spanning the reconstructed mid-18th century geography of Lower Manhattan (Figures 4.1 and 4.4 from the former glacial kame known as Catimuts Hill, across the lower sub-basin of the Collect Pond and terminating at Kalkhoek Promontory). As depicted, Stage IV is the last time that the natural topography and glacial aged landscape dictated the course of human settlement prior to large-scale re-landscaping during the 19th century. It is also the most accurate representation of the city setting during the peak period of the Pearl Street tannery complexes. As the stages progress the natural landscape is systematically displaced by the humanly engineered landscape. The Collect Pond serves as a barometer of these changes.

Stage I (Terminal Pleistocene/Early Holocene: to 10,000 BP) identifies the early post-glacial environment (Figure 4.7a). The MCC Tunnel preserved deeply stratified and steeply bedded gravels and sands that could only have been derived from extensive melt-water drainage. The Hudson was the trunk stream for a series of minor drainage lines that migrated in a southeasterly direction and created the trough that evolved historically into Lispenard's Meadows. The glacial kames that are noted in 19th century geological and geographic accounts of New York City (Cozzens 1843; Gratacap 1909; Hobbs 1905) attest to the survival of these steep-sided gravel-sand hills well into the 19th century. While Archaic-period Native American sites could have been located in the vicinity of Pearl Street, the stratigraphy, as well as regional geo-archaeological successions, argue for strong dynamism of northeastern drainages in the early post-glacial period (Schuldenrein 2003; Thieme 2003), thus minimising the survival potential for Early Holocene (Lower and Middle) Archaic sites in the flood belts of major stream valleys.

Stage II (Middle-Late Holocene: to 2000 BP) is the earliest period for which geological evidence is recorded, albeit indirectly, in the study area (Figure 4.7a). The Unit IV paleosol illustrated in the Collect Pond profile (Figure 4.5) began to form over 1,500 years ago, and surfaces progressively stabilised by soil-forming processes. Since the paleosol caps a regionally extensive surface, it would have extended across the terrace bisected by the former channel incising the antecedent Lispenard's Meadows. The historical breadth of the meadow and the coarseness of the channel fill indicate that a braided stream would have migrated across the valley. However, the Late Holocene channel would have been considerably narrowed in response to the 10–15 metre rise in sea level and the expansion of near-shore environments to which the stream trended. The stream was also sustained by active springs that discharged through the local aquifer. It is probable that considerable late prehistoric Native American activity (Terminal Archaic-Woodland: c. 2000–500 BP) would have occurred along the channel banks. Evidence was not encountered at the MCC Tunnel because of the limited window afforded by the excavation. However, site expectation is high in such a setting.

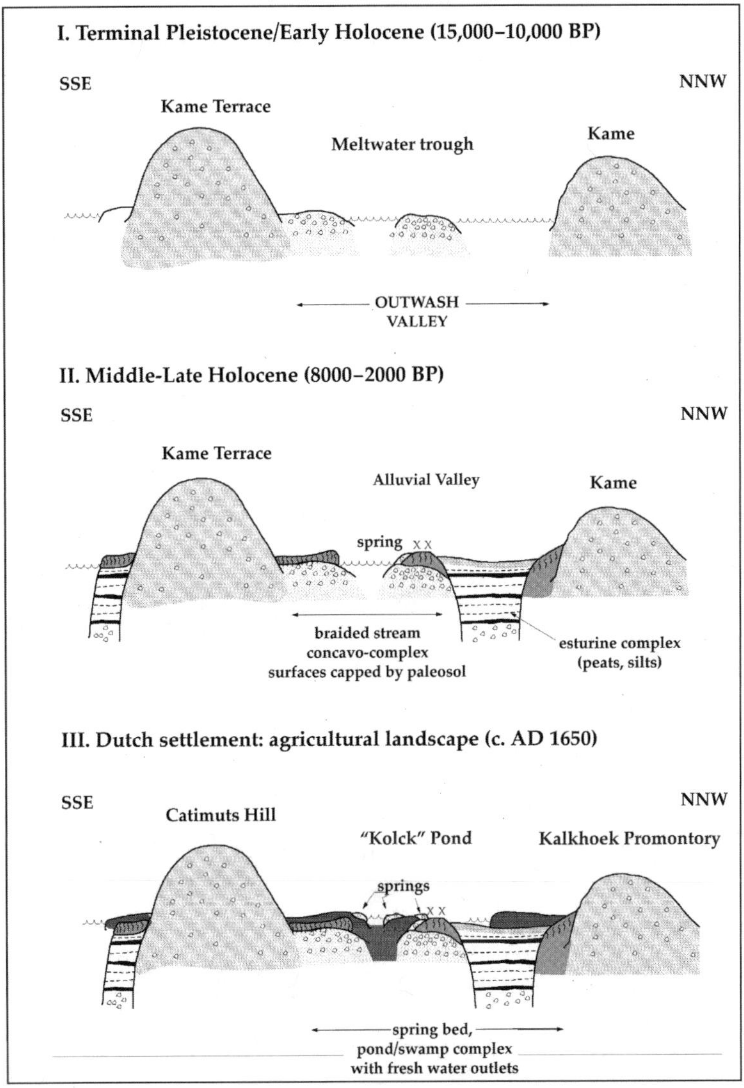

Figure 4.7a Diachronic model of land use and occupation, Lower Manhattan (15,000 BP–AD 1650).

By *Stage III* (AD 1650-Dutch settlement), historical maps depict extensive agricultural lands (Figure 4.7a). Lispenard's Meadows functioned as farm, pasture and swamp lands depending on the permeability and composition of the soils (see Castello Plan of 1660; Lyne Plan of 1728). At this time the natural waters that drained the antecedent Collect were marshalled for agricultural purposes. Who but the Dutch

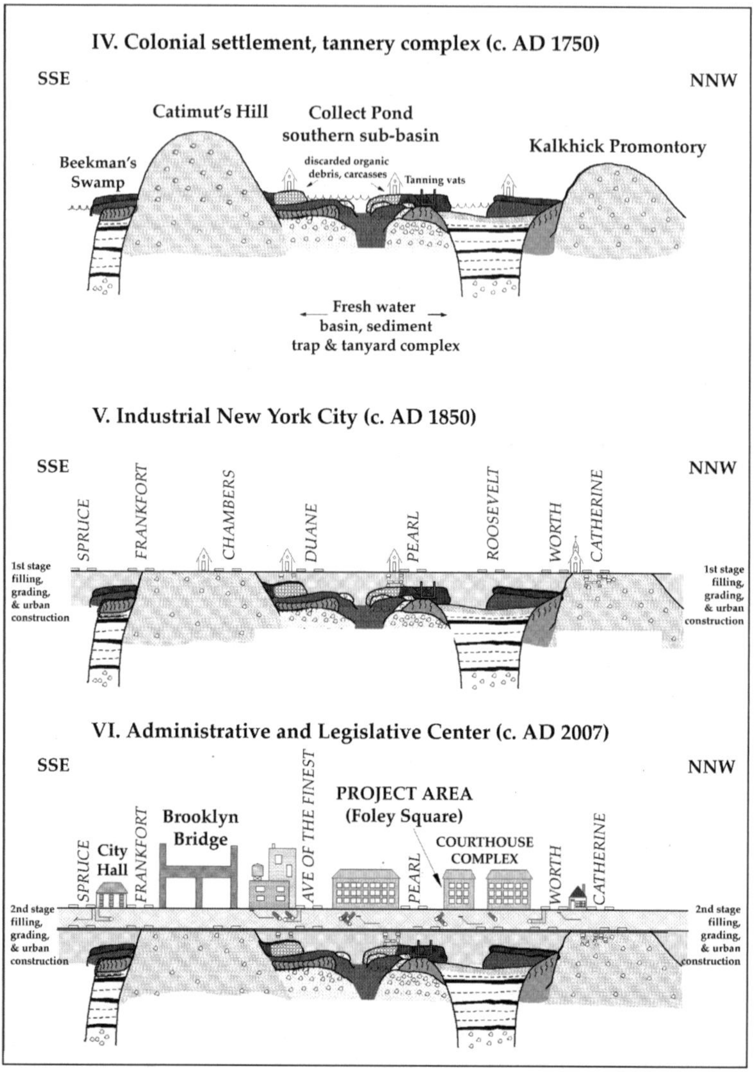

Figure 4.7b Diachronic model of land use and occupation, Lower Manhattan (AD 1750–present).

agriculturalists could have been so keenly aware of delicate hydrological balances related to land use (i.e. overgrazing, destructive farming practices, accelerated erosion), given their heritage? Indications are that there were at least two water bodies originally, nourished by springs and runoff from the surrounding kames (Figure 4.4). By the time of peak Dutch settlement the southwestern hill had been named Catimuts

Hill. The topographic gradients across the transect began to diminish as a result of limited landscaping and the aforementioned sedimentation of the floodplain.

In *Stage IV* (AD 1750), the Collect Pond became the central fixture of the economic landscape (Figure 4.7b). The entire expanse between Catimuts Hill and Kalkhoek Promontory was part of the tannery infrastructure. Both the Eastern Outlet and the Collect Pond served distinctive functions in the tanning process. The pond was a swamp that was probably utilised for long-term hide soaking, while the Eastern Outlet was the source of running water and served to remove chemicals and rinse the hides. Springs feeding both softened the hide fabric. It is probable that vats lined the sill separating the Collect's basins and that sets of vats were emplaced everywhere along the swamp margin of the southern Collect. Extensive debris, animal carcasses and tanyard waste were probably discarded everywhere along the swamp margins. Processing sheds would also have been built along the sill and the higher ground flanking the Collect and the Eastern Outlet.

Stage V (AD 1850) registers the abandonment of the tanneries and burial of the Collect and the sub-basins associated with the tannery infrastructure (Figure 4.7b). A clay liner at the base of Unit I (in Figure 4.5) (separates the overlying landfill from tanyard debris. Extensive grading was the dominant regional activity over the course of this period, as the area around Pearl Street was transformed from a centre of the leather trade to an urban residential hub. The kames were levelled to provide fill for the pond and more generally supported the surfaces of the grid of streets and sidewalks that were eventually built.

There was a second major phase of land filling after AD 1900. This was *Stage VI* (Contemporary), a period marking a functional transition in the life of the Pearl Street area. The district evolved to its contemporary role as an administrative and legislative centre. The excavated fill records more complex artefact assemblages of the 20th century, including utility debris and large-scale construction rubble (see sequence in Figure 4.6). The transformation of the prehistoric environment was completed.

THE COLLECT AS CULTURAL LANDMARK

The model of landscape change presented above ties the remembered role of the Collect Pond in the history of Lower Manhattan to its physical imprint in the ground. It is one thing to imagine the remains of the pond below the complex of courthouses that now (at least partially) cover the space where it once was, but quite another to see its actual remnants under many layers of fill and utility trenches. The archaeological investigations of both the 18th century African Burial Ground that once flanked the western edge of the Collect Pond of and the 19th century Five Points neighbourhood that grew up on top, and east, of the filled pond have successfully revealed pieces of the city's history that were in the first case forgotten and in the second seriously misunderstood. The African Burial Ground project, in particular, connected New Yorkers to the reality of slavery in the city's past, a fact that had been all but overlooked. The physical remains – in this case, burials – provided a powerful connection to that past and a rallying point for the local community, which became involved in the conduct of the project, in the disposition of the human remains and the memorialisation of the site (LaRoche and Blakey 1997). The Collect Pond is part of that history. It was the central feature in the landscape where Africans and African Americans were 'allowed' to bury their dead. To reveal its physical presence and explain its evolution and demise adds to an understanding of how the Collect functioned as a boundary between the inside and the outside, and was ultimately absorbed into the city's grid, not coincidentally at the very same time the African Burial Ground was eliminated. Just as the human remains connect us to the reality of an unremembered past, the buried remains of the Collect connect us to the landscape of exclusion.

For the earliest New Yorkers the landscape of the Collect – be it the stream and basin complex of fresh water that stretched from one side of the island to the other, or the impoundments that were created by Dutch agricultural practices, possible climatic changes or some combination of both – was a boundary. It separated wilderness from city, rural from urban, safe from unsafe. Simon Schama (1995: 7–9) claims that wilderness is created with words. It is not wilderness until we name it and even then it is often structured from our own perceptions, contrivances and visions. The landscape of the Collect was such a wilderness. For the urbanites who sought out the Collect for an afternoon of skating or picnicking, it was a benign wilderness, a pleasant place to escape the rigours of city living. For the farmers who faced the threat of Indian attack, it was a dangerous wilderness and they protected themselves by creating a buffer zone of African American farmers who were offered land and partial manumission in return. The Collect district was also wilderness in the sense of being uncivilised. More rowdy (i.e. uncivilised) behaviour was tolerated outside the bounds of the genteel city than inside. One could even get stronger drink and any number of other dubious pleasures in the taverns near the Collect than were available in the city proper (Neville 1994: 44–46).

But we know from the historical record that at the same time as the Collect was valued as 'wilderness', it was exploited for industry. From the late 17th century tanneries lined the eastern shore and southeastern outlet, and by the end of the 18th century they, and other noxious industries, had so contaminated the pond that it was covered over. The people who worked in the industrial landscape saw something very different than the well-dressed revellers who only occasionally visited. The workers – and industries – took advantage of the Collect for its productive potential. It supplied the water for tanning and brewing and even for drinking until it was no longer potable. The slaughterhouse worker who mapped his world – the streets to the east of the Collect – included the slaughterhouse, his workplace, but he also included the African Zion Church, probably the church he attended, and the Tea Water Pump, the source of drinking water for the neighbourhood.

In addition to attracting industrial workers, the streets first developed on the eastern edge of the pond – Orange Street, which became Baxter, and Magazine, which became Pearl – at the turn of the 19th century also attracted artisans who couldn't afford to buy property in more established parts of town. Several features that belonged to such artisans were investigated during the Five Points project (see Yamin 2000: Chapter 5). Tobias Hoffman, a baker, for instance, brought his family to Pearl Street in about 1798. He had his bake oven on one lot and his house next door. What was most surprising about the Hoffmans' household goods was their quality, detail and sophistication. In spite of living in the midst of foul-smelling industrial establishments and within a block or so of the pond that was being condemned as a sink of filth at the time, they served meals on elegant Chinese porcelain and drank from decoratively etched glasses. Clearly, for them, the condition of the pond was less important than the opportunity to own property and set up business.

For others, however, the opportunity to convert the pond to saleable land was extremely important, and it is not inconceivable that complaints about its condition were simply an excuse. The pond, after all, could have been cleaned up. Instead, it was filled and carved up into building lots, which eventually became part of the Five Points neighbourhood. Again, the lure of affordable housing attracted residents and immigrants – first from Eastern Europe and Ireland and later from Italy and China – who put up with overcrowded and miserably unsanitary conditions in the tenements that lined the streets. This, too, is part of the legacy of the Collect. One of the common complaints about Five Points was the dampness of the ground, presumed to be caused by the fact it was built on top of the old pond. Dickens described 'lanes and alleys paved with mud knee-deep' (Dickens 1842: 36), and the implication was that only the less than human would tolerate such conditions.

The Collect's association with Five Points is much better known than its association with the African Burial Ground, but when the burial ground was declared a national landmark in 1993 the Collect figured prominently in the article and accompanying bird's-eye view that appeared in the *New York Times*. The Collect appears in the lower left-hand corner of the image while the African Burial Ground is in the middle and is shown abutting the palisade that separated it from the city until at least the 1760s (Neville 1994, Figure 1). It was this image that seems to have been the impetus for Neville's study, *Overlooking the Collect: Between Topography and Memory in the Landscape of Lower Manhattan*. In it he tried to make the pond visible:

> The idyllic Collect, the horrific Collect, and the suppressed Collect have all left their traces.... A phantom historical district underlies a broad swath of the city, and signals its presence if we know how to look for it. (Neville 1994: 91)

Neville's search was through documents and maps, but neither he nor anyone else had looked for the physical remains of the Collect Pond. The reconstruction presented here is centred on the physical development of the pond and land use in its environs. A construction project made it possible to research the Collect in a way that had never been done previously. A systematic reconstruction of events was structured on the accounts of historic geographers, maps and scientific geomorphic and archaeological investigation. The model illustrated here demonstrates that the Collect had its beginnings as a post-glacial landscape feature; human intervention made it into a socio-economic hub; and it eventually faded in the wake of urban expansion. It was idyllic and horrific and then it was gone, but only from view.

CONCLUDING THOUGHTS

The Collect re-emerges in various stories of the past, the most recent one having to do with the African Burial Ground. It was clearly a centrepiece in the cultural fabric of early Manhattan and cannot be completely suppressed. Even Pearl Street and the 4 metres of fill interlaced with utilities that underlies it could not completely suppress the physical remnants of the old Collect and its catchment. The tanners who made use of the stagnant waters of the pond and the running water of the Eastern Outlet are not forgotten. Fragments of the facilities they built to take advantage of the stagnant and running water endure and remind us of their presence, just as the network of utility trenches that cut through them reminds us of the infrastructure that supports the modern city.

Curiously, Maya Lin's sculpture in front of the courthouse to which the MCC Tunnel connects consists of four blocks of granite, each containing an inverted water fountain. Viewing circles on two sides of each block make it possible to see through all four cubes simultaneously – to look through the water to the other side. It is as if she were capturing above ground what was once below and is still there for those who want to remember. Just as Robertson's painting (Figure 4.2) contrasts the natural world (the Collect in the foreground) with the civilized world (the settled city in the distance), Lin's sculptures 'reflect her concern with the notion of place, point of view and the contrast between artificial and natural, public and private' (GSA flyer describing 'Sounding Stones' by Maya Lin). Quite unintentionally, it seems, the present landscape speaks to the past – unless, of course, Lin had done more research than we know.

'Made land', it would seem, cannot hold its secrets forever. In January 2006, fragments of old sea walls were unearthed beneath the relatively recent (1960s) made land that is Battery Park, just to the south of where the World Trade Center stood. And, of course, the destruction of the Trade Center revealed its underpinnings – the bathtub that had been carved into historical fill and the bedrock below to hold back the groundwater that would otherwise have undermined its foundations. Cultural landscapes in cities undergo dramatic changes, some gradual, some all too sudden, but in changing they add to the complexity that makes cities what they are. The depth of the stratigraphic record revealed in the MCC Tunnel visually connected the present streetscape to the landforms that pre-dated Dutch occupation. Specifically, the layers revealed the impact of human agency on the landscape in one part of New York City over an almost 400-year-long period. In a more general sense, they revealed how cities remake themselves, and while much is destroyed in the process much is also absorbed and reinterpreted. In a physical sense, the Collect is gone, the African Burial Ground is gone, Five Points is gone and the World Trade Center is gone, but all of them belong to the layered cultural landscape of Lower Manhattan.

ACKNOWLEDGEMENTS

We wish to thank graphic artists Sarah Ruch, Nick Dubroff and Mark Smith for their careful and creative graphics without which this paper would make much less sense. We are also grateful to Maria L. Schleidt-Peñalva for her contributions to the original study and to Joan Geismar for her insights into the evolution of Foley Square. And finally, we are grateful to the editors of this volume for inviting us to contribute and for putting up with, and constructively commenting on, our delayed submissions.

REFERENCES

Andrews, W.L. 1893. *The Bradford Map: The City of New York at the time of the Granting of the Montogomerie Charter*. New York: De Vinne Press.

Asbury, H.H. 1928. *The Gangs of New York: An Informal History of the Underworld.* New York: Alfred A. Knopf.

Baron, W.R. 1988 'Historical climates of the northeastern United States: seventeenth through nineteenth centuries,' in G.P. Nicholas (ed) *Holocene Human Ecology in Northeasten North America*. New York: Plenum Press, pp. 29–46.
Blakey, M.L. and L. Rankin-Hill (eds) 2004. *Skeletal Biology Final Report, Vols. I and II*. Prepared for the United States General Services Administration, Northeastern and Caribbean Region.
Brown, H.C. 1913. *Old New York*. New York: Lent & Graff Company.
Burroughs, W.J. 1992. *Weather Cycles: Real or Imaginary?* New York: Cambridge.
Cantwell, A.M. and D. diZerega Wall 2001. *Unearthing Gotham: The Archaeology of New York City*. New Haven: Yale University Press.
Castello Plan. 1660. Redraft of the Castello Plan, New Amsterdam in 1660. John Wolcott Adams. Map Division, New York Public Library.
Cheek, D.C. (ed) 2006. *The History and Archeology of the Secular Uses of 290 Broadway*. Ms. Report prepared for the General Services Administration.
Cozzens, I. 1843. *A Geological History of Manhattan or New York island Together with a Map of the Island, and a Suite of Sections, Tables, and Colu*mns. New York: W.E. Dean.
Dickens, C. 1842. *American Notes, A Journey*. New York: Harper and Brothers.
Fisher, L.E. 1986. *The Tanners*. Colonial American Craftsmen Series. Boston: David R. Godine.
Foster, G.G. 1990. *New York by Gas-Light*. Originally published in 1850. Edited and with an introduction by Stuart Blumin. Berkeley: University of California Press.
Fullerton, D.S. (ed) 1992. Quaternary Map of the Hudson 4×6 Quadrangle, United States and Canada. Map I-1420 (NK-18). Miscellaneous Investigations Series. Reston, Virginia: U.D. Department of the Interior, U.S. Geological Survey.
Geismar, J. 1993. Reconstruction of Foley Square: Historical and Archaeological Report. Prepared for the City of New York Parks and Recreation. New York: Coe Lee Robinson Roesch, Inc. (submitted to the City of New York Department of Parks and Recreation).
Gratacap, L.P. 1909. *Geology of the City of New York with Numerous Illustrations and Maps*. 3rd ed. New York: Henry Holt and Company.
Harris, E. 1993. *Principles of Archaeological Stratigraphy*. 2nd ed. London: Academic Press.
Hayden, D. 1995. *The Power of Place: Urban Landscapes as Public History*. Cambridge: The MIT Press.
Hobbs, William H. 1905. The Configuration of the Rock Floor of Greater New York. *United States Geographical Survey Bulletin* 270: 1–93.
Holliday, Vance T. 2004. *Soils in Archaeological Research*. Oxford: Oxford University Press.
Ingle, M., J. Howson and E.S. Rutsch 1990. *A Stage IA Cultural Resources Survey of the Proposed Foley Square Project, Manhattan, New York*. New York: Historic Conservation and Interpretation, Inc. (report prepared for Edwards and Kelcey Engineers, Inc. Manuscript on file with Landmarks Preservation Commission).
James, B.B. and J.F. Jameson 1913. *Journal of Jasper Danckaerts 1679–1680*. New York: Charles Scribner's Sons.
Knight, E.H. 1876. *Knight's American Mechanical Dictionary*. New York: Hurd and Houghton.
LaRoche, C.J. and M.L. Blakey 1997. Seizing Intellectual Power: The Dialogue at the New York African Burial Ground. *Historical Archaeology* 31(3): 84–106.
Lyne, J. 1728. *A Plan of the City of New York from an Actual Survey Made by James Lyne*. Orrin and Vanderhoven, New York. Map Division, New York Public Library.
Maerschalck, F. 1754. *Plan of the City of New York from an Actual Survey. Anno Domino MDCCLV by Francis Maerschalck, City Surveyor*. Lithographed by G. Hayward for D.T. Valentines Manual. Map Division, New York Public Library.
Milne, C. 2000. 'Slaughterhouses in Which I Labored': Industry, Labor and the Land from Colonial Times to 1830. In R. Yamin (ed) *Tales of Five Points: Working-Class Life*

in Nineteenth-Century New York, Vol. 1. A Narrative History and Archeology of Block 60. West Chester: John Milner Associates, pp. 15–36.

Negendank, J.F.W. 2004. The Holocene: Considerations with Regard to its Climate and Climate Archives. In H. Fischer, T. Kumke, G. Lohmann, G. Floser, H. Miller, H. von Storch and J.F.W. Negendank (eds) *The Climate in Historical Times* (Berlin: Springer-Verlag, pp. 1–12.

Neville, C.P. 1994. *Overlooking the Collect: Between Topography and Memory in the Landscape of Lower Manhattan*. Unpublished thesis submitted in partial fulfilment of the requirements for MSc in Historic Preservation, Columbia University.

Newman, W., D. Thurber, H. Zeiss, A. Rokach and L. Musich 1969. Late Quaternary Geology of the Hudson River Estuary: A Preliminary Report. *Transactions of the New York Academy of Sciences* 31: 548–570.

Perry, W.R., J. Howson and B.A.Bianco (eds) 2006. *New York African Burial Ground Archeology Final Report*, Vols. I-IV. Prepared by Howard University, Washington, DC for the United States General Services Administration, Northeastern and Caribbean Region.

Ratzer, B. 1854 [1767]. Plan of the City of New York. Surveyed in 1767. New York: D.T. Valentine. Map Division, New York Public Library.

Rothschild, N. 1990. *New York City Neighborhoods: The 18th Century*. New York: Academic Press, Inc.

Schama, S. 1995. *Landscape and Memory*. New York: Alfred A. Knopf.

Schuberth, C.J. 1968. *The Geology of New York City and Environs*. Garden City, NY: The Natural History Press.

Schuldenrein, J. 2003. Landscape Change, Human Occupation and Archaeolgical Site Preservation at the Glacial Margin: Geoarchaeological Perspectives from the Sandts Eddy Site (36Nm12), Middle Delaware Valley, Pennsylvania. In D.L. Cremeens and J.P. Hart (eds) *Geoarchaeology of Landscapes in the Glaciated Northeast*, New York State Museum Bulletin 497, The University of the State of New York, Albany, pp. 181–210.

Schuldenrein, J. 2004. *Geoarchaeological Investigations at Thomas Paine Park and Foley Square*. Report prepared for J. Geismar.

Sheppard, W.J. 1989. *A Good Summer's Tan: A Report on the Excavation of the Royer Tannery site (36Fr325), 1987–1988*. Waynesboro, PA: Renfrew Museum and Park.

Stuiver, M., P.J. Reimer, E. Bard, J.W. Beck, G.S. Burr, K.A. Hughen, B. Kromer, G. McCormac, J. van der Plicht and M. Spurk 1998 INT-CAL 98. Radiocarbon Age Calibration 24,000–0 cal B.P. *Radiocarbon* 40: 1041–1083.

Taylor, B. and J. Roberts 1797. *A New and Accurate Plan of the City of New York in the State of New York in North America*.

Thieme, D.W. 2003. Archaeological Site Formation in Glaciated Settings, New Jersey and Southern New York. In D.L. Cremeens and J.P. Hart (eds) *Geoarchaeology of Landscapes in the Glaciated Northeast*, New York State Museum Bulletin 497, The University of the State of New York, Albany, pp. 163–180.

Viele, E.L. 1874. *Topographical Atlas of the city of New York Including the Annexed Territory. Showing Original Water Courses and Made Land. Prepared under the Director of Egbert L. Viele, Civil and Topographical Engineer*. Map Division, New York Public Library.

Vingboom, Joan. 1639. *Joan Vingboom's Map, 1639. Edward Van Winkle, New York, 1916.* Map Division, New York Public Library.

Yamin, R. (ed) 2000. *Tales of Five Points: Working-Class Life in Nineteenth-Century New York.* Philadelphia: John Milner Associates, Inc. (Report prepared for Edwards and Kelcey and the General Services Administration).

Yamin, R. 2001. From Tanning to Tea: The Evolution of a Neighborhood. *Historical Archaeology* (special edition 'Becoming New York: The Five Points Neighborhood', ed. R. Yamin) 35(3): 6–15.

Yamin, R. and K. Metheny (eds) 1996. *Landscape Archaeology, Reading and Interpreting the American Historical Landscape*. Knoxville: University of Tennessee Press.

Yamin, R., J. Schuldenrein and M.L. Schleidt-Peñalva. 1995. *Archeological and Geoarcheological Investigations Associated with the Construction of the Metropolitan Corrections Center Tunnel Under Pearl Street, Foley Square, New York*. Prepared by John Milner Associates, Inc. for Edwards and Kelcey Engineers, Inc. and General Services Administration, Region 2.

CHAPTER 5

Cultural Landscapes, Communities and World Heritage: In Pursuit of the Local in the Tsodilo Hills, Botswana

Susan O. Keitumetse, Geoffrey Matlapeng and Leseka Monamo

INTRODUCTION

Almost all heritage sites and monuments in Africa lie within wider landscapes in which local communities live. Accordingly, heritage management, like broader political efforts to achieve community participation, often aims to encourage local participation in decision-making. Recognising local cultural values attached to the physical heritage is a central means through which sustainable approaches in African heritage management can be pursued. For the United Nations Educational, Scientific and Cultural Organisation (UNESCO), local values are significant as a means of achieving a satisfactory intellectual, emotional, moral and spiritual existence, allowing communities to define their futures in an integrated manner (Edroma 2003; UNESCO 2003). United Nations Conference on Environment and Development (UNCED) 'Agenda 21' guidelines define sustainable development as an approach that emphasises 'giving communities a large measure of participation in the sustainable management and protection of the local ... resources in order to enhance their productive capacity' (Robinson 1993: 33).

The UNESCO World Heritage list, particularly in its section on 'Cultural Landscapes' – which by virtue of their scale, dynamism, multiple types of ownership and significance are usually much more embedded in community and local life – is for Africa an increasingly important arena for local, national and supra-national values to meet, and is increasingly framed as concerned with sustainable tourism and sustainable heritage. While the sustainable use of cultural heritage is an important aim, the question arises: what might 'sustainable' approaches to heritage management entail in practice in the diverse contemporary social, historical and prehistoric landscapes of Africa?

The production of heritage can take place in several different contexts, including archaeological excavations, historical documents or the placement of values and meanings on the physical heritage (Keitumetse 2005a) – buildings, landscapes or even the natural environment (Lipe 1984). It is also common for archaeological heritage to be produced and valorised through research (Bednarik 1994; cf. Carver 1996), and academic and professional discourses are to some extent responsible for what is promoted as heritage in many varied situations around the world. In particular, the application and translation in local contexts of concepts such as 'world heritage' highlights and promotes certain 'heritages' as more valued than others (Keitumetse 2005b). Alongside such heritage 'production', however, the consumption of heritage also involves social processes, from national ideologies to individual experiences, as local communities attach their own values and meanings to the heritage (Keitumetse 2006). Archaeological heritage remains unpopular among local communities in Botswana (Keitumetse 2002) and in other African countries (Bednarik 1994; Schmidt 1995; Schmidt and Patterson 1995; Drewal 1996; Fontein 2000; Ndoro 2001; Matsikure 2002). This requires archaeologists and other heritage professionals to look more carefully at the nature of communities' relationships with official approaches to heritage management, especially because the cultural values attached to monuments by officials may be perceived very differently by local communities (Ndoro 2001; Matsikure 2002; Keitumetse 2005a). Indeed, by focusing only upon 'official' values we can compromise the potential for local communities to use archaeological heritage in education, cultural or economic life. In most parts of Africa, current approaches to heritage management can serve to hold back communities from acknowledging archaeological heritage as cultural heritage.

In this wider context, this chapter explores how the cultural landscapes of 'world heritage' are understood and experienced by the local communities for whom they are home as well as heritage. By addressing the issues pertaining to the engagement of local communities as stakeholders of heritage, with particular reference to the Tsodilo Hills World Heritage Site in Botswana, we suggest that a range of local communities' perceptions be considered and acknowledged as important data in understanding the complexities of social landscapes. Conclusions drawn from such data should then be compared with the impulses of official (both national and international) heritage managers, a crucial element in developing sustainability in the management of world heritage landscapes.

THE LANDSCAPE OF THE TSODILO HILLS

The Tsodilo Hills are located in the Kalahari Desert in the Ngamiland District of Botswana. The site lies about 250 kilometres from Maun, the tourism capital of northern Botswana. It is approximately 50 kilometres west of the panhandle of the Okavango Delta (Figure 5.1). Rising 410 metres above the surrounding country, the Male hill is the highest point in the desert region and, at roughly 1,395 metres above sea level, has the highest peak in Botswana (Brook et al. 1992). The site can be reached by a charter plane – there is an airstrip at Tsodilo, which is a settlement of approximately 200 people, or by road on the Trans-Kalahari Highway, leaving at either Sepopa or Nxamasere villages.

The most striking and famous aspect of the Tsodilo Hills landscape is its rock art (Figure 5.3). The Tsodilo Hills are a major geological landmark in Botswana, comprising four quartzite hills, locally known as *mosadi* (female), *monna* (male), *ngwana* (child) and *ngwana-wa-ngwana* (grandchild). These rock formations are covered by over 4,500 paintings on 400 rock panels, and many carvings concentrated in the site's three main hills, or *inselbergs*[1] – one of the highest concentrations of rock art in the world. The art appears on isolated rock panels and

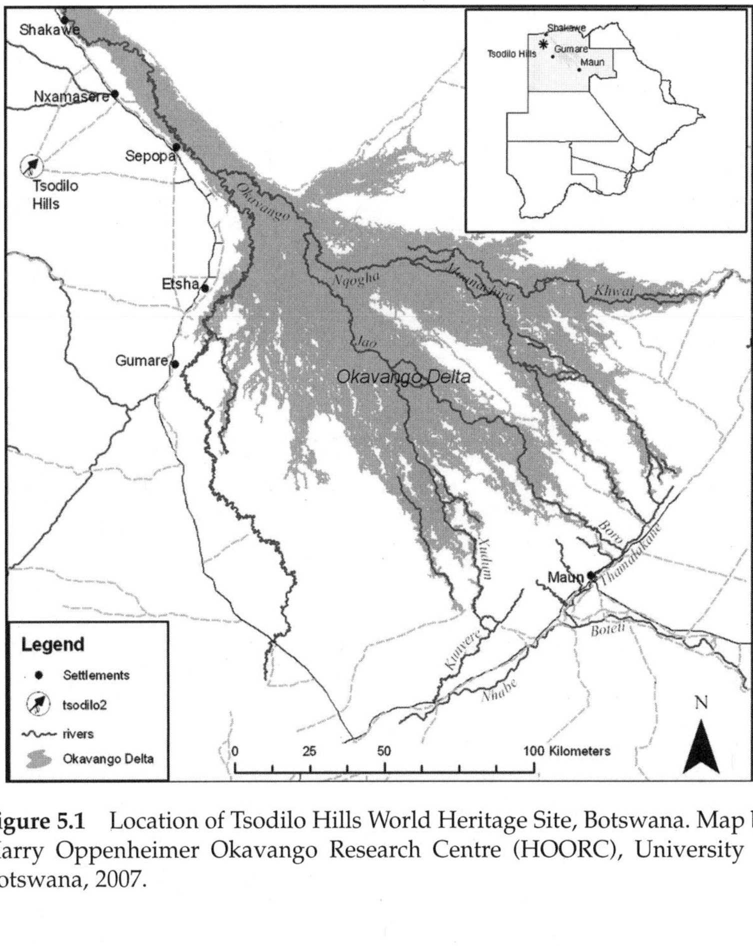

Figure 5.1 Location of Tsodilo Hills World Heritage Site, Botswana. Map by Harry Oppenheimer Okavango Research Centre (HOORC), University of Botswana, 2007.

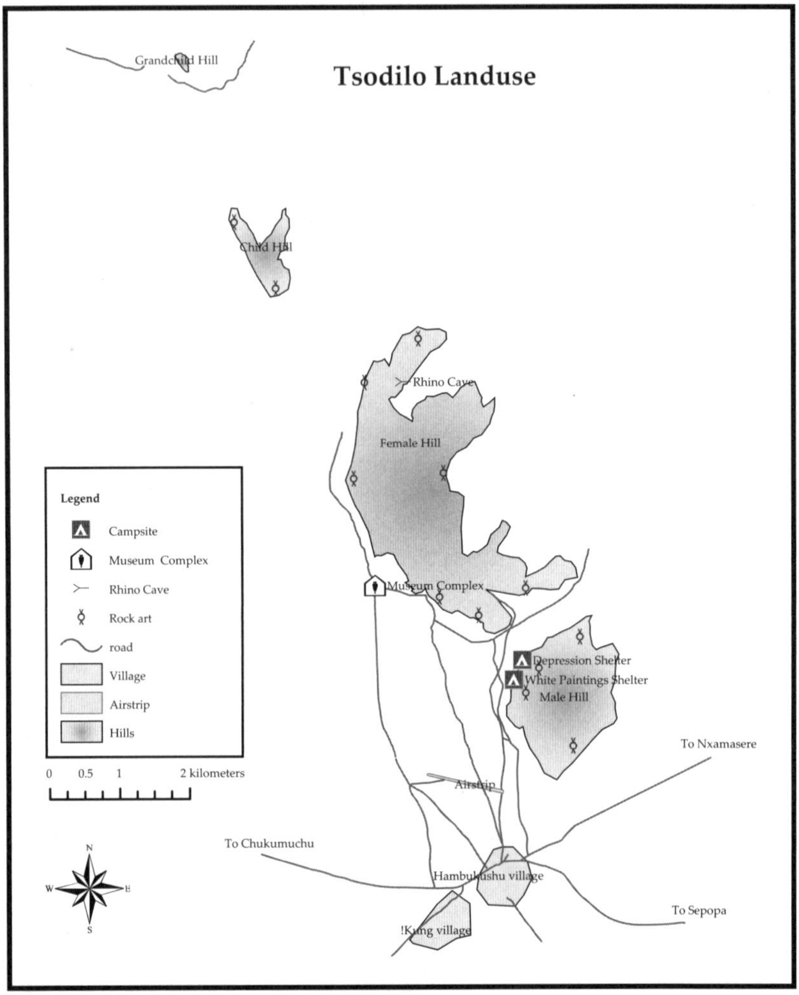

Figure 5.2 Tsodilo Hills and other geographical features mentioned in the text. Data supplied by Department of Surveys, government of Botswana, 2000 and modified at HOORC-UB, 2007.

in relatively unsheltered overhangs, compared with the more sheltered locations of other paintings in southern Africa such as those at Matopo, Zimbabwe and Drankensburg, South Africa (Robbins et al. 1989; Lewis-Williams 1990; Campbell 1994).

The Tsodilo rock art was first recorded in the early 20th century (Passarge 1907: 97), and the landscape was the subject of an intensive programme of survey and recording by the Botswana National Museum during the 1980s and 1990s (Brook et al. 1992; Campbell

Figure 5.3 Some of the rock art at Tsodilo Hills World Heritage Site. *Photograph from Susan Keitumetse 2003.*

1994; De Maret 1995). Included are white ochre paintings, red ochre paintings and engravings; relative dates have been provided for both the red and white paintings. Some of the red paintings have been dated to AD 600–1200, while some of the white paintings have been dated to the 19th century. The red ochre paintings mainly consist of wild animals such as eland, giraffe, zebra, rhinoceros and elephants, with geometric design depictions. The white paintings, made from chalk-like calcrete, depict very few animals but 'include more domestic species, especially cattle' (Campbell et al. 1994), and in all cases they are superimposed onto the red ones, leading some researchers to conclude that the white paintings are of a significantly more recent date than the latter. Some of the white paintings have been interpreted as riders on horses, and this has been used to suggest a 19th century date for the paintings because the Griquas[2], famous for their horses, are known to have passed through the area during the 1850s. In general, the Tsodilo rock art paintings are more schematic and less naturalistic in depiction (Robbins et al. 1989; Robbins 1990) than those in Namibia, Zimbabwe and South Africa. Because of its appearance in unsheltered areas, the art in Tsodilo Hills has been interpreted as having been executed mainly for aesthetic, as opposed to spiritual, purposes (Campbell 1994; Botswana National Museum 2000).

However, the Tsodilo landscape holds a wide range of archaeological evidence for human settlement over a much longer time span than that of the art. Evidence of human activity in caves, rock shelters, and at seasonal camps and other settlements range from the Middle Stone Age (c. 100,000–50,000 years ago) to the Late Stone Age (first century AD), and to the Iron Age sites at Divuyu and Nqoma, where there is evidence of episodes of permanent settlement from the 7th and 8th centuries AD (Robbins et al. 1989; Robbins 1990; Brook et al. 1992), to the material remains of hunter-gatherer societies up to the recent past (Campbell 1994; Keitumetse 2005a). Archaeological evidence from Tsodilo indicates that the sites at Nqoma and Divuyu were abandoned around AD 1200, possibly due to local and regional climate fluctuations (Miller 1996). Two lacustrine carbonate samples from a former lakebed at Tsodilo yielded a date of c. 17,500 BP, indicating the presence of standing water, in contrast with the present environmental conditions (Brook et al. 1992).

UNIVERSAL VALUES AND WORLD HERITAGE AT TSODILO

The landscape described in the previous section was inscribed on the World Heritage list in 2001. This was the culmination of a series of designations and plans for Tsodilo. The Tsodilo Hills had been declared a national monument in 1927, and had been protected under the Botswana Government's Monument and Relics Act in 1970 (since been re-enacted as the *Monument and Relics Act 2001*, see Republic of Botswana 2001). The landscape was the subject of an official management plan – *Tsodilo Hills Management Plan: Scheme for Implementation* – in 1994 (Campbell 1994), and in 2000 a dossier – *Tsodilo, Mountain of the Gods* – outlined Tsodilo's case for World Heritage status (Botswana Government 2000, cf. Ecosurv 2005).

As with all World Heritage Sites, the inscription of Tsodilo followed the bureaucratic procedures of assessment against pre-defined criteria. The procedures took into account a range of factors and knowledge that largely stem from academic, or at least expert discourse and knowledge. It was recommended that the landscape met three of the ten UNESCO criteria for outstanding universal value in cultural heritage: namely representing 'a masterpiece of human creative genius' (criterion i), bearing 'a unique or at least exceptional testimony to a cultural tradition or to a civilisation that is living or which has disappeared' (criterion iii), and bearing 'a unique or at least exceptional testimony to a cultural tradition or to a civilisation that is living or which has disappeared' (criterion vi) (see UNESCO 1996). The Report of the 25th session of World Heritage Committee described how Tsodilo fitted these criteria:

> ***Criterion (i)***: For many thousands of years the rocky outcrops of Tsodilo in the harsh landscape of the Kalahari Desert have been visited

and settled by humans, who have left rich traces of their presence in the form of outstanding rock art.

Criterion (iii): Tsodilo is a site that has witnessed visits and settlement by successive human communities for many millennia.

Criterion (vi): The Tsodilo outcrops have immense symbolic and religious significance for the human communities who continue to survive in this hostile environment. (UNESCO 2001)

Thus, the Committee highlighted the remarkable concentration of rock art at the site, the range of evidence for human occupation in the area (which includes significant Early Stone Age, Middle Stone Age and Late Stone Age remains), and the significance of the site for the communities living in Tsodilo today. However, these criteria, like the management plans and historical and archaeological understanding of the region that lies behind them, were produced through academic and expert cultures, and their influence upon subsequent approaches to the management of the Tsodilo cultural landscape has served to exclude the ideas and values of local communities. One problem has been that the language in which the planning and recommendations has been conducted is very specialised, technical, scientific and un approachable to all but experts. For instance, the hills' description as 'inselbergs' is scientifically accurate, but is hardly a word that captures the spirit of the place as interpreted and described in legends by the Hambukushu and San communities living in the Tsodilo landscape (discussed below). Indeed, it is clear that the expert knowledge of the archaeological landscape of Tsodilo that was brought together in the process of the inscription of World Heritage status has at no stage been significantly communicated to a non-specialist audience, whether in Tsodilo or further afield (Keitumetse 2005b). As a result, local communities do not place value upon the archaeology in the landscape, because they do not 'know' about it in comparable ways to archaeologists – just as the processes of heritage management and designation have not engaged with local knowledge or understandings.

We want to suggest here that highlighting the overlaps and tensions between the different ways of knowing a landscape such as Tsodilo can be creative and constructive as well as challenging. In the rest of this chapter, we hope to show that by considering the human landscape of the newly defined Tsodilo Hills World Heritage Site, we can explore an interface between 'universal' and local values.

LOCAL VALUES: COMMUNITY PERSPECTIVES AT TSODILO

Today, the entire area of the World Heritage Site is owned by the Government of Botswana. However, for official management purposes, the core zone has been leased from the Tawana Land Board by the Botswana National Museum, which bears the official responsibility of managing the whole site as per the World Heritage requirements (Botswana National Museum 2000). The buffer zone is under the full control of the Tawana Land Board, which oversees land distribution in the Ngami District of Botswana. In this buffer area, communities' involvement in the management of the site is permitted and encouraged at an official level, through the activities of the Botswana National Museum.

Substantial infrastructural developments have taken place in Tsodilo since 1998 and continue today (Figure 5.2). Most of these infrastructural developments were implemented in preparation for nomination on the UNESCO World Heritage list. These include upgrading of roads that were previously sandy and difficult to navigate (Figures 5.1 and 5.2). In addition, a site museum and two permanent exhibitions were installed. Part of the museum area is used as a craft shop for curios made by members of the Hambukushu and San communities in Tsodilo. However, most of the tourism that takes place in the area is not systematic (Carman and Keitumetse 2005; Keitumetse 2005b). Part of the core zone was fenced at some stage. However, some members of the local communities vandalised part of the fence as a sign of protest against what appeared to them as the appropriation of their cattle-grazing area (Monamo 2004; Carman and Keitumetse 2005; Keitumetse 2005b).

Tsodilo is inhabited by two communities: the Bantu Hambukushu and the !Kung San, both of which comprise the descendants and relatives of the people who lived at and used the Tsodilo Hills from around AD 1860 (Campbell et al. 1994). These two ethnic groups are the principal custodians of the Tsodilo Hills. Both groups have traditional beliefs associated with the site. The San regard one painting of an eland in the Rhino trail as a rain spirit, whereas the Hambukushu believe that the rock engravings are the prints of the first cattle to arrive in Tsodilo when the rocks were still wet (Interviews by Monamo and Keitumetse 2003). The Hambukushu, the !Kung San and other communities from neighbouring settlements, such as Sepopa, Chukumuchu and Nxamasere (Figure 5.1), visit and use the site for spiritual, cultural, recreational and religious purposes. For example, members of the Zionist church occasionally use part of the site for religious worship and also use the waterhole in Tsodilo for religious purposes.

As part of the management plan of the Tsodilo site the Hambukushu and the San have been granted restricted hunting and gathering rights in the buffer zone (Campbell 1994; De Maret 1995; Botswana National Museum 2000). This has altered their way of interacting with the site. As previously mentioned, community initiative is mainly encouraged in the buffer zone but this does not mean that communities have no interest in the core zone. This is particularly true of the San, who were relocated to the buffer area of the site in preparation for the World Heritage listing. However, both the San and the Hambukushu believe that the heritage management of the rock art at the site builds upon earlier activities of their communities. They guided tourists around and ensured no damage was done to the heritage. Kgosi (chief) Samochau of the Hambukushu described how, long before the site museum was established, he discovered that some tourists had begun to tap the rock panels with hammers, and he reported such incidents to the Ngamiland District local authorities (Community meeting June 2003; Monamo 2004).

Both the San and the Hambukushu have traditional beliefs associated with the World Heritage Site. A popular legend attached to the site by both ethnic groups is featured in a permanent exhibition in the site museum and reads as thus:

> Tsodilo, which means Damp Earth, was according to oral traditions, once a family … a husband, mother and two children. When the wife divorced the husband she moved away with both children. Whilst the older child stayed near her mother, the younger child returned to the father. Hence the positions and names of the hills: Male, Female and Child (Keitumetse 2005a; Nthoi 2007). (Permanent exhibition entitled 'Welcome to Tsodilo', Botswana National Museum, Permanent Exhibition 2001)

Some of the San report that a hidden cave in the Male hill was in the past used as a ritual site where spiritual help was solicited during hunting expeditions (Monamo interviews 2000, see Campbell et al. 1994). A range of alternative interpretations of, and beliefs about, the rock art on the site exist. One myth, relating to creation and reproduction, suggests that the rock carvings at a site known as *Mangole* (knees) on the Female hill (Figure 5.2) was the place where the first sexual act took place after God sent an angel to drop people down to earth (Monamo 2004), and where God demonstrated how to perform sex by carving out three valves, resembling female genitals on the hills. In contrast, the Hambukushu believe that rock engravings on the site are the prints of the first cattle to arrive in Tsodilo when the rocks were still wet. Such intangible cultural meanings are constantly attached to the prehistoric and historical landscapes by local communities, and are closely bound up with the creation of contemporary cultural identities and a sense of place that constitute their perception of heritage at a cultural landscape such as that of Tsodilo.

However, the World Heritage status has also resulted in increased tourism activities that are gradually having impacts on local communities' perception of the Tsodilo landscape and heritage. The most significant impact is the increase in the production of crafts for sale to tourists and other visitors.

LOCAL MEETS UNIVERSAL: COMMUNITY ATTITUDES TO WORLD HERITAGE AT TSODILO

How, then, do 'official' approaches to heritage at the site – those of the Botswana government and the criteria used by UNESCO in the process of inscription – relate in practice to how local people value the rock art heritage, and why have they not easily translated into a local appreciation of the archaeological landscape as lived cultural heritage? The principal contrast between these two sets of values is that local communities live with heritage in their everyday lives in the landscape, rather than viewing the landscape in global academic terms. At times the 'universal' cultural values and meanings placed on monuments by national governments, academics and international organisations such as UNESCO coincide with those of local communities, while at other times they contradict them (Keitumetse 2005a). Community perceptions of and attitudes towards the landscape will always respond and adapt to new values imported from outside. But there is currently no mechanism in place through which World Heritage criteria can adapt to community views – a fact most sharply demonstrated by the relocation of the San from the core zone to the buffer zone in Tsodilo. Furthermore, no mechanism exists for government implementation to take stock of local community values. While government policy documents might emphasise the relevance of involving local communities in managing resources, in the case of Botswana this occurs only at the natural resources management level, as contained in the country's 1990 tourism policy together with the National Conservation Strategy of 1991. Lack of significant involvement of local communities affects the construction of landscape as a mental and emotional construct in that communities are then compelled to view and value the landscape only in economic terms, i.e. the amount of money produced from craft production (Keitumetse 2005a; Nthoi 2007).

In order to explore the changing perceptions of the cultural landscape of Tsodilo by the Hambukushu and the San in response to its new status as a World Heritage Site, Keitumetse conducted ethnographic research (informal interviews, participation observation and attendance of community meetings) in 2003 (interviews reproduced below conducted by Keitumetse unless stated otherwise). At least one member from each household in Tsodilo was interviewed. Although not discussed in this chapter, tourists were also interviewed after the tour (Keitumetse 2005). In order to gain insights into the attitudes of resident communities towards their heritage in Tsodilo the research focused upon the changing use of the site by local communities, communities' perceptions of other users (officials, tourists), and how notions of the site as a World Heritage Site informed residents' ideas about it. The research also considered the reactions of official management towards local communities' perception of the heritage, and whether local responses to the site management represented a potential threat to the heritage.

A major indication of communities' lack of participation in the management of heritage in Tsodilo is their disengagement from, and at times discomfort with, the idea of 'world heritage'. Thus, in responses to the question 'To whom does the heritage belong?', most respondents were happy to associate the heritage with the nation (people of Botswana), but not beyond that. At the time of the research, the idea of 'world' heritage had not been absorbed into local perceptions of the landscape (Keitumetse 2005: Appendix 8):

> Heritage that belongs to all is not a good point. It is Tsodilo people's culture. They say it is for all people. But we have not been explained to what exactly that means (Female, born 1975).
>
> We don't know, Tsodilo became everyone's. They never explained thoroughly what this means.... My life has not changed much since then (Male, born 1975).
>
> They found us here. They have their own cultures overseas so it can't be everyone's culture. It is for Batswana only (Male, year of birth not provided).
>
> It belongs to all Batswana, not tourists (Male, born 1961).
>
> It is not only Tsodilo heritage but for all Batswana except foreigners (Male, born 1980).
>
> Tsodilo heritage belongs to all Batswana. As for foreigners, it is not theirs, they just come to visit and that's all (Female, born 1956).

The continual arrival of visitors and tourists to the site to view the rock art is now part of the everyday experience of living at Tsodilo – it is now a characteristic of the landscape. Both the Hambukushu and the San perceive paintings as the most meaningful component of the site heritage. When asked 'Why do tourists visit?' some members of the communities responded that:

> I know of paintings in the site. I know that tourists come specifically for these. I don't know about other things [that might attract them] (Female, Hambukushu, 27 years old).
>
> Tourists come for paintings only (Male, San, 67 years old).
>
> The paintings are the main attractions for the tourists (Male, Hambukushu, 31 years old).
>
> Paintings bring tourists who then buy my crafts (Female, San, born 1956).

In these cases both communities and officials attach a shared value to the rock art heritage in Tsodilo. The likelihood of resident communities advocating the protection and preservation of the rock art as cultural heritage is high, and is closely connected with the fact that it is seen in practice by local communities as having economic value in that it attracts tourists. This is expressed in the following statements:

> Paintings are the cultural heritage made by God. We found them here. They are durable (Female, Hambukushu, 46 years old).

> Tsodilo should be protected so that the pictures could be saved (Male, San, 55 years old).

As Monamo (2004) discovered during her research in Tsodilo, a profound knowledge of the perceived value of the rock art heritage is evident in the debates that members of the two communities engage in regarding the authorship of the paintings. In most of southern Africa, rock art authorship is attributed to the San (Masao 1979; Lewis-Williams 1990; Hall and Frans 1994). However, some members of the Hambukushu community at times argue that the paintings represent the work of a supernatural being:

> There is no individual who can paint on the rocks and then the paintings survive throughout the years without fading. Only god can do that (Male, Hambukushu, year of birth not provided).
>
> If the San say that their great grandfathers painted on the Tsodilo Hills why are they not able to do that today? (Male, Hambukushu, 31 years old).

In contrast, members of the San community perceive the paintings as a form of cultural capital that can be used to support contemporary political agendas. They believe that their predecessors painted the rocks to legitimise ownership of the Tsodilo by the San. As pointed out by some interviewees from the San community, failure to replicate the paintings should not be used to deny them a direct link to their heritage in the present (Monamo interviews 2004):

> Our predecessors found it necessary to leave paintings as evidence of our [the San's] existence in Tsodilo so that no one could take our land. We might not be able to replicate like our predecessors did but that does not mean that our predecessors did not paint the hills (Male, born 1973).
>
> This is the work of our forefathers and they left the art so that their children can learn artistic concepts expressed on the hills. My father told me that, the hills are dangerous to those who do not respect them. Although I do not live in Tsodilo anymore I teach my children about the hills and still fear the hills (Male, year of birth not provided).

Such contestations of authorship illustrate a significant level of attachment to the rock art heritage by both communities at Tsodilo, and demonstrate how the two communities value and perceive such heritage. This profound knowledge and attachment manifests itself as a form of participation that can ensure the sustainability of the rock art heritage even where other socio-economic aspects compete with it. An example of an alternative type of heritage that competes with cultural heritage in Tsodilo is cattle herding by the local communities. Although several respondents indicate that the fencing of the Tsodilo site interferes with grazing land for their cattle, they still emphasise that they would like the paintings to be conserved. This indicates an understanding and appreciation of the rock art component of heritage. The responses below illustrate the compromises that community members are willing to make in order to balance cattle-grazing and rock art conservation:

> We have long tried to say that the fence border is not good for our livestock. It is too close to our borehole ... the museum people are not co-operating. We are not against fencing. *We don't want to be treated just like paintings on the rocks...* (Male, Hambukushu, community meeting, 10th June 2003).
>
> They should fence the hills and exclude the airstrip [the airstrip is closer to communities' settlements hence within the sphere of cattle grazing] (Male, Hambukushu, community meeting, 10th June 2003).

Such responses indicate that due to the comprehensive knowledge and understanding that communities have about the value and meaning of the rock art heritage, community members can appreciate approaches to its conservation even when they impact on other valuable resources such as cattle. It is clear from comments such as 'We don't want to be treated just like paintings on the rocks' that local communities are searching for significant involvement in the management of a heritage that is found in a landscape which they perceive as intertwined with their socio-historical life. Relocation of some members of the community in order to pave the way for management of the core zone appears to imply that the local communities have to be managed just as the physical heritage is.

In contrast, the knowledge and theories that communities have about the rock art are not extended to the excavated and other types of archaeological heritage in Tsodilo. Community members who were asked about the archaeological heritage, in particular, did not demonstrate the same understanding and appreciation of its value as they did with the rock art heritage. As outlined earlier, at the UNESCO level Tsodilo is valued for its archaeological significance, which is perceptible only to academics and intellectuals. Due to a lack of education by official managers, local communities continue to have limited knowledge about the scientific, hence conventional, significance of the archaeological heritage in Tsodilo (Keitumetse 2005a). As evident from responses from some residents, craft heritage is more valued than the archaeological heritage, which nonetheless is recognised at an official level as the primary value of the Tsodilo site:

> I think our crafts can bring more money than archaeological remains (Female, born 1980, Hambukushu).
>
> I don't know if the pottery in the museum was excavated in Tsodilo. I think it comes from 'the river' [*Ba-noka* (People of the river) are people from wetter areas in the region (Okavango), south of Tsodilo, in Shakawe] (Figure 5.1) (Female, born 1956, Mbukushu).

Communities' indifference towards knowledge about both the excavated and underground archaeological heritage can be attributed to a tendency by archaeologists to transport excavated material to the national museum in the capital city, as well as overseas and far away from their original cultural landscapes. This has also been aggravated by a lack of a site museum until the late 1990s. Educating communities about the importance and relevance of the archaeological heritage can result in communities cultivating a balanced appreciation of the various forms of 'heritages' found in sites such as Tsodilo. Communities have to be accorded an opportunity to become active stakeholders in the management of the site, rather than elements of the site that have to be managed by officials.

As highlighted earlier, the Botswana National Museum has secured a space for a craft shop within the site museum to assist in the marketing of community crafts to tourists. This is a very helpful gesture by the officials, but it is constantly cited by them as a significant form of resident communities' participation in the management of the heritage on the site (cf. Carman and Keitumetse 2005; Keitumetse 2005). However, although this approach benefits communities financially, it appears to have limited impact on communities' perception of the excavated and deposited archaeological heritage in particular. Responses have indicated that communities visit the site largely to engage in the sale of crafts rather than to engage with the overall archaeological heritage within the Tsodilo landscape:

> I have always sold the crafts long before the museum was established (Male, San, born 1975).
>
> We were still selling our crafts to tourists long before the museum (Female, San, born 1957).
>
> We used to sell traditional jewellery to *makgoa* [white people, tourists] even without the museum initiative (Female, Hambukushu, 50 years).
>
> Tourists still came without the museum. We have always sold our products (Female, Hambukushu, born 1980).

In the process, the museum curio shop enhances craft tourism, but not local communities' attitudes towards the significance of the archaeological heritage in the Tsodilo landscape. In addition, since the curio shop is located within the core zone, where part of the Tsodilo community was relocated from, it is questionable as to how the government intends it to be useful for the community rather than just a standard part of the visitor experience. Both local communities sold their curios to visitors long before the museum was established. It appears therefore that the museum was built to satisfy a conventional approach to heritage management, which has a tendency to view landscapes such as Tsodilo as primarily heritage areas and only secondly as a settlement area. This in turn disconnects local communities from interacting with all components of heritage in a landscape that has since acquired World Heritage status.

Although communities understand the spiritual and political relevance of the rock art heritage, it appears that this understanding does not guarantee them meaningful participation at the level of official management of the site. Some of the statements gathered during fieldwork indicate a degree of cultivated indifference on the part of communities towards official site management strategies. Community members seem not to consider themselves as having control over approaches to Tsodilo heritage management. Undertones of alienation were expressed when community members were asked the question 'Should tourists pay?' (Keitumetse 2005: Appendix 8):

> The question of whether the tourists should pay depends on the museum not us (Male, Hambukushu, 32 years).
>
> Tourists don't pay to go to the site. I don't know if it is okay or not. The museum knows if they should pay or not. I don't know (Male, San, 32 years).
>
> I just see them [tourists] go in there but I never talk to them. I can't take them in because I don't speak their language … I want them to pay the guides not the government (Female, Hambukushu, 44 years).
>
> We are fine, what they do in the museum is theirs and we are not part of it (Male, Hambukushu, 54 years).

These statements indicate that to a great extent resident communities do not perceive themselves to be significant stakeholders in the management of the heritage in the Tsodilo landscape. Indifference to decisions concerning whether or not tourists should pay mirrors the helplessness of communities in the face of relatively new and powerful stakeholders such as the Botswana government officials and UNESCO. Such alienating heritage management approaches serve to dissociate less visible components of cultural heritage, such as buried archaeological remains, from community heritage.

The general indifference among the Tsodilo communities towards official heritage approaches means that community members are more likely to view managing the site as a task for the government. As one of the respondents said:

> We don't control the heritage. The government controls it. I find that this is okay except that they do not make tourists pay … but it could have been better if the community also had a share in the control (Male, Hambukushu, 44 years).

It is evident that these communities feel excluded from a heritage that they used to interact with before zoning and fencing was implemented by the national museum within the Tsodilo landscape. Here, approaches that develop from a privileging of 'universal' values perpetuates communities' dependence on the government to manage heritage resources and establish strategies through which communities could benefit. The Tsodilo communities expect officials to identify heritage management strategies that are of financial benefit to them, like other government programmes for low-income communities and Remote Area Dwellers, which have been running long before the establishment of the museum:

> The museum does not help us at all, only the government does. I focus on making crafts to sell to tourists (Male, Hambukushu, 66 years).
>
> There is no benefit. Even when residents show tourists around, it is up to the tourists to decide the price which can be a meagre sum at times. Some tourists don't need any help therefore residents lose (Male, Hambukushu, 37 years).

There is even a lack of clarity on the fact that the national museum is a government body:

> The site is controlled by the museum at the moment but I would like it to be controlled by the government (Female, 46 years).

Nevertheless, many among the Tsodilo communities express significant interest in being involved in managing heritage to benefit the area that they perceive as culturally belonging to them. Examination of the responses for the question 'To whom should tourists pay?' (Keitumetse 2005a) illustrates that communities prefer to be actively involved in the management of the heritage:

> I want tourists to pay but only if the money is used to develop our village [settlement] (Female, 30 years).
>
> Tourists should not pay because the money will go to the government and become useless for Tsodilo residents (Male, 23 years).
>
> Money should be paid to the museum to hold and monitor for the community (Male, 37 years).

The responses documented above indicate that where heritage management remains focused upon universal, rather than local values, local communities are likely to participate in a passive, rather than an active, manner. To develop alternatives, actively and constantly engaging local communities is not enough: ways of recognising those values attached to heritage landscapes by local communities, and incorporating them into the practices of heritage management, must be found. By seeing World Heritage sites as contemporary landscapes, we can start to acknowledge how the heritage is lived and valued at a local level.

CONCLUSIONS

Advocacy for community participation in Western contexts is often based on the assertion that 'the public' are taxpayers and are automatically entitled to participate in the management of heritage resources (Schaafsma 1989; Faulkner 2000; Merriman 2003). However, the example of Tsodilo reminds us that unlike conventional models of community involvement or engagement, we must acknowledge how communities represent a central part of the landscape in which a particular type of heritage is found. Local communities' historical, as well as social worldviews of ownership and belonging are tied in to landscapes such as the Tsodilo Hills. When such landscapes are exposed to the processes of international heritage management, as through World Heritage status, too often the involvement of local communities is secondary to the preservation of the physical resource and the development of broad-scale economic and cultural benefits. At Tsodilo, although in theory communities are acknowledged as important stakeholders, their responses indicate that there are subtle but profound dissatisfactions that gradually nurture local communities' indifference towards the heritage. While world heritage is often framed in terms of 'sustainable development', in which government and international agencies aim to be 'responsible and accountable for ensuring that their policies, programmes ... encourage and support activities that are economically and ecologically sustainable both in the short and longer terms' (WCED 1987: 312), a more radical approach that understands how people themselves form part of heritage landscapes is required. As the implementation of another management plan gets underway at Tsodilo (Ecosurv 2005), it is clear that in order to achieve sustainable use of this cultural landscape it must be recognised that local communities have always been part of such landscapes, long before ideas of universal value and world heritage were put into practice.

NOTES

1. 'Inselberg' is a German term used in geography to refer to 'an isolated hill of circumdenudation e.g. a steep-sided, isolated residual hill, common in semi-arid and savannah lands, rising from a plain which is, in many cases, monotonously flat' (Clark 1998: 2002–2003).
2. People of mixed Khoisan and European descent found in parts of the present-day South Africa.

REFERENCES

Bednarik, R.G. 1994. Archaeology: Empiricist Determinism or Cultural Synthesis? *South African Archaeological Bulletin* 49: 96–99.

Botswana National Museum 2000. *Tsodilo, Mountain of the Gods: World Heritage Nomination Dossier*. Gaborone: Botswana National Museum (31 May 2000).

Brook, G.A., K.A. Haberyan and S. De Filippis 1992. Evidence of a Shallow Lake at Tsodilo Hills, Botswana, 17,500 to 15,000 yr BP: Further Confirmation of a Widespread Late Pleistocene Humid Period in the Kalahari Desert. In K. Heine (ed) *Palaeoecology of Africa and the Surrounding Islands*. Rotterdam: Balkerma, pp. 165–175.

Campbell, A.C. 1994. *Tsodilo Hills Management Plan: Scheme for Implementation*. Gaborone: Botswana National Museum and Art Gallery.

Campbell, A.C. 1998. Archaeology in Botswana: Origins and Growth. In P. Lane, L.A. Reid and A. Segobye (eds) *Ditswa Mmung: Archaeology of Botswana*: Gaborone: Pula Press, pp. 24–42.

Campbell, A.C., J. Denbow and E. Wilmsen 1994. Paintings Like Engravings: Rock Art at Tsodilo. In T.A. Dowson and D. Lewis-Williams (eds) *Contested Images: Diversity in Southern African Rock Art*. Witwatersrand: Witwatersrand University Press, pp. 131–158.

Carman, J. and S. Keitumetse 2005. Talking About Heritage and Tourism. *Society for American Archaeologists Archaeological Record* 3(5): 39–41.

Carver, M. 1996. On Archaeological Value. *Antiquity* 70: 40–56.

Clark, A.N. 1998. *The Penguin Dictionary of Geography*. 2nd ed. London: Penguin

De Maret, P. 1995. *Evaluation of the Tsodilo Hills Management Plan and its Implementation*. Paris: UNESCO/National Museum and Art Gallery [now Botswana National Museum].

Drewal, H.J. 1996. Past as Prologues: Empowering Africa's Cultural Institutions. In P.R. Schmidt and R. McIntosh (eds) *Plundering Africa's Past*. Bloomington and Indianapolis: Indiana University Press, pp. 110–124.

Ecosurv 2005. *Tsodilo Hills World Heritage Site: Integrated Management Plan*. Gaborone: Ecosurv.

Edroma, E.L. 2003. Linking Universal and Local Values for the Sustainable Management of World Heritage Sites. In UNESCO *World Heritage Papers 13, Linking Universal and Local Values: Managing a Sustainable Future for World Heritage. A conference organized by the Netherlands National Commission for UNESCO, in Collaboration with the Netherlands Ministry of Education, Culture and Science, 22–24 May 2003*. Paris: UNESCO, pp. 36–41.

Faulkner, N. 2000. Archaeology from Below. *Public Archaeology* 1: 21–33.

Fontein, J. 2000. *UNESCO, Heritage and Africa: An Anthropological Critique of World Heritage*. Edinburgh: Centre of African Studies Occasional Paper No. 80.

Hall, S. and P. Frans 1994. Expressions of Fertility in the Rock Art of Bantu-Speaking Agriculturalists.*African Archaeological Review* 12: 171–203.

Keitumetse, S.O. 2002. *Living and Archaeological Sites in Botswana: Value and Perception in Cultural Heritage Management*. Unpublished M.Phil. thesis, Department of Archaeology, University of Cambridge.

Keitumetse, S.O. 2005a. *Sustainable Development and Archaeological Heritage Management: Local Community Participation and Monument Tourism in Botswana*. Unpublished Ph.D. thesis, Department of Archaeology, University of Cambridge.

Keitumetse, S.O. 2005b. Living and Archaeological Sites in Botswana: Values and Perception in Cultural Heritage Management. *Pula: Botswana Journal of African Studies* 19(1): 37–47.

Keitumetse, S. 2006. UNESCO Convention on Intangible Heritage: Practical Implications for Heritage Management Approaches in Africa: *South African Archaeological Bulletin* 61(184): 166–171.

Lewis-Williams, J.D. 1987. A Dream of Eland: An Unexplored Component of San Shamanism and Rock Art. *World Archaeology* 19: 165–177.

Lewis-Williams, J.D. 1990. *Discovering Southern African Rock Art*. Cape Town: University of Cape Town Press.

Lipe, W.D. 1984. Value and Meaning in Cultural Resources. In H.F. Cleere (ed) *Approaches to the Archaeological Heritage*. Cambridge: Cambridge University Press, pp. 1–10.

Masao, F.T. 1979. *The Late Stone Age and the Rock Paintings of Central Tanzania*. Wiesbasen: Franz Steiner.

Matsikure, J. 2002. The Past Through the Eyes of the Present: The Management of Rock Art Sites in Zimbabwe. Unpublished M.Phil. thesis, Department of Archaeology, University of Bergen, Norway.

Merriman, N. 2003. *Public Archaeology*. London: Routledge.

Miller, D.E. 1996. *The Tsodilo Jewellery: Metal Work from Northern Botswana*. Cape Town: University of Cape Town Press.

Monamo, L. 2004. The Relevance of Tsodilo to !Kung and Hambukushu. Unpublished B.A. thesis, Department of History (Archaeology unit), University of Botswana.

Ndoro, W. 2001. *Your Monument Our Shrine: The Preservation of Great Zimbabwe*. Uppsala: Uppsala University, Institutionen för Arkeologi och Antik Historia (Studies in African Archaeology 19).

Nthoi, O. 2007. Investigating the Impact of World Heritage Status on Local Communities' Traditional Production of Crafts: Tsodilo Hills World Heritage Sites. Unpublished research report submitted in fulfilment for course *GEC 431: Introduction to Wetlands Research*. Harry Oppenheimer Okavango Research Centre, University of Botswana: Maun.

Passarge, S. 1907. *Die Buschmänner der Kalahari*. Berlin: Dieter Reimer.

Republic of Botswana 2001. *Monument of Relics Act No. 12*. Gaborone: Government of Botswana (Ministry of Home Affairs).

Republic of Botswana. *Botswana Population and Housing Census 2001*. Gaborone: Government of Botswana.

Robbins, L.H. 1990. The Depression Site: A Stone Age Sequence in the Northwest Kalahari Desert, Botswana. *National Geographic Research* 6: 329–338.

Robbins, L.H. and A.C. Campbell. 1989. The Depression Rock Shelter, Tsodilo Hills. *Botswana Notes and Records* 20(3): 1–3.

Robinson, N.A. (ed) 1993. *Agenda 21 and the UNCED Proceedings*. Third Series: International Protection of the Environment, vol. 6. New York: Oceana Publications.

Schaafsma, C.F. 1989. Significant until Proven Otherwise: Problems versus Representative Samples. In H.F. Cleere (ed) *Archaeological Heritage Management in the Modern World*. London: Unwin (One World Archaeology 9), pp. 38–51.

Schmidt, P. 1995. Using Archaeology to Remake History in Africa. In P.R. Schmidt and T.C. Patterson (eds) *Making Alternative Histories: The Practice of Archaeology and History in Non-Western Settings*. New Mexico: School of American Research Press, pp. 119–147.

Schmidt, P.R. and T.C. Patterson 1995. Introduction: From Constructing to Making Alternative Histories. In P.R. Schmidt and T.C. Patterson (eds) *Making Alternative Histories: The Practice of Archaeology and History in Non-Western Settings*. New Mexico: School of American Research Press, pp. 1–24.

UNESCO 1996. *Operational Guidelines for the Implementation of the World Heritage Convention*. Paris: UNESCO.

UNESCO 1972. The General Conference of the United Nations Educational, Scientific and Cultural Organization (UNESCO) meeting, Paris, 17th October to 21st November 1972, 17th session. *Convention for the Protection of the World Cultural and Natural Heritage, adopted 1972*.

UNESCO 2001. Report of the 25th Session of the World Heritage Committee, Helsinki, Finland. Paris: UNESCO. Archived at http://whc.unesco.org/archive/repcom01.htm (Consulted 8 June 2007).

UNESCO 2003. The General Conference of the United Nations Educational, Scientific and Cultural Organization (UNESCO) meeting, Paris, 29 September to 17th October 2003, 32nd session. *Convention for the Safeguarding of the Intangible Cultural Heritage, adopted 17th October 2003*. Paris: France.

WCED (World Commission on Environment and Development) 1987. *Our Common Future*. Oxford: Oxford University Press.

CHAPTER 6

Common Culture: The Archaeology of Landscape Character in Europe

Sam Turner and Graham Fairclough

INTRODUCTION

This chapter describes a relatively new way of doing landscape archaeology that focuses on the historic dimension of the present-day landscape. It is both an applied and a research approach, closely connected to archaeological resource and cultural heritage management, but also (through spatial planning) to the creation of future landscape. By 'creation' we mean both the physical and perceptual senses of the word, covering both the material and ideational construction of landscape. The approach we discuss regards landscape as a matter of perception but it also recognises that material things (the 'stuff' that archaeologists call material culture) underlie and support this perception. It is thus aligned with the new European Landscape Convention (hereafter ELC) (Council of Europe 2000), and underlying the following discussion is a concern with how landscape is being used in the ongoing creation of European cultural identities. It uses landscape as a tool in developing a 'common culture' – a heritage that is shared by all, existing wherever people see it and experience it.

We shall refer to this type of landscape archaeology by its English name, 'Historic Landscape Characterisation' (hereafter HLC) (Herring 1998; Fairclough 1999; Clark et al. 2004; Macinnes 2004; Turner 2006a). Similar approaches exist or are developing under different names in other western European countries, usually in the context of the ELC or organisations such as the European Association of Archaeologists and the *Europae Archaeologiae Consilium* (Fairclough and Rippon 2002) and transnational European Union (EU) funding programmes such as European Pathways to the Cultural Landscape (www.pcl-eu.de, Clark et al. 2003) and COST A27 (Landmarks 2007).

We hope to show that HLC is a valuable and in some ways innovative research tool. HLC was initially designed with two aims: to help other disciplines involved with landscape management and planning to incorporate landscape's historic and archaeological dimension into real-world decision-making; and to use landscape approaches to find new ways of looking after historic and archaeological resources. Central to the HLC approach is the treatment of the material remains of the past as part of present-day perception. This 'presentness' of the landscape requires different objectives. It requires that we acknowledge landscape as perception – an intellectual and emotional construct and a personal and social way of being in the world. HLC is a political approach because it places landscape at the centre of individual and collective identity. It promotes interdisciplinary approaches. HLC creates characterisations – maps of negotiable perceptions, rather than objective 'truths' in any straightforward way – that are informed by the processes of history in each place, and that seek to capture some of landscape's time-depth (Macinnes and Wickham-Jones 1992). HLC is applied archaeology conducted in the present, its social purpose being to help people to understand the historic dimensions of the things they value in the world around them, to recognise and facilitate the creation of new narratives and perceptions that attach to their landscape and to shape development and change as part of the process of creating future landscape – the key aspiration of the ELC.

THE EUROPEAN LANDSCAPE CONVENTION

How does the ELC define landscape? Such definitions have always been problematic, and have often been an obstacle to interdisciplinary dialogue or even to dialogue within disciplines. The ELC's very concise definition is therefore a large step forward, persuasive in its simplicity and powerful in its inclusivity:

> an area, as perceived by people, whose character is the result of the action and interaction of natural and/or human factors. (Council of Europe 2000: Article 1)

Or, in the parallel and equally 'authentic' French-language version:

> *une partie de territoire telle que perçue par les populations, dont le caractère résulte de l'action de facteurs naturels et/ou humains et de leurs interrelations* (Council of Europe 2000: Article 1).

This is a high-level, umbrella definition that seeks to enable different landscape disciplines to read into it their own concerns. Landscape archaeologists, for example, might focus on scale ('area'; '*une partie de territoire*'), human agency, and change and the passage of time (action and interaction). But the ELC also has at its core the idea that landscape is an issue of perception, through which people make sense of the world around them using 'ways of seeing'. These might follow aesthetic rules (for example), but might equally reflect personal memory or associations, or be strictly practical and functional; they can be collective or personal.

HLC recognises that landscape is contested. As many archaeologists have observed, landscape is an arena for the expression of power and the construction of identity in the present just as it was in the past (Hall 2006: 204; Tilley 2006: 7). The ways in which archaeologists assign value to landscapes, and how they react to proposed changes, are bound up with these contestations. Are traditional monument-based paradigms appropriate for landscape, or do archaeologists need to adopt a new, more appropriate, ethos or 'aesthetic' in the face of continuing landscape change (Fairclough 2006a)?

The ELC also underlines the ubiquity of landscape. Landscape is urban as well as rural, maritime as well as terrestrial, and 'everyday' and ordinary, or even 'degraded' and damaged, as well as special or outstanding. Accordingly, the ELC does not claim that certain places are more important than others in an absolute sense. It insists on the democracy of landscape through the idea of landscape as a physical manifestation of human rights – a 'common culture', as we have suggested above. Finally, the ELC is forward-looking (Council of Europe 2002: 5–6) – concerned with how people will create and construct landscape tomorrow. 'Landscape protection' is only one of three instruments the ELC proposes for looking after landscape, and not the most important. Landscape management (upkeep focused on continuing and modifying the *processes* of landscape creation) and landscape planning (including enhancement, change and creation) are both seen to play bigger roles (Sarlöv Herlin 2004, Selman 2006). Both instruments need access to knowledge and ideas about why the landscape today looks and feels as it does due to its histories (Figure 6.1). It is all the more important, therefore, for landscape archaeology to be one of the disciplines helping to construct future landscapes that contain the past as well as being new.

DIVERSITY, PERCEPTION, APPLICATION AND SCALE

What are the implications of the approach to landscape put forward by the ELC for landscape archaeology? In this section, we want to suggest that drawing upon the ELC has implications for four aspects of landscape archaeology: *diversity, perception, application* and *scale*. These implications form the basis of the HLC approach.

Figure 6.1 Holbeton, Devon (UK). Most of the hedge-banks in the photograph were first built in the Middle Ages. Such boundaries are highly characteristic of Devon landscapes.

Diversity

Landscape archaeology comprises a diverse set of approaches (cf. Muir 1999), but it is only one of many ways of envisioning landscape. Landscape is studied in landscape architecture (LE:NOTRE 2002–2006), cultural historical geography (Cosgrove and Daniels 1988; Matless 1998; Olwig 2002, 2004), cultural history (e.g. Schama 1995) and landscape ecology (Jongman and Pungetti 2004). But landscape is also a current theme in many other disciplines. For instance, in environmental psychology the potential influence of a 'good' quality landscape – that is, the recuperative or restorative powers of 'nature' – upon physical and mental health of those living in it have been studied (Mabey 2005; Ottoson and Grahn 2005). At the same time, there is a growing recognition in Europe that landscape is not only an area for aesthetic appreciation by trained specialists in particular areas, but a social and cultural matter concerning democratic rights and the common good as part of heritage or *patrimoine*. For example, the emphasis on landscape as cultural in many senses is beginning to change the rhetoric and aspirations of the IUCN (World Conservation Union) 'Protected Areas' approach, which was originally designed to protect nature and biodiversity but is increasingly being modified to recognise that biodiversity cannot be maintained without people and without establishing a strong culturally based sustainable management of landscape (Brown et al. 2005).

No single discipline can encompass landscape's wide diversity, but different disciplines often operate in isolation. Indeed, these interdisciplinary dimensions are part of one of landscape's essential qualities – diversity. Landscape archaeologies must find ways of contributing across disciplinary boundaries, situating their approaches between those disciplines that deal mostly with landscapes as concrete entities and those interpreting landscapes as ways of seeing. Especially pressing are interdisciplinary debates on shaping the landscapes of the future, particular in relation to spatial planning and practical management (Palang and Fry 2003; Scazzosi 2002, 2004).

Perception

The second issue raised by the ELC is that of alternative perceptions of landscape, and the importance of recognising multiple interpretations and realities, at both social (collective) and individual (personal) levels. We might picture four people looking at the same field: a farmer sees the different grasses and can tell what the soil and drainage are like and how good the arable potential might be; a walker from the nearby city appreciates the beautiful flowers growing in the hedges; an ecologist realises they are non-native species, feels very unhappy about them and suggests they be removed. Meanwhile, the archaeologist looking at the same field may not understand the agronomics or the ecology but would see the grassed-over ridge and furrow earthworks of medieval ploughing as an important part of the field's history.

Alternative perceptions of landscape are more than different ways of 'seeing': they are expressions of reciprocating social interaction and representations that are mediated through the materiality of the landscape. The four people imagined above look at the same things, but interpret them differently: they are in a recursive relationship with the landscape in which many social and material factors are implicated.

Application

Landscape archaeology provides research tools that can examine why contemporary landscapes look as they do, how alternative perceptions of landscape are created and to what degree humanly induced change is part of landscape's character. But it must also be an applied field. Its application is not restricted to resource management, spatial planning or environmental management, but extends into socially critical issues of the lived environment, sense of place, identity and how people see the world. Landscape archaeology involves both conventional archaeological research and heritage management, and also draws archaeological practice into debates about the relationship of people to nature, to their environment and to each other, and discussion about identity and ownership at local, regional, national and European scales.

Scales of Analysis

The fourth issue relates to archaeological data and scales of analysis. Traditionally, most landscape archaeology has collected detailed data about features in particular territories – a parish, for example – and tried to understand how people in the past understood, lived in, shaped and were shaped by their landscape. Because archaeologists regard all things as material culture (just as modern and contemporary historians see all documents and oral accounts as historical sources), such work gives rise to extremely detailed data about individual sites

Figure 6.2 Agricultural landscape near Budingen, Hessen (Germany), showing long-standing nucleated settlements in a modern version of open field farming, a highly distinctive regional type across central Europe.

of many different periods. This data could be mapped into detailed images of spatial and temporal origins, but these images provide only a limited sense of archaeological 'landscape'. Landscape approaches can work on much wider scales, across which it would be too time-consuming to make detailed records of great swathes of land (Figure 6.3). In doing so, they can perceive the spaces between features and sites. Data-heavy, locally grounded approaches can be difficult for other people to access, and often fit poorly with how the character of landscape is perceived.

HLC, then, has developed to address together these four aspects – diversity, perception, application and broad scales of analysis. It represents a research tool that can assist the archaeological interpretation of past landscapes (Turner 2003a, 2006b; Rippon 2004; Martin and Satchell forthcoming), and that seeks to acknowledge that people in past societies had multiple viewpoints too – seeing things in different ways depending upon where and who they were. HLC can help to reveal processes of long-term change in the landscape, and shows the connections between people at various times in the past (Turner 2006b) and people in the present and future (e.g. Fairclough 2006b). When we create an HLC, we make a representation using analogies based on archaeological and historical research into specific sites and

Figure 6.3 A typical 'fieldscape' palimpsest in the UK, in this case on the Isle of Man.

monuments or areas. The research informs our perception of the rest of the landscape, and our ideas about what we value inform what we research. In turn, HLC can inform our planning for the future – not just the future for a few enthusiasts or professionals, but the future of everyone who sees, visits, remembers or lives in a landscape. This future-oriented aspect offers a new type of importance and relevance to 'doing' archaeology.

THE DEVELOPMENT OF THE HLC APPROACH IN ENGLAND

In practical terms, HLC represents the physical landscape of fields, boundaries, roads, waterways, buildings and so forth, with reference to their development over time (McNab and Lambrick 1999: 54). Although the method has only become widespread in its present form in the last decade or so, in England it has much earlier roots.

In his seminal study *Domesday Book and Beyond: Three Essays in the Early History of England* (1897), the historian F.W. Maitland recognised the historical significance of the physical differences between English landscape regions that had been noted by commentators since the 16th century (Edelen 1994: 217) – in his case, the contrast between dispersed settlements and small fields on the Devon/Somerset border and nucleated villages and former common fields in the south Midlands. By the mid-20th century, archaeologists and geographers were also showing the surviving landscape to be the result of past human activity. O.G.S. Crawford, following Maitland's description of the Ordnance Survey map as a 'marvellous palimpsest' (Maitland 1897: 15), described the broader landscape as a palimpsest: 'a document … written on and erased over and over again … it is the business of the field archaeologist to decipher [it]' (Crawford 1953: 51). Crawford was principally interested in understanding discrete 'archaeological' features (Crawford and Keiller 1928) and his approach gave rise to another type of landscape archaeology, the tradition of field survey developed in Britain by generations of Royal Commission surveyors (Bowden 1999).

In the decades after 1950 a generation of scholars were inspired by the work of W.G. Hoskins (Johnson 2005) and the Leicester School of Local History (Tranter et al. 1999) to bring together historical and archaeological approaches in a rich understanding of the historic landscape. They realised that the countryside of rural England presented a complex picture of loss, maintenance and creation – some villages with medieval origins were still flourishing as living communities, for example, whereas others had been reduced to earthworks or crop marks (e.g. Hoskins 1955; Beresford and Hurst 1971). Fifty years after

Maitland, H.P.R. Finberg combined information from several early historic maps and documentary sources to produce a composite plan of pre-modern landscape resources around Tavistock Abbey in Devon (Finberg 1951, Figure 1). We can regard his plan as a distant ancestor of the maps produced in many recent HLC projects.

The landscape archaeology we are discussing in this chapter has roots in all these traditions, as practiced in recent decades by scholars such as Mick Aston (1985), Oliver Rackham (1986) and Steven Rippon (2001). Like them, HLC practitioners see landscape archaeology as a comprehensive field of study. It embraces complete landscapes and represents the historicity of landscape in various ways, often through the idea of places through time, an approach to archaeology that has something in common with *Annales* history (Morris 2000, 24–25).

By the late 1980s in the UK, the growing appreciation of the historic value of everyday landscapes suggested that they were not receiving adequate attention under the current conservation paradigm (Fairclough et al. 1999, 69–70). Early attempts to broaden the scope of archaeological resource management focused largely on the 'setting' of sites and monuments, designed landscapes such as parks and gardens and the land-use context of monuments (e.g. Darvill 1987). It was clear that a more ambitious approach was needed for the wider historic landscape. Landscape archaeologists looked to other disciplines for ideas. One place in which they found inspiration was in landscape architects' procedures of 'assessing' landscape character defined largely by visual or aesthetic criteria. Robust methods were finally being developed for assessing landscape in visual terms at the end of the 1980s (Countryside Commission 1987, 1993) but most applications of these methods did not do justice to the time-depth and archaeological diversity of the European landscape, despite attempts to emphasise the need for landscape characterisation to take the historic dimension more fully into account (e.g. Countryside Commission 1994a, 1996; Countryside Agency and Scottish Natural Heritage 2002). There was thus a need to develop HLC as a separate but related method, whilst creating a common language by borrowing from these other approaches a generalised, broad-brush, area-based approach, and a concern for the whole, not the parts: in short, a unifying rather than a reductionist approach.

English Heritage had already begun to evaluate and compare different methods for understanding and valuing the historic landscape in 1993–1994. This work concluded that no single method of historic landscape assessment existed that was fit for purpose in the planning and management sphere (Fairclough et al. 1999). Some approaches suggested relying on experts' ability to identify areas of landscape according to the historical and archaeological characteristics they considered of particular significance (e.g. Cadw et al. 1998: xxiii–xxvi).

Others proposed that 'relict cultural landscapes ... conceived of as articulated sets of monuments that exist in repetitious patterns of integration and association' could be assessed against criteria such as 'integrity and articulation', 'diversity and structure', 'pattern and repetition' or 'completeness'; clusters of important *monuments* within landscapes would then define entire 'relict cultural landscapes'. This is a problematic approach because it takes the individual archaeological site or 'monument' as its basic unit of analysis, not the best of starting points for understanding the landscape as 'a seamless canvas extending out in all directions' (Darvill et al. 1993: 566).

The seamless quality of the historic landscape was more effectively recognised in other techniques which sought to present interpretations of the whole landscape's historic character, and assess and compare methods for identifying and mapping it (e.g. Masters 1999). The Cornwall Archaeological Unit was able to undertake such work on Bodmin Moor in 1993, which was subsequently extended to cover the whole county (CCC 1994, 1996; Countryside Commission 1994b; Herring 1998) and which became the prototype for the national HLC programme in England.

Most other local government archaeological teams in England have since adopted the general principles of Cornwall's HLC (Figure 6.4), although there has also been significant methodological development (Aldred and Fairclough 2003), and 'historic characterisation' more generally is now an accepted tool in heritage management and in other spheres (e.g. English Heritage 2005, Fairclough 2006c). The take-up of HLC was helped greatly by the UK government's planning guidance on the historic environment published in September 1994 as PPG15. Whilst mainly about historic buildings, two sections were included about the wider historic landscape:

> ... local planning authorities ... should take account of the historical dimension of the landscape as a whole rather than concentrate on selected areas. (PPG15 1994: 2.26)

and

> Suitable approaches to the identification of the components and character of the wider historic landscape are being developed by the Countryside Commission ... and English Heritage.... The whole of the landscape, to varying degrees and in different ways, is an archaeological and historic artefact, the product of complex historic processes and past land-use ... Much of its value lies in its complexity, regional diversity and local distinctiveness ... (PPG15 1994: 6.40)

This was a significant milestone. It ensured that the idea that the whole landscape had interest, significance and value (now accepted through the ELC) began to be enshrined in the UK policy framework.

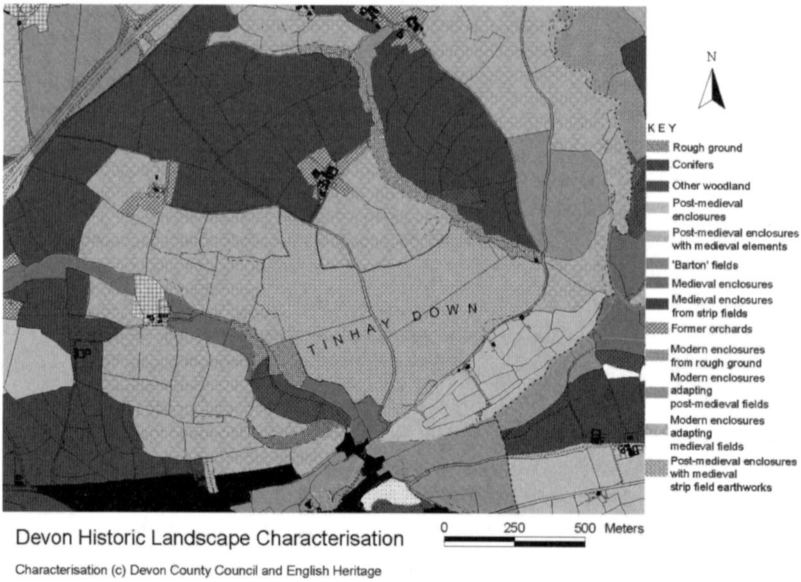

Figure 6.4 A landscape view generated from the Devon HLC database: Tinhay Down, west Devon. *This map is reproduced from the Ordnance Survey material with the permission of Ordnance Survey on behalf of the Controller of Her Majesty's Stationery Office © Crown Copyright. Unauthorised reproduction infringes Crown Copyright and may lead to prosecution or civil proceedings. 100019783.2006.*

At around the same time, in a coastal wetland landscape context, Steven Rippon developed a closely related methodology using regressive map analysis. He used successive historic maps to unpick the landscape's development and morphological analysis at a detailed scale to produce a discussion of how the historic landscape of a particular area came to look the way it did. In the *Gwent Levels*, his work was primarily concerned with 'the last cultural layer', that is, the present pattern of fields, roads and settlement. This can be termed the 'historic landscape' (Rippon 1996: 1). Rippon's work, like most traditional 'landscape archaeology', retained the idea that 'our' landscape is the uppermost of a stratified layering of past landscapes which survive to varying degrees. To some extent, this contrasted with the HLC view of landscape as a present-day construct containing the remains, or simply the memory and associations, of past activities. Rippon has since suggested that other techniques of landscape archaeology can broaden out the frame of reference for HLC to encompass not only this last cultural layer, but all those elements of the landscape, such as earthworks or prehistoric monuments, that contribute to historic character (Rippon 2004).

THE HLC METHOD IN ENGLAND: PRACTICAL ISSUES AND IMPLEMENTATION

The HLC method employs generalising mapping techniques similar to those used commonly in other academic disciplines, for example in geology (to show soil type) or ecology (to map habitats). Peter Herring has explained the basis of the method developed in Cornwall as follows:

> Closer examination [of the landscape] reveals that particular groupings and patterns of components which recur throughout the county can be seen to have been determined by similar histories. Cornwall's historic landscape can, therefore, be characterised, mapped and described, using a finite number of categories or types of 'historic landscape character'. (Herring 1998: 11)

In the Cornwall HLC, 18 broad 'types' of historic landscape character were identified, covering aspects of historic landscape such as rough ground (e.g. unenclosed grazing), enclosed land (fields of prehistoric, medieval, post-medieval and modern character), industrial areas, settlements, parks and gardens (Herring 1999: 21). Later HLC projects have adopted this system but they have also subdivided these broad types into more detail while following the basic concept established in Cornwall and ensuring that historic processes as well as form were being taken into account (e.g. Darlington 2002; Dyson-Bruce 2002; Fairclough et al. 2002; Bannister and Wills 2004; Fairclough and Wigley 2005; Turner 2005). In different areas of the country, different HLC 'types' will be appropriate because of differing landscape histories, or because the landscape characterisation has been designed to be used at larger or smaller scales. The method can therefore be very flexible, and can be used at more local scales than the 'county scale' so far adopted by HLC in England (Herring 1998: 20–21; Turner 2003a).

HLC projects use a range of sources to inform their mapping, although the main sources need to be uniformly available across the study area. In Britain, the most commonly used sources are modern and historic Ordnance Survey mapping at 1:25,000 scale or more detailed, and vertical air photography. These sources depict field boundaries, roads, waterways and settlement plans that can all be used to make interpretations about landscape history. Non-comprehensive data sets can also be used to inform the process where available and if spatially based – for example, archaeological field surveys of extensive visible earthwork remains or tree surveys (Clare and Bunce 2006).

As with all archaeological work, HLC raises methodological problems (Fairclough 2002a, 2002b; Aldred and Fairclough 2003; Fairclough and Macinnes 2003). For example, the categorisation of landscapes into different 'types' relies on a researcher's ability to place areas in the 'correct' categories with some consistency. The HLC mapping process does not simply involve identifying particular physical features, but relies on the interpretation of combinations of features as a reflection of certain landscape histories. The majority of the English landscape is covered with agricultural land, but categorising fields into different historical types/periods based on their morphology alone can be fraught with difficulties. Some HLC projects aim for more transparency in this regard and seek a sort of 'objectivity' that interprets the landscape according to quantifiable criteria, such as the straightness or curviness of boundary features (e.g. Aldred 2001). Used on their own, however, such approaches can overlook subtle variations between different historic trajectories. In an attempt to counteract these problems, other projects have used more local case studies from published archaeological and historical research to identify a range of HLC types that can provide analogies for areas lacking detailed research.

Another potentially difficult task is to give due weight in HLC analysis to the features from several different periods that contribute to overall character, part of the issue of time-depth. Any type of landscape (e.g. one created by post-medieval enclosure) will normally include elements of other kinds of landscape (e.g. prehistoric enclosures; although see Herring 1998: 106–109). Whether (and to what degree) these are perceived as contributing to landscape character depends on the judgement and priorities of the person(s) perceiving the area as landscape. This is one of the many areas where 'landscape' is contested.

Various techniques have been used to help mitigate these problems, but GIS in tandem with supporting explanatory text provides a reasonable solution (Herring 1999: 22). Using a GIS gives considerable flexibility and allows each of the character areas, or 'polygons' (i.e. coherent blocks sharing the same historical development), to be given more than one descriptive characteristic (Wills 1999: 38–39). More recent HLCs have sought to represent broad changes in historic character through time by designing a database that allows a sequence of character types to be recorded for each 'polygon' (Turner 2005, Figure 6.5). The most recent county HLC pro-jects in England add an additional layer of sophistication by assessing whether a 'previous' landscape character in the current landscape is highly accessible (easily legible to all) or only slightly accessible (legible only to those with special knowledge). In this way HLC can add greater understanding of the long-term trends that have shaped our perceptions of the present.

The representation of the historic landscape provided by HLC has several advantages over other techniques (Turner 2003b). The coverage of conventional archaeological databases like the UK's Sites and

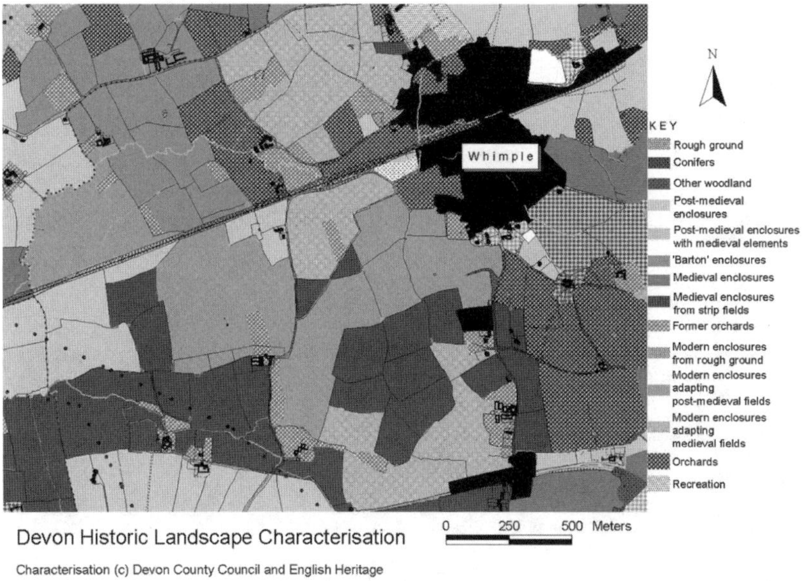

Figure 6.5 A landscape view generated from the Devon HLC database: Whimple, east Devon. *This map is reproduced from the Ordnance Survey material with the permission of Ordnance Survey on behalf of the Controller of Her Majesty's Stationery Office © Crown Copyright. Unauthorised reproduction infringes Crown Copyright and may lead to prosecution or civil proceedings. 100019783.2006.*

Monument Records (SMRs), for example, tends not to be even but to vary across geographical and administrative regions. SMRs in any case map only what is known (or worse, what has been chosen to be recorded), and reflect the imbalances of research. Traditionally, these imbalances related to particular individuals' research 'patches' or interests; more recently, development-led excavations or SMR-based predictive modelling and evaluation have led to 'hot spots' and biases.

SMRs also mainly concern individual archaeological sites. Plotting of SMR entries on maps in order to present a picture of the historic landscape therefore presents an imbalanced picture – for example, often showing medieval field boundaries that have been destroyed, but ignoring those that remain (Johnson 1999: 121). The act of displaying such a database entry as a point on a map (so-called 'point' data) also limits how the feature can be referred to as part of a historic landscape (Herring 1998: 9, Figure 10). HLC is a significant improvement because it allows the historic landscape to be given archaeological significance on a wider scale. Its maps allow a 'break-out from the site-based myopia' of the past (Herring and Johnson 1997: 54) (Figure 6.6).

Figure 6.6 Prehistoric rock art site at Lordenshaws, Northumberland (UK). Managing the site as a discrete monument risks cutting it out of its landscape context.

HLC BEYOND ENGLAND

Whilst HLC *sensu stricto* was initiated in an English context, similar principles have has been adopted in other European countries, and different ways of achieving its aims, methods and applications have been developed. A diversity of approaches to understanding the 'present historic' landscape across Europe is part and parcel of the wide diversity of landscapes in Europe (Figure 6.7). The ELC recognises and celebrates this: there is a single European landscape just as there is an 'English' landscape, but both are characterised by their diversity.

Some of this diversity reflects bio-geographic or climatic differences (e.g. between Atlantic, Boreal or Mediterranean zones). Much of it, however, reflects cultural differences which extend beyond the character of landscape to differences in how landscape is studied or perceived, to different attitudes to change and modernity and to different emphasis upon certain disciplines or sectoral interests. Different cultures of landscape archaeology experiment with different metaphors, aesthetics or ideologies. For example, the metaphor of landscape as biography is particularly emphasised in the Netherlands (Hidding et al. 2001); the journey or the pathway was chosen as a way of seeing landscape by the European landscape network EPCL (Clark et al. 2003); landscape is seen partly as a reflection of community and neighbourhood in Wales (Gwyn 2002); and in Germany landscape as mentality

Figure 6.7 The pastoral landscape of the Limousin at Chéronnac, Haute Vienne (France). The post and wire fences, established in the 20th century, are mostly temporary.

and myth often comes to the fore (Ermischer 2004). What unites these approaches, however, is the distinctively archaeological view of landscape as a perception of the world that is informed by the distant as well as the contemporary past. These approaches can raise context and character above value and significance – in short, they recognise the importance of the local and the ordinary, the small things that in combination make landscapes most relevant for most people.

Again, scale is a critical factor. A European or global view puts the diversity of western European approaches to landscape into a wider context, and highlights their similarities (e.g. Fairclough and Rippon 2002 or Ucko and Layton 1999). A number of broadly similar European approaches, along the lines of HLC, are developing. Only a few pointers can be given here to methods and approaches in other Western European countries, but there are now a number of published reviews such as (Fairclough and Rippon 2002; Scazzosi 2002; Doukellis and Mendoni 2004; Kelm 2005; Ruiz del Árbol and Orejas 2005; Meier 2006; Fairclough and Grau Møller forthcoming). HLC is often adapted to suit local needs. The Scottish method of 'Historic Land-use Assessment' (Dixon et al. 1999; Dixon and Hingley 2002) adds a stronger and more separated layer showing 'relict' landscape in order to capture the deep layers of visible time-depth of the Scottish uplands, with their well-preserved landscape-scale remains of abandoned prehistoric and medieval land use. This mirrors the way that English HLC includes interpretations of still visible 'previous' historic landscape characters in any given area. In Wales, HLC has been drawn into a more narrative-based approach, with a special emphasis on community and place (Gwyn 2002). Characterisation in Wales is also very strongly linked to other aspects of landscapes through the 'Landmap' database, designed to inform land management (CPAT n.d.), although GIS-based characterisation closer to English models is also being adopted in some areas, such as Flintshire (Bill Britnall pers. comm.). The English method of HLC has been used for three counties of Ireland as part of the EU 'Pathways to the Cultural Landscape' project. Work is currently in hand elsewhere, and there are parallel methods too, notably an approach that uses the distinctively Irish territorial pattern of historic 'townlands' as a foundation for rapidly creating HLC-type understanding (Cooney et al. 1998, 2001, 2002; Environmental Resource Management and ERA-Maptec 2000). Elsewhere, the exportability of HLC, and the extent to which its methods need to be modified within its overall philosophy, has been tested in various research contexts, for example in Sweden (Nord 2006a, 2006b).

But it would be a mistake to see HLC, even in broad terms, as the only approach to landscape in heritage management. In both Denmark and Flanders (Belgium), for example, the lack of a landscape scale in archaeological databases and the 'historic' gap in landscape assessment (both starting points in England for developing HLC) have been filled by the creation of online digital atlases.

In Denmark, the *Digital Atlas on Cultural Environments* focuses on the historical period since 1000 AD, and all environments in the landscape are registered and assessed according to their preservation value. Selection is on the basis of discrete defined *'kulturmiljø'* (cultural settings), initially usually defined around focal points such as historic villages, but now focusing on areas of greatest survival – for example, pre-18th century field patterns. What is significant here is the use of an apparently traditional, almost site-based, approach to raise awareness of the historic character of the whole territory: the 'white areas' become areas that pose questions (DAKD n.d.; Grau Møller pers. comm.; Schou 2001; Stoumann 2002).

In a similar way, in Belgium the *Landscape Atlas of Flanders* (2001) has changed perceptions of landscape – and therefore arguably is changing the landscape itself, since landscape is 'as perceived by people'. This is a region where significant modern change, especially peri-urbanisation, is often thought to have obscured or removed most of the character derived from earlier activities. The Flanders atlas selected its areas according to the legibility of 'relict' features. In practice this involved defining areas whose historic character predominantly related to the period before the 1850s and 1860s (Antrop 2004, 2005; *Landscape Atlas of Flanders* 2001). In consequence, it strengthened recognition that the past survives in the present, but also drew attention to the other blank areas. This raised a number of questions: if the relict features are not predominantly pre-mid-19th century, what are they? Can they be characterised as being more historic (i.e. having greater time-depth) if the threshold of legibility (which after all is conditioned by knowledge and the ability to recognise the past) is lowered?

We might ask, in any case, why historic 'value' cannot be ascribed to landscape elements of the late 19th century, early 20th century or even the later 20th century (Bradley et al. 2004, Fairclough forthcoming). 'Historic' can include even the most recent periods such as the 1980s or 1990s, which are also the result of past cultural processes and of social interaction with nature, part of the trajectory along which the present-day landscape has reached us. Not recognising this has in the past been one of the reasons why the sort of landscape assessment (and therefore landscape management and protection) carried out by other disciplines has not included sufficient historic depth; it is as if researchers believed that the past stopped a hundred years ago, and has no connection with the present. The word 'historic', like 'traditional', especially when applied to landscape, requires careful thought and use; they are not unproblematic terms.

CONCLUSIONS

The HLC method of landscape archaeology is distinguished by a concern for how the past and its remains contribute to people's contemporary perception of landscape at a variety of scales and to a variety of degrees (dependent on the knowledge, understanding and interest of an individual beholder). Because it recognises that archaeology studies the contemporary landscape, HLC also offers its insights as a contribution to a broader inter-disciplinary view of the world. HLC studies *the past within the present*, and the effect of past human agency on the present day: a perspective that other landscape disciplines do not offer (Figure 6.8).

HLC thus contributes a sense of time-depth, and of change in the past, to landscape studies. These counteract the tendencies of some landscape disciplines, against all evidence, to regard things or processes older than about a century as simply 'traditional'. Landscape archaeology can use these methods to understand present-day change, and to comment on and even shape future change. We are used to the idea that archaeological resource management must be grounded in an essentially protective, conservationist mindset. HLC shows that this need not be the case, by highlighting the idea that change is a central

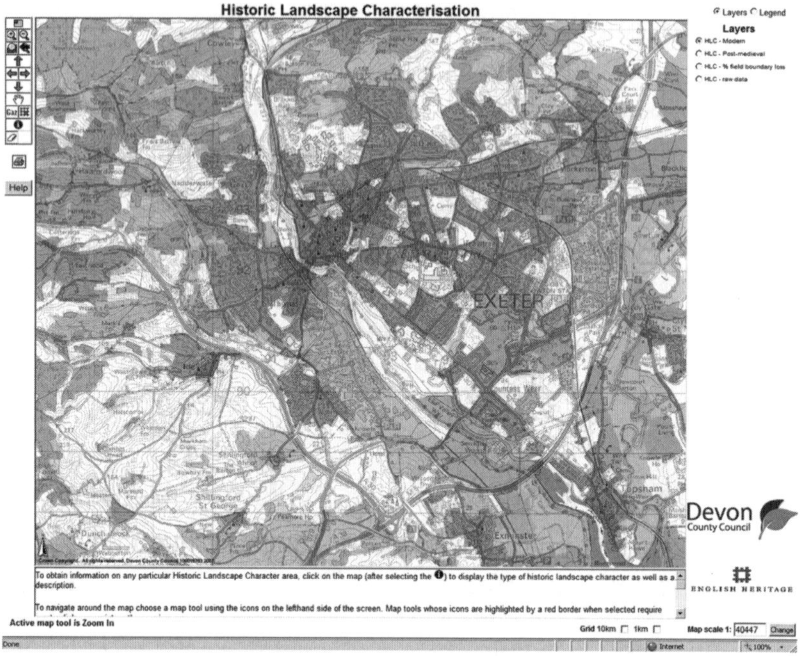

Figure 6.8 Devon HLC: publicly accessible web-based GIS version. *This map is reproduced from the Ordnance Survey material with the permission of Ordnance Survey on behalf of the Controller of Her Majesty's Stationery Office © Crown Copyright. Unauthorised reproduction infringes Crown Copyright and may lead to prosecution or civil proceedings. 100019783.2006.*

attribute of historic landscape character. An area of land that is not changing is unlikely to be anyone's landscape, merely a museum piece.

There are already many examples in the UK of how HLC is coming to be used both to help shape the landscape that is bequeathed to the future – to create future heritage, so to speak – and to help people to enjoy and appreciate it in the present day (see www.english-heritage.org.uk/characterisation). Applications directed towards research have already been mentioned, and this chapter is not the place for detailed description of practical landscape management applications. These are discussed elsewhere, but include supporting the archaeological contribution to landscape management through the EU agricultural support regime; informing strategic spatial planning; and providing information and prediction for the operation of Environmental Impact Assessments and development control through the local government-based spatial-planning process (Fairclough 2002b, 2003; Clark et al. 2004; English Heritage 2005). In addition, of course, HLC can be used to extract some of landscape's economic benefits – through tourism, for instance – and social value – through outreach, community involvement or volunteer land management. These might be seen as the 'instrumental' values of the archaeological landscape.

HLC offers a way to bring landscape archaeology into constructive contact with other landscape disciplines through a common language that other practitioners, such as landscape architects and ecologists, can more easily understand. HLC also draws landscape into the realm of archaeological resource management, from which it was noticeably absent in the UK until the early 1990s, and it helps to balance the tendency of other types of landscape archaeology to study past environments only for favoured historical periods rather than the full continuum of time.

Finally, HLC is an approach that brings archaeology into closer contact with questions of participation, inclusion and ownership. What could be more relevant for a discipline that studies people and their actions than to become closely involved, as the ELC says, with how people define and decide what happens next to their landscapes? Yet here – and this is almost as equally true of other action-oriented landscape disciplines such as landscape architecture – it could be said that something is missing. If HLC encourages and facilitates an inclusive approach to public perceptions of landscape – and thus, one hopes, highlights the past's fundamental relevance to contemporary society – then where can this be seen to be working in practice? There are some examples, such as the Devon 'Oral History' approach (Riley and Harvey 2005), but they are few in number, and rarely rise above collecting recent memories loosely linked to a site-based conception of place; so far they come close to capturing some public views of 'monuments' but not of landscape.

A second area where there is a need for new ideas and practices is that of objectives and purposes. The now-traditional values-based approach to heritage management, based on an agenda of preserving the fabric of monuments that are seen to have survived from the past, should not be uncritically transferred to the field of landscape. Landscape is not simply a larger type of monument; rather, it transcends such approaches. To help 'landscape' play its role in contemporary society, heritage managers and archaeologists effectively need to review the appropriateness of their inherited assumptions about *protection* (Fairclough 2006b). If landscape is 'common culture' – heritage that is accessible and shared – then it must be alive; coming to terms with change is therefore essential. Having accepted that, archaeology will be well placed to forge new landscapes for the future as well as the past.

REFERENCES

Aldred, A. 2001. *Somerset and Exmoor National Park Historic Landscape Characterisation Project, 1999–2000*, unpublished report (2 vols.). Taunton: Somerset County Council.

Aldred, A. and G. Fairclough 2003. *Historic Landscape Characterisation: Taking Stock of the Method*. Taunton: English Heritage and Somerset County Council. Available http://www.english-heritage.org.uk/characterisation (Consulted 7 July 2006).

Antrop, M. 2004. Landscape Change and the Urbanisation Process in Europe. *Landscape and Urban Planning* 67: 9–26.

Antrop, M. 2005. Why Landscapes of the Past Are Important for the Future. *Landscape and Urban Planning* 70: 21–34.

Aston, M. 1985. *Interpreting the Landscape*. London: Batsford.

Bannister, N.R. with P.M. Wills 2004. The Surrey Historic Landscape Characterisation Project. In J. Cotton, G. Crocker and A. Graham (eds) *Aspects of Archaeology and History in Surrey: Towards a Research Framework for the County*. Guildford: Surrey Archaeological Society, pp. 119–132.

Beresford, M. and J. Hurst 1971. *Deserted Medieval Villages*. London: Lutterworth Press.

Bowden, M. 1999. *Unravelling the Landscape: An Inquisitive Approach to Archaeology*. Stroud: Tempus.

Bradley, A., V. Buchli, G. Fairclough, J. Miller and J. Schofield 2004. *Change and Creation, Historic Landscape Character 1950–2000*. London: English Heritage and WS Atkins. http://www.changeandcreation.org.uk (Consulted 7 July 2006).

Brown, J., N. Mitchell and M. Beresford (eds) 2005. *The Protected Landscape Approach, Linking Nature, Culture and Community*. Cambridge (UK) and Gland (CH): IUCN.

Cadw, Countryside Council for Wales and ICOMOS-UK 1998. *Register of Landscapes of Outstanding Historic Interest in Wales*. Cardiff: Cadw.

CCC 1994. Cornwall HLC 'types' maps (unpublished historic landscape characterisation maps, held by Cornwall Historic Environment Service). Truro: Cornwall County Council.

CCC 1996. *Cornwall Landscape Assessment 1994*. Truro: Cornwall County Council.

Clare, T. and R. Bunce 2006. The Potential for Using Trees to Help Define Historic Landscape Character Zones: A Case Study in the English Lake District. *Landscape and Urban Planning* 74: 34–45.

Clark, J., J. Darlington and G. Fairclough (eds) 2003. *Pathways to Europe's Landscape*. Heide: EPCL/EU.

Clark, J., J. Darlington and G. Fairclough 2004. *Using Historic Landscape Characterisation*. London: English Heritage and Lancashire County Council.

Cooney, G., T. Condit and E. Byrnes 1998. *Archaeological Landscapes in Ireland: A Pilot Study*. Dublin: University College Dublin.

Cooney, G., T. Condit and E. Byrnes 2001. *The Archaeological Landscapes Project: Final Report (Draft for Comment)*. Dublin: University College Dublin (for Heritage Council, Ireland).

Cooney, G., T. Condit and E. Byrnes 2002. The Archaeological Landscape Project: An Approach to Cultural Landscapes in Ireland. In G. Fairclough and S. Rippon (eds) *Europe's Cultural Landscape: Archaeologists and the Management of Change*. Brussels and London: Europae Archaeologiae Consilium and English Heritage (EAC Occasional Paper no 2), pp. 39–48.

Cosgrove, D. and S. Daniels (eds) 1988. *The Iconography of Landscape*. Cambridge: Cambridge University Press (Cambridge Studies in Historical Geography).

Council of Europe 1995. *Recommendation R(95)9 of the Committee of Ministers to Member States on the Integrated Conservation of Cultural Landscape Areas as Part of Landscape Policies*. Strasbourg: Council of Europe.

Council of Europe 2000. *European Landscape Convention*. Florence: Council of Europe (European Treaty Series No. 176). http://conventions.coe.int/Treaty/en/Treaties/Html/176.htm (Consulted 12 July 2006).

Council of Europe 2002. *The European Landscape Convention* (*Naturopa* Issue 98). Strasbourg: Council of Europe. http://www.coe.int/T/E/Cultural_Co-operation/Environment/Resources/Naturopa_Magazine/index.asp (Consulted 17 July 2006).

Countryside Commission 1987. *Landscape Assessment: A Countryside Commission Approach, CCD18*. Cheltenham: Countryside Commission.

Countryside Commission 1993. *Landscape Assessment Guidance*, CCP423. Cheltenham: Countryside Commission.

Countryside Commission 1994a. *Views from the Past – Historic Landscape Character in the English Countryside*. Cheltenham: Countryside Commission (Consultation Report).

Countryside Commission 1994b. *The Landscape of Bodmin Moor*. Cheltenham: Countryside Commission.

Countryside Commission 1996. *Views from the Past – Historic Landscape Character in the English Countryside*, CCW4. Cheltenham: Countryside Commission.

Countryside Agency and Scottish Natural Heritage 2002. *Landscape Character Assessment: Guidance for England and Scotland*. Cheltenham: Countryside Commission. http://www.countryside.gov.uk/LAR/Landscape/CC/index.asp (Consulted 17 July 2006).

CPAT n.d. Historic Landscape in Wales. http://www.cpat.org.uk/projects/longer/histland/histland.htm (Consulted 6 July 2006).

Crawford, O. 1953. *Archaeology in the Field*. London: Phoenix House.

Crawford, O. and A. Keiller 1928. *Wessex from the Air*. Oxford: Clarendon.

DAKD n.d. Digitalt Atlas over Kulturmiljøer i Danmark. http://www.humaniora.sdu.dk/kulturmiljoe/index.html (Consulted 6 July 2006).

Darlington, J. 2002. Mapping Lancashire's Historic Landscape: The Lancashire HLC Programme. In G. Fairclough and S. Rippon (eds) *Europe's Cultural Landscape: Archaeologists and the Management of Change*. Brussels and London: Europae Archaeologiae Consilium and English Heritage (EAC Occasional Paper no 2), pp. 97–105.

Darvill, T. 1987. *Ancient Monuments in the Countryside*. London: English Heritage.

Darvill, T., C. Gerrard and B. Startin 1993. Identifying and Protecting Historic Landscapes. *Antiquity* 67: 563–574.

Dixon, P., L. Dyson-Bruce and J. Stevenson 1999. *Historic Landuse Assessment (HLA): Development and Potential of a Technique for Assessing Historic Landscape Patterns, Report of the Pilot Project 1996–98*. Edinburgh: Historic Scotland/RCHMS.

Dixon, P. and R. Hingley 2002. Historic Land-Use Assessment in Scotland. In G. Fairclough and S. Rippon (eds) *Europe's Cultural Landscape: archaeologists and the Management of Change*. Brussels and London: Europae Archaeologiae Consilium and English Heritage (EAC Occasional Paper no 2), pp. 85–88.

Doukellis, P and L.G. Mendoni (eds) 2004. *Perception and Evaluation of Cultural Landscapes*. Athens: National Hellenic Research Foundation.

Dyson-Bruce, L. 2002. Historic Landscape Assessment – the East of England Experience. In G. Burenhult (ed) *Archaeological Informatics: Pushing the Envelope, Procs. 29th CAA Conference, Gotland, April 2001*. Oxford: Archaeopress (British Archaeological Reports International Series 1016), pp. 35–42.

Edelen, G. (ed) 1994. *The Description of England by William Harrison* [1587]. Washington and New York: Folger Library and Dover Publications.

English Heritage 2005. Characterisation. *Conservation Bulletin* 47 (themed issue).

Environmental Resources Management/ERA-Maptec Ltd 2000 (for the Heritage Council). *Pilot Study on Landscape Characterisation in County Clare*. Oxford/Dublin.

Ermischer, G. 2004. Mental Landscape: Landscape as Idea and Concept. *Landscape Research* 29(4): 371–383.

Fairclough, G.J. 1999. Historic Landscape and the Future. In G. Fairclough, G. Lambrick and A. McNab (eds) *Yesterday's World, Tomorrow's Landscape: The English Heritage Landscape Project 1992–94*. London: English Heritage, pp. 60–67.

Fairclough, G.J. 2002a. Cultural Landscape, Computers and Characterization. In G. Burenhult (ed) *Archaeological Informatics: Pushing the Envelope, Procs. 29th CAA Conference, Gotland, April 2001*. Oxford: Archaeopress (British Archaeological Reports International Series 1016), pp. 277–294.

Fairclough, G.J. 2002b. Cultural Landscape and Spatial Planning: England's Historic Landscape Characterisation Programme. In L.M. Green and P.T. Bidwell (eds) *Heritage of the North Sea Region: Conservation and Interpretation: Papers Presented at the Historic Environment of the North Sea InterReg IIC Conference 29–31 March 2001, South Shields*. Shaftesbury: Donhead, pp. 123–149.

Fairclough, G.J. 2003. Cultural Landscape, Sustainability and Living with Change? In J.M. Teutonico and F. Matero (eds) *Managing Change: Sustainable approaches to the Conservation of the Built Environment, Procs. of the 4th Annual US/ICOMOS International Symposium, Philadelphia, Pennsylvania, April 2001*. Los Angeles: Getty Conservation Institute, pp. 23–46.

Fairclough, G.J. 2006a. Our Place in the Landscape? An Archaeologist's Ideology of Landscape Perception and Management. In T. Meier (ed) *Landscape Ideologies*. Budapest: Archaeolingua, pp. 177–197.

Fairclough, G.J. 2006b. A New Landscape for Cultural Heritage Management: Characterisation as a Management Tool. In L. Lozny (ed) *Landscapes under Pressure: Theory and Practice of Cultural Heritage Research and Preservation*. New York: Springer, pp. 55–74.

Fairclough, G.J. 2006c. From Assessment to Characterisation. In J. Hunter and I. Ralston (eds) *Archaeological Resource Management in the UK*, 2nd ed. Stroud: Sutton, pp. 250–270.

Fairclough, G.J. 2006d. Large Scale, Long Duration and Broad Perceptions: Scale Issues in Historic Landscape Characterisation. In G. Lock and B. Molyneaux (eds) *Confronting Scale in Archaeology: Issues of Theory and Practice*. Dordrecht: Kluwer Academic Publishers, 203–215.

Fairclough, G.J. forthcoming . The Contemporary and Future Landscape: Change and Creation in the Later 20th Century. In L. McAtackney, M. Palus and A. Piccini (eds) *Contemporary and Historical Archaeology in Theory*. Oxford: Archaeopress (Studies in Contemporary and Historical Archaeology).

Fairclough, G.J. and L. Macinnes 2003. *Understanding Historic Landscape Character*, Topic Paper 5. Landscape Character Assessment Guidance for England and Scotland, Countryside Agency, Scottish Natural Heritage, Historic Scotland and English Heritage. http://www.ccnetwork.org.uk/lca_topic_5_sum.htm (Consulted 17 July 2006).

Fairclough, G.J. and P. Grau Møller forthcoming. *Managing and Protecting Landscape in Europe: A Summary of National Approaches by COST A27*.

Fairclough, G.J. and S.J. Rippon (eds) 2002. *Europe's Cultural Landscape: Archaeologists and the Management of Change*. Brussels and London: Europae Archaeologiae Consilium and English Heritage (EAC Occasional Paper no 2).

Fairclough, G.J. and A. Wigley 2005. Historic Landscape Characterisation: An English Approach to Landscape Understanding and the Management of Change. In M. Ruiz del Árbol and A. Orejas (eds) *Landscapes as Cultural Heritage in the European Research: Proceedings of COST A27 Workshop, Madrid 2004*. Madrid: CSIC, pp. 87–106.

Fairclough, G.J., G. Lambrick and D. Hopkins 2002. Historic Landscape Characterisation in England and a Hampshire Case Study. In G. Fairclough and S. Rippon (eds) *Europe's Cultural Landscape: Archaeologists and the Management of Change*. Brussels and

London: Europae Archaeologiae Consilium and English Heritage (EAC Occasional Paper no 2), pp. 69–83.

Fairclough, G.J., G. Lambrick and A. McNab (eds) 1999. *Yesterday's World, Tomorrow's Landscape: The English Heritage Landscape Project 1992–94*. London: English Heritage.

Finberg, H.P.R. 1951. *Tavistock Abbey: A Study in the Social and Economic History of Devon*. Cambridge: Cambridge University Press.

Gwyn, D. 2002. Associative Landscape in a Welsh Context. In G. Fairclough and S. Rippon (eds) *Europe's Cultural Landscape: Archaeologists and the Management of Change*. Brussels and London: Europae Archaeologiae Consilium and English Heritage (EAC Occasional Paper no 2), pp. 187–192.

Hall, M. 2006. Identity, Memory and Countermemory: The Archaeology of an Urban Landscape. *Journal of Material Culture* 11(1/2): 189–209.

Herring, P. 1998. *Cornwall's Historic Landscape: Presenting a Method of Historic Landscape Character Assessment*. Truro: Cornwall Archaeological Unit.

Herring, P. 1999. Cornwall: How the Historic Landscape Characterisation Methodology Was Developed. In G. Fairclough (ed) *Historic Landscape Characterisation: 'The State of the Art'*. London: English Heritage, pp. 15–32.

Herring, P. and N. Johnson 1997. Historic Landscape Character Mapping in Cornwall. In K. Barker and T. Darvill (eds) *Making English Landscapes*. Oxford: Oxbow Monograph 93, pp. 70–91.

Hidding, M., J. Kolen and T. Spek 2001. De biografie van het landschap. Ontwerp voor een inter en multidisciplinaire benadering can de landschapsgeshiedenis en het culturele erfgoed. In J.H.F. Bloemers et al. (eds) *Bodemarchief in Behoud en Ontwikkeling. De Conceptuele Grondslagen*. Den Haag: NOW, pp. 7–109.

Hoskins, W.G. 1955. *The Making of the English Landscape*. London: Hodder and Stoughton.

Johnson, M.H. 2005. On the Particularism of English Landscape Archaeology. *International Journal of Historical Archaeology* 9(2): 111–122.

Johnson, N. 1999. Context, Meaning and Consequences: Using the Map in Cornwall. In G. Fairclough (ed) *Historic Landscape Characterisation: 'The State of the Art'*. London: English Heritage, pp. 117–122.

Jongman R. and G. Pungetti 2004. *Ecological Networks and Greenways*. Cambridge Studies in Landscape Ecology. Cambridge: Cambridge University Press.

Kelm, R. (ed) 2005. *Frühe Kulturlandschaften in Europa: Forschung, Erhaltung und Nutzung*, Papers from the 3rd and 4th Albersdorfer Kolloquium Albersdorfer Orschungen zur Archäologie und Umweltgeschichte, Band 3. Heide: ÄOZA and Boyens.

Landmarks 2007. Landmarks: Understanding Pre-industrial Structures in Rural and Mining Landscapes. http://www.soc.staffs.ac.uk/jdw1/costa27home.html (Consulted 8 June 2007).

Landscape Atlas of Flanders 2001. *Landscape Atlas of Flanders*. Bruxelles: Ministry of the Flemish Community, Support Center GIS-Flanders. http://geo-vlaanderen.gisvlaanderen.be/geo-vlaanderen/landschapsatlas (Consulted 6 July 2006).

LE:NOTRE, 2002–2006. Le Notre: Thematic Network Project in Landscape Architecture. http://www.le-notre.org (Consulted 19 July 2006).

Mabey, R. 2005. *Nature Cure*. London: Chatto and Windus/Pimlico.

Macinnes, L. 2004. Historic Landscape Characterization. In K. Bishop and A. Phillips (eds) *Countryside Planning: New Approaches to Management and Conservation*. London: Earthscan, pp. 155–169.

Macinnes, L. and C.R. Wickham-Jones 1992. Time-Depth in the Countryside: Archaeology and the Environment. In L. Macinnes and C.R. Wickham-Jones (eds) *All Natural Things: Archaeology and the Green Debate*. Oxford: Oxbow (Oxbow Monograph 21), pp. 1–13.

McNab, A. and G. Lambrick 1999. Conclusions and Recommendations. In G. Fairclough, G. Lambrick and A. McNab (eds) *Yesterday's World, Tomorrow's Landscape: The English Heritage Landscape Project 1992–94*. London: English Heritage, pp. 54–59.

Maitland, F.W. 1897. *Domesday Book and Beyond: Three Essays in the Early History of England*. Cambridge: Cambridge University Press.
Martin, E. and M. Satchell forthcoming. *'Wheare Most Enclosures Be' – Historic Fields in East Anglia*. Norwich: East Anglian Archaeology Reports.
Masters, P. 1999. County Durham. In G.J. Fairclough, G. Lambrick and A. McNab (eds) *Yesterday's World, Tomorrow's Landscape: The English Heritage Landscape Project 1992–94*. London: English Heritage, pp. 43–47.
Matless, D. 1998. *Landscape and Englishness*. London: Reaktion.
Meier, T. (ed) 2006. *Landscape Ideologies*. Budapest: Archaeolingua.
Morris, I. 2000. *Archaeology as Cultural History*. Oxford: Blackwell.
Muir, R. 1999. *Approaches to Landscape*. London: Macmillan.
Nord, J.M. 2006a. Förhistoriska vägval och dagens väglandskap på Bjärehalvön i Nordvästra Skåne. In M. Qviström (ed) *Gångna landskap: möten mellan väghistoria och landskapshistoria*. Alnarp: Institutionen för Landskapsplanering, Rapport 06:1, pp. 43–50. http://www2.lpal.slu.se/skyltfonstret/Mattias/Rapport%2006_1.pdf (accessed 17 July 2006).
Nord, J.M. 2006b. *Landscape as Heritage: The Bjäre Peninsula. I: Utskrift 8*. Halmstad: Hallands länsmuseer.
Olwig, K.R. 2002. *Landscape, Nature, and the Body Politic: From Britain's Renaissance to America's New World*. Madison: University of Wisconsin Press.
Olwig, K.R. 2004. 'This is not a landscape': Circulating Reference and Land Shaping. In H. Palang, H. Sooväli, M. Antrop and G. Setten (eds) *European Rural Landscapes: Persistence and Change in a Globalising Environment*. Dordrecht: Kluwer Academic Publishers, pp. 41–65.
Ottosson, J. and P. Grahn 2005. A Comparison of Leisure time Spent in a Garden with Leisure Time Spent Indoors: On Measures of Restoration in Residents in Geriatric Care. *Landscape Research* 30(1): 23–55.
Palang, H. and G. Fry (eds) 2003. Land*scape Interfaces: Cultural Heritage in Changing Landscapes*. Dordrecht: Kluwer Academic Publishers.
PPG15 1994. *Planning Policy Guidance Note 15: Planning And The Historic Environment*. London: Office of the Deputy Prime Minister. http://www.odpm.gov.uk/index.asp?id=1144041 (Consulted 5 April 2006).
Rackham, O. 1986. *The History of the Countryside*. London: Dent.
Riley, M. and D. Harvey 2005. Landscape Archaeology, Heritage and the Community in Devon: An Oral History Approach. *International Journal of Heritage Studies* 11(4): 269–288.
Rippon, S. 1996. *The Gwent Levels: The Evolution of a Wetland Landscape*. York: Council for British Archaeology.
Rippon, S. 2001. *The Transformation of Coastal Wetlands: Exploitation and Management of Marshland Landscapes in North West Europe during the Roman and Medieval Periods*. Oxford: Oxford University Press.
Rippon, S. 2004. *Historic Landscape Analysis: Deciphering the Countryside*. York: Council for British Archaeology.
Ruiz del Árbol, M. and A. Orejas (eds) 2005. *Landscapes as Cultural Heritage in the European Research: Proceedings of the Open Workshop, Madrid 29th October 2004*. Madrid: CSIC.
Sarlöv Herlin, I. 2004. New Challenges in the Field of Spatial Planning: Landscapes. *Landscape Research* 29(4): 399–411.
Scazzosi, L. (ed) 2002. *Leggere il paesaggio. Confronti internazionali*. Rome: Gengemi.
Scazzosi L. 2004. Reading and Assessing the Landscape as Cultural and Historical Heritage. *Landscape Research* 29(4): 335–355.
Schama, S. 1995. *Landscape and Memory*. London: Harper Collins.
Schou, A. 2001. *Cultural Heritage in Planning: Identifying Valuable Cutural Environments*. Copenhagen: Danish Forest and Nature Agency (Ministry of Environment).

Available at http://www2.skovognatur.dk/Udgivelser/2001/87-7279-298-1/default_eng.htm (Consulted 17 July 2006).

Selman, P. 2006. *Planning at Landscape Scale*. London: Routledge.

Stoumann, I. 2002. Cultural History in Spatial Planning in Denmark. In G. Fairclough and S. Rippon (eds) *Europe's Cultural Landscape: Archaeologists and the Management of Change*. Brussels and London: Europae Archaeologiae Consilium and English Heritage (EAC Occasional Paper no 2), pp. 61–68.

Tilley, C. 2006. Introduction: Identity, Place, Landscape and Heritage. *Journal of Material Culture* 11(1/2): 7–32.

Tranter, M., K. Hawker, J. Rowley and M. Thompson (eds) 1999. *English Local History: The Leicester Approach, A Departmental Bibliography and History, 1948–1998*. Leicester: Friends of the Department of English Local History.

Turner, S. 2003a. Making a Christian Landscape: Early Medieval Cornwall. In M. Carver (ed) *The Cross Goes North: Processes of Conversion in Northern Europe, AD 300–1300*. Woodbridge: Boydell and Brewer, pp. 171–194.

Turner, S. 2003b. Christianity and the Landscape of Early Medieval South-West Britain. Unpublished Ph.D thesis, University of York.

Turner, S. 2005. The Devon Historic Landscape Characterisation: Methodology, Results and Preliminary Analysis. Unpublished report, Devon County Council Archaeology Service. Exeter: Devon County Council/English Heritage.

Turner, S. 2006a. Historic Landscape Characterisation: A Landscape Archaeology for Research, Management and Planning. *Landscape Research* 31(4): 385–398.

Turner, S. 2006b. *Making a Christian Landscape*. Exeter: University of Exeter Press.

Ucko, P.J. and R. Layton (eds) 1999. *The Archaeology and Anthropology of Landscape: Shaping Your Landscape*. London: Routledge (One World Archaeology 30).

Wills, J. 1999. Cotswolds AONB: Characterisation, Classification and GIS. In G.J. Fairclough (ed) *Historic Landscape Characterisation: 'The State of the Art'*. London: English Heritage, pp. 33–40.

CHAPTER 7

Landscape Archaeology and 'Community Areas' in the Archaeology of Central Europe

Martin Kuna and Dagmar Dreslerová

> Landscape is why people climb up to a viewing tower.
> — Jiří Sádlo, quoted by Storch 2002: 9

The terminology of 20th century European archaeologists included the terms 'space', 'environment', 'settlement' and 'landscape'. These terms are closely interrelated: they can be treated as synonyms, their meanings can overlap or they can be defined as quite separate concepts. Their meanings change according to the traditions of use in various regional and period-based communities of archaeologists, as well as under the influence of shifting disciplinary paradigms. Our focus in this chapter is on how approaches to 'landscape' have changed in the archaeology of central Europe, drawing upon archaeological material from the Czech Republic, Poland and Germany. We explore the changing approaches to 'settlement' in central European archaeology, introducing the concepts of *Landesaufnahme* and *Siedlungskammer*, and Neustupný and Kuna's notion of 'community areas'. We shall then discuss how the idea of 'community areas' can be used to trace the relationships between identity and landscape in central European prehistory through two case studies, examining an Iron Age industrial zone and Bronze Age burial mounds in landscape perspective. By doing so, we hope to demonstrate the distinctive way in which the archaeology of landscapes has been approached in the study of the later prehistory of central Europe.

SPACE, SETTLEMENT AND LANDSCAPE IN CENTRAL EUROPEAN ARCHAEOLOGY

The archaeological landscapes of central Europe range from Upper Palaeolithic sites at the foot of the South Moravian hills in the eastern Czech Republic to the changing landscapes of the Neolithic and Neolithic (Sherratt 1981); Corded Ware and Bell Beaker sites; Celtic hill forts, medieval field boundaries and deserted villages; 19th century industrial landscapes; and the remains of 20th century collectivisation and industrialisation programmes, opencast coal mining, factories and motorways. A variety of spatial approaches, with a diversity of methods and aims, have been employed in the region.

The earliest attempts to study archaeological material in relation to geographical space in central Europe developed in the culture historical approaches of the early 20th century. The German prehistorian Gustav Kossina (1858–1931) was one of its foremost proponents, with his *Siedlungsarchäologie* (settlement archaeology), which focused on the ethnic dimension of spatial distribution of archaeological sites (Kossina 1911). The culture historical approach developed ever more detailed typologies and chronologies of archaeological cultures. Geography and landscape were used to identify individual social and political groups in the archaeological record, and to highlight migrations from 'homelands' that were analogous to conventional historical accounts of geopolitical events. Similar approaches were common in other European countries at that time and even later, such as the school developed by Kossina's student Józef Kostrzewski at Poznań University in Poland (Kostrzewski 1949), and many archaeologists in Czechoslovakia, e.g. Jaroslav Böhm (1937, 1941) and Jan Filip (1956). Occasionally, culture historical approaches led to nationalistic excesses and political manipulations: most notoriously in the appropriation of Kossina's attempts to define the primacy of German culture in Europe by the Nazis (e.g. Mähling 1944). But such perspectives were not necessarily an integral part of culture history. The culture historical paradigm has survived across central European archaeology in a moderated form until today, and the influence of culture historical interpretive frameworks continues to be significant.

The gradual dissolution of Kossina's concept of settlement archaeology began in central Europe during the 1940s and 1950s, with the work of German scholars such as Ernst Wahle (1941) and Hans J. Eggers (1950). These scholars argued that ethnic reasons alone did not explain the spatial organisation of archaeological cultures and that a range of other explanations (socio-economic, political, ritual) were required. Later, Herbert Jankuhn (1955, 1977) revised Kossina's approaches, and his work produced a school of thought with a new theoretical framework. Jankuhn retained the term *Siedlungsarchäologie*, but sought to examine the relationships of past settlements with the natural environment, demography and socio-economic relationships, rather than just engaging in 'ethnic' interpretations. Spatial relationships between archaeological sites, the natural environment and remnants of settlement activities were emphasised in studies that made use of a range of methodologies: the analysis of geological and soil maps, palynological and climatic data, and the density of settlement units in space.

Slightly later, between the 1950s and 1960s, similar approaches to settlement and archaeology emerged in the United States. A significant asset of the American school was contained within the concept of 'settlement pattern' – the layout of past settlements, containing evidence of their socio-economic systems (Willey 1953; Chang 1968). The term 'settlement pattern', whether adopted directly from Willey and Chang, or from those influenced by their work, percolated into European archaeology after the 1960s in innumerable partial meanings and contexts. The extent of intellectual exchanges between central European and American schools of settlement archaeology during the 1950s is unclear. References to central European approaches are not however to be found in the classic statements of either American (e.g. Chang 1968) or German versions of settlement archaeology (e.g. Jankuhn 1977) during the 1960s and 1970s. The processes through which ideas of settlement archaeology developed therefore remains an interesting target for future research into the history of thought in world archaeology.

Central European approaches to landscape were, however, significantly influenced by the spatial archaeology formulated in Britain by David Clarke. Clarke drew upon the New Geography to develop both methods and theoretical frameworks for the archaeological study of spatial relationships (Clarke 1977). The spatial organisation of archaeological data was studied at various levels – from a single assemblage to an archaeological culture. Spatial archaeology enriched the discipline with several valuable methods and models, such as site-catchment analysis, flow-off curves, etc., but as a general theoretical model it nevertheless gradually became unfashionable during the 1980s. This occurred mainly under the pressure and influence exerted by another approach, also originating in Britain, which did not bow to the exact methodology and economic interpretations of spatial archaeology, but preferred a hermeneutic approach to archaeological material – symbolic archaeology, later to be known as post-processual archaeology.

However, as British scholars were moving away from the term, in Czech archaeology spatial archaeology was revitalised in the 1990s with the work of Evžen Neustupný (1991, 1998a). As with earlier British approaches to spatial archaeology, Neustupný argued that space forms an important aspect of the archaeological evidence, which should be analysed next to the aspects of its (artefactual) form. In a sense, Neustupný's spatial archaeology is a conceptual antinomy and supplement of the typology of artefacts – that is, archaeology that studies the formal characteristics of the data. General spatial characteristics and spatial units of archaeological sources have been formulated in Neustupný's theory of community areas.

In central Europe, landscape archaeology was also developed by avocational archaeologists – for example, in the development of research on deserted medieval villages and medieval field systems by Czech physician Ervín Černý (1973, 1979). Such approaches developed a concept of landscape as revealing an object of study outside the 'site' (which in central European prehistory were often simply rich concentrations of artefacts). It was clear that the areas between the distinct sites (settlements and burial grounds) also held archaeological potential, although often of a different nature from those found on more visible 'sites'. As in Britain, so in central Europe this was particularly significant for the recognition of landscape features such as field boundaries, enclosures of various functions, remnants of quarrying and processing raw materials, etc. These artefacts and ecofacts were often difficult to date, resulting in their long-lasting neglect by previous archaeological schools in central Europe, which presumed chronology to be the main aim of archaeological knowledge.

Within the new conceptual framework, however, the concept of landscape (in the sense mentioned above) brought a new insight to archaeology. Firstly, it showed that the geographical space was used continually and not just at the selected loci, and that this continuity of use is reflected in the spatial continuity of archaeological data (components). Secondly, it showed that spatial continuity could be interpreted as the continuity of complementary functions – most settlements, for example, must have been associated with arable fields, paths, burial grounds and pastures. Consequently, the understanding of past communities has become more complex than had so far been realised. Thirdly, archaeology's capability of studying large objects of anthropogenic origin altered the concept of the surrounding material world as a whole: it has become clear that man was not only influenced by his environment, but often also impacted upon it and re-created it. Rather than thinking of humans adapting to the environment, the emphasis shifted to human interaction with the environment (Zvelebil et al. 1993; Zvelebil 1994), and its adaptation to human uses (Neustupný in prep.). Finally, the fact that archaeological features are still part of the landscape illustrates a further characteristic of the landscape: continuity in time. Individual people and their settlements can relatively easily appear and disappear, while the landscape, although transformed by man and nature, cannot. All processes in the landscape relate to the state of the previous period and previous generations of its inhabitants – in this sense landscape has a memory. Erasure of that memory can be achieved only by drastic means, such as the surface mining of brown coal in northwestern Bohemia during the communist period, or the building of the Three Gorges Dam in China today.

In contemporary archaeology, landscape is often understood not only as a concept complementary to 'sites', but also as an overarching approach to multiple scales of analysis – from site to region. This is also how the term landscape has been used in the German landscape archaeology (*Landschaftsarchäologie*) since the mid-1990s (Lüning 1997; Gramsch 2003). According to Andreas Zimmerman (2002; Zimmermann et al. 2004), landscape archaeology represents the research of the spatial organisation of human activity above the level of individual sites or communities, while emphasising the structure and function of landscape components (understood generally as settlement components). In his two seminal studies on the prehistoric settlement (mainly Neolithic, excavated in the opencast mining areas) in the Rhineland, Zimmermann and his colleagues provide evidence for aspects of landscape archaeology in two spheres: settlement history reconstructing the development of the area under study to the scale of individual villages and generations of their inhabitants, and the study of settlement structure on various scales. As exemplary questions of landscape archaeology, in this sense the author considers settlement density, demographic trends, circulation of raw materials, hierarchy of settlement units and similar.

We suggest here that the term 'landscape' can incorporate one crucial element which is not considered by the contemporary German model of *Landschaftsarchäologie*, and which is often missed in settlement and environmental archaeologies in general: unlike the 'natural environment', 'landscape' is indivisible from human beings and their cognitive activities. Human beings not only interact with the environment through materials, but they also perceive and interpret it, and the part of the environment which can be perceived and given meaning by people is incorporated within the idea of 'landscape'. The idea of landscape, then, can be used to capture how the physical environment is manipulated by people in a symbolic as well as a practical way. We naturally assume that this manipulation existed in the past, too, and should have left traces in the archaeological record. Therefore, we see the aim of landscape archaeology as being a process of reconstruction of landscape elements or larger parts of landscape in the past that works upon the identifiable and interpretable imprint of human activity in both its practical and symbolic aspects.

SETTLEMENT AND LANDSCAPE ARCHAEOLOGY IN CENTRAL EUROPE

In this section, we examine how archaeologies of settlement and landscape have developed in central Europe, introducing several influential concepts (*Landesaufnahme*, microregions, community area theory) and methods (analytical survey, predictive modelling). However, before introducing these concepts, it is important to underline some of the particular influences upon the development of central European settlement and landscape archaeologies in this region – since the character of the central European archaeological record has shaped how archaeologists in the region have developed and used ideas of landscape.

Today, landscape archaeology in central Europe presents a number of challenges, not only because of the influence of culture history or Jankuhn's settlement archaeology on research in the region, but also due to the ongoing destruction of the archaeological landscape through intensive arable agriculture, which continues a process that began with medieval and post-medieval clearance and leaves just small areas of pasture and woodland in which elements of past landscapes survive above ground. Apart from hill forts or medieval field systems, above-ground earthwork features – whether from the prehistoric or medieval periods – are rare in the central European landscape. Another specific problem is the low frequency of artefacts made of durable materials such as flint. Whereas surface collection in southern England can recover thousands of flint tools and *débitage* during a short field survey project, but very few sherds of prehistoric ceramics (Shennan 1985), the situation in central Bohemia is quite the opposite. For example, during the British-Czech Ancient Landscape Reconstruction (ALRB) project in Bohemia approximately 30,000 fragments of prehistoric pottery were found during the five collection campaigns, contrasting with only approximately 600 items of flint, including Palaeolithic material and post-medieval strike-a-lights, and even fewer other durable items (fragments of stone axes, glass or metal objects) (Kuna 1994, 1998, 2000).

Comparison of settlement densities in various prehistoric periods is not without its pitfalls, too. As already mentioned, the probability of finding a prehistoric settlement depends directly on defining subsurface archaeological features. For example, the period of Corded Ware (Late Eneolithic in the Czech terminology, 2900–2500 cal. BC) is not characterised by the presence of sub-surface features such as houses or storage pits, and therefore not a single settlement site is known in the Czech Republic from this period – whereas very many Corded Ware cemeteries survive. This poses particular challenges for landscape archaeology. One answer is to maximise the exploitation of the evidence available. For example, a detailed analysis of residual pottery fragments, which previously had not received analytical attention, has recently been undertaken. The results of this research surprisingly revealed that there were more periods represented, often almost their complete spectrum, on most prehistoric settlement sites than had been recorded by standard excavation methodology (Kuna 2002). This example is a warning against premature conclusions concerning the density of prehistoric settlements on the basis of standard field data.

The central European archaeological resource also presents particular advantages, however. For example, in some periods ceramics from prehistoric settlements and cemeteries are so abundant and typologically distinctive that assemblages can be sorted into very fine-grained phases. As reflected in the analysis of Neolithic settlements in both Germany and Bohemia, by using this data the development of individual settlements and microregions can be detailed down to almost individual generations of their inhabitants. Such clarity in prehistory can only be surpassed by some of the dendrochronologically dated sites of the Alpine region.

Landesaufnahme

The first concept that we want to introduce, that of *archäologische Landesaufnahme* ('archaeological area scan'), has played an important role in the development of central European landscape archaeology since the 1960s. *Landesaufnahme* involves the systematic collection of all archaeological data within the maximum possible chronological span from the region under study. All the possible resources are used: museum collections, sites and monument records and published archaeological literature are combined with new data generated through systematic field surveys. The cartographic visualisation of these data creates an integrated overview of the landscape which serves as a source for theoretical research and effective heritage management.

The term *archäologische Landesaufnahme* was first used by Karl Heinz Jakob in 1908, and was further developed between the wars by other German archaeologists (Alfred Tode, Albert Kiekebusch, Karl Kersten). It was Herbert Jankuhn (1973), however, who defined it as a principal method of settlement archaeology. In 1978, it was adopted in Poland by an ambitious project, *Archeologiczne Zdjęcie Polski* (Archaeological Record of Poland), for mapping the whole state territory (Mazurowski 1980) that continues to this day. Combining fieldwalking and documentary research on the standing monuments, it has now covered approximately 75% of the country and has registered some 290,000 archaeological sites (Barford et al. 2000: 73; Barford 2001: 27).

Unlike the focus of previous culture historical approaches, the *Landesaufnahme* allowed a regional (rather than a supra-regional) and diachronic (rather than chronologically selective) approach to the archaeological resource. In regions where archaeological evidence was sufficient, settlement archaeology, using the methodology of *Landesaufnahme*, achieved very complex reconstructions of the past landscape (cf. Jankuhn 1977; Kuna et al. 2004: 454). However, *Landesaufnahme* and its applications were characterised by an excessive empiricism and over-confidence in the surviving data patterns. The collection of 'all' the data from any area is obviously impossible. *Landesaufnahme* as a method does not sufficiently consider the character of research as a never-ending and iterative process, in which sources themselves (their kind, quantity and availability) are inseparable from research questions and simultaneously change with them. Therefore, no research sources can ever be assembled either in completeness, or in advance for future research. For this reason, current archaeology has replaced the methodology of 'archaeological scan' with methods of archaeological sampling. It is no longer believed that archaeology can obtain a 'complete' data set once forever and for any future research question – it is rather believed that the method of data collection and the data itself always reflect the research question posed at the beginning. Different questions, therefore, deserve different field methods, which bring data sets of different characteristics (Kuna et al. 2004).

Siedlungskammer

While regional approaches developed in many traditions of archaeology during the 1970s, inspired especially by the New Geography (Coones 1985), in central Europe studies of *Siedlungskammer* (microregions) became a popular form of settlement archaeology, especially as an alternative to previous culture historical studies. It formed part of a general shift in the region away from typological and chronological questions towards a focus on the social and economic dimensions of settlement. The term 'microregion' appears to have first been used in the late 1960s in Poland by Witold Hensel (Barford 2001: 22). A similar approach had already been applied in the Czech Republic during the excavations of Neolithic settlements in Bylany area near Kutná Hora. Within this framework Bohumil Soudský (1966) formulated, for example, the model of cyclic (shifting) agriculture. This was a perfect example of an effective exploitation of microregional data, and although it is no longer generally accepted as a useful method, its stimulating effect cannot be doubted.

The term 'microregion' gained in importance in the Czech Republic, Poland and both German states during the 1980s, since it was during that decade that extensive rescue excavations were undertaken in areas of brown coal surface mining in these countries – especially in Bohemia in the Czech Republic (Velímský 1987), in the Cottbus area in Niederlausitz, eastern Germany (Wetzel 1987), and in the Aldenhovener Platte in Rhineland, Germany (Schwellnuß 1987). With square kilometres of cultural landscape disappearing almost instantaneously it was clear that archaeological excavation could not document the whole region with equal quality. The methodological answer therefore was to change to the systematic research of landscape samples, or microregions. The available capacities of research institutions concentrated on microregions, usually defined as basins of small streams with fewer than a dozen settlement areas (prehistoric community areas). This in turn promoted studies concerning the mobility of settlement areas and settlement 'rotation' (Smrž 1981, 1987, 1994; Smejtek 1994), size of the settlement areas and settlement concentration (Kuna 1991) or the dynamics of settlement processes in the early medieval period (Meduna and Černá 1991).

The methodology used in the coal mining areas was consequently applied to other excavations and field survey projects, later on including the use of geographical information systems (GIS) and statistical analysis. Questions of settlement area dynamics, size and the density of prehistoric communities began to be investigated (Kuna 1991, 1997), as well as the location of activity areas in relation to local geomorphology and other landscape features (Kuna and Adelsbergerová 1995). In the mid-1990s an economic model of a prehistoric microregion was developed (Dreslerová 1995, 1996, 2002), and a method of prediction of prehistoric settlement sites was tested (Dreslerová 1998).

Similarly in Germany, the 1970s archaeological excavations in the brown coal district of Aldenhovener Platte led to a new focus on settlement archaeology (Lüning 1997; Zimmermann 2002). Extensive excavations were also undertaken in the southeastern Swabian Alb (Knipper et al. 2005) and in Hessen, in the central part of western Germany (Ebersbach and Schade 2005). These projects focused on settlement patterns from the level of individual houses to complete regions, demographic estimations, quantification of the settlement economy (mainly the size of arable fields and pastures) and mapping communication networks as demonstrated by the distribution of lithics.

Community Area Theory

Like the field research of *Siedlungskammer*, Evžen Neustupný's 'community area theory' (1986, 1991, 1998a) sought to give a deeper theoretical background to this research by presenting a general model of the settlement and social structure of prehistoric agricultural societies (Table 7.1). Neustupný suggested that the regularities observed in the organisation of archaeological evidence, including regular distances between sites and continuity in occupation, often resulted from 'settlement areas' in the past. In this model, the remains of settlement activities of individual prehistoric communities accumulated within the original 'community areas'. Community areas were composed of activity areas with various functions. Neustupný's suggestion that components of various functions and dates with different archaeological visibility resulted from archaeological formation processes was

Table 7.1 The basic concepts of the community area theory. *After E. Neustupný 1986, 1998a, 1998b, 2001*

Living culture	*Dead (Archaeological) culture*
activity area (e.g. residential, funerary, production area)	component (residential, funerary, production component)
community area	settlement area
supra-community area (ritual, production area)	extra-settlement area (hill forts, rondels, mines)
sphere (world) of otherness	landscape

closely related to his wider 'theory of archaeological transformations' (Neustupný 1998a; Kuna 2001). Community area theory highlighted how some aspects of the past use of the landscape – such as fields, pastures, areas of wood resources or boundaries – are less archaeologically visible. Settlement space was understood as a continuum, rather than as a set of isolated points.

Community area theory is important mainly for the general perception of past settlement activities and the corresponding archaeological evidence. It is an approach that makes it possible to work with results of modern non-destructive methods in a way that conventional site-based archaeologies cannot. Neustupný has recently extended community area theory to explore the landscape dimensions of interaction between communities through the idea of *sféra jinosti* ('sphere of otherness'), which focuses especially upon the exchange of material culture and kin relations between different communities in the past (Neustupný 1998a, 1998b, 2001). The idea of the *sféra jinosti* highlights how activity areas may relate not just to a single community, but to relationships between communities. It has been applied to sites such as hill forts or industrial zones, for example in the Loděnice area discussed below (Neustupný and Venclová 1996; Venclová 2001); recently the idea of analogical 'ritual zones' has been considered (Waldhauser 2001).

Community area theory introduced a distinctive approach to landscape, which was seen as a 'spatially unconfined, but richly structured area, where the relations of otherness were in motion' (Neustupný 2001: 17). Landscape in this sense is not a geographical unit, but the relic of a past social world. Landscape, that is to say, is a part of the dead (archaeological) culture and comprises settlement areas and supra-settlement areas.

Analytical Survey

The large excavations of the 1980s inspired further projects in central Europe. Programmes of intensive field survey were developed in the Rhineland (Schwellnuß 1987), Bohemia (Meduna and Černá 1991) and outside the area of brown coal mining (Kuna 1994), and came to include not only surface artefact collection but also intensive aerial photography, which had been forbidden during the Communist period. In the 1990s similar surveys were also developed in other former Communist central European countries such as the Czech Republic, eastern Germany, Poland and Hungary, and were characterised by a new focus on landscape, rather than on sites in isolation. Analytical field survey used arbitrary spatial units which were covered with standard intensity; the results were quantified and analysed by various statistical methods.

Such analytical surveys approached the continual occurrence of archaeological data in the landscape as deriving from past activity areas that shifted location over time, although their movements were usually constrained by the boundaries of a community area, creating an extensive palimpsest of variously dated overlapping remains. Hence, concentrations of finds ('sites') might be seen not as the exclusive loci of past activities but as 'epicentres' of activity areas, where accumulations of remains mostly relate to more archaeological phases or cultures (Kuna 1998, 2000). Archaeological 'sites' are not identical with 'original' functional units, but are a result of secondary processes, of archaeological transformations. Analytical methods of field survey enable a quantitative analysis of the data collected and the application of sophisticated mathematical procedures. In particular, multivariate analysis of data (factor analysis, etc.) and its combination with geographical information systems (GIS) have proven to be very effective (Neustupný 1996).

Predictive Modelling

The increasing use of GIS since the 1990s has revived interest in location analysis and archaeological predictive modelling in central Europe, topics that can contribute to archaeological research as well as being applied to archaeological heritage management. By facilitating predictive modelling, GIS allowed archaeologists to assess the potential archaeological 'risk' of large industrial landscape interventions. In countries such as Holland and Denmark, predictive modelling has become part of the common heritage protection strategy, creating 'indicative maps of archaeological value' (Deeben and Hallewas 2003: 109). As a method, archaeological prediction is based upon an assumption of regular relationships between the use of the land and the characteristics of its geography. This assumption does not necessarily mean geographical or ecological determinism, because the relationships under attention need not be only those with practical functions (such as the distance from a water source), but could also be those with social or symbolic value: their relation to the landscape may more easily change in time and between various cultures.

TWO CASE STUDIES

A number of shifts in central European landscape archaeologies have occurred in recent years. Previous approaches to archaeological sites as corresponding to past activity foci have been replaced by an awareness of how they are the products of formation processes. Site-focused approaches have been replaced by an understanding that the areas between sites can be significant – that a lot of significant data lies beyond sites. Where previously it was believed that a concise list of sites could represent the full complexity of past activities (this is where the concept of *Landesaufnahme* comes from), a shift towards a more appropriate theory and a more flexible field methodology is represented by the community area theory and the methodology of analytical field survey, as well as predictive modelling.

Two case studies drawn from the archaeology of the Czech Republic will serve to demonstrate the potential of contemporary landscape approaches: the study of an Iron Age zone of industrial production and the use of predictive modelling in the study of the construction of burial mounds during the Middle Bronze Age to Hallstatt periods (c. 1600–400 cal. BC) in South Bohemia (Figure 7.1).

The Landscape of an Iron Age Industrial Zone

The potential of the combination of community area theory and analytical field survey is clearly illustrated by the Loděnice project, which was carried out in the early 1990s (Neustupný and Venclová 1996; Venclová 2001). Field survey during this project concentrated on the microregion along the Loděnice and Bakov streams in the western part of central Bohemia, where nucleated Iron Age (La Tène) settlement and the remains of iron smelting and production of decorative arm rings made from sapropelite (a shiny, black shale related to cannel coal) have been identified. This area is also well known in Europe for the excavation of a *Viereckschanze* (rectangular enclosure) dating from the 2nd–1st centuries BC at Mšecké Žehrovice, which produced a famous Celtic carved ragstone head. Two relatively innovative elements were used in this project: an analytical fieldwalking survey of surface ecofact scatters, and the combination of multivariate mathematical analysis with GIS.

The ecofacts collected on the surface from an area of c. 11 square kilometres included concretions of pelosiderite, a potential raw material for a production of iron; both smelting and smithing slag; sapropelite both

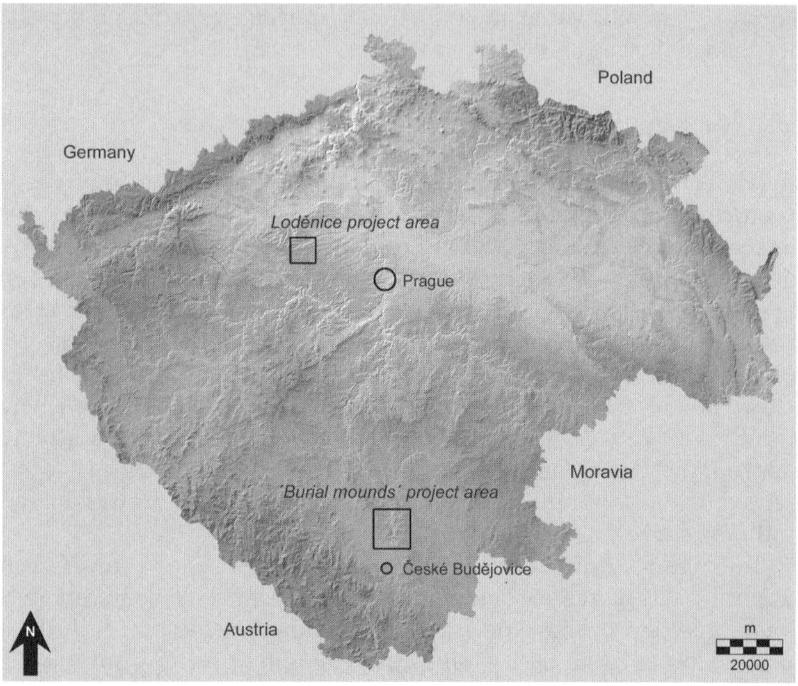

Figure 7.1 Map of Bohemia in the western part of the Czech Republic. The locations of the two projects discussed as case studies are marked by rectangles: Loděnice (north) and Hluboká nad Vltavou (south). The boxes mark the Czech capital Prague (north) and the regional center of České Budějovice (south).

as raw material and as waste from the production of sapropelite arm rings (Figure 7.2); as well as prehistoric and medieval pottery and fragments of arm rings. The surface patterning of the La Tène period ecofacts provided a completely new understanding of the prehistoric industrial activities. Traditional field methods such as excavations and survey aimed at identifying sites could not have detected prehistoric industrial waste on such a scale. Since the remains of these activities occur only in the plough soil and often spread over very large areas, their patterning cannot be recognised other than by analytical fieldwalking survey covering large parts of the landscape.

On the basis of the surface ecofact scatters, the term 'absolute quantities' was used to describe items that usually do not decay and disappear in the course of archaeological transformations and, therefore, their number in the archaeological record can correspond to the original number of the once-deposited items. This applies, of course, only to durable materials, in this case to iron slag pieces and items of sapropelite, with which we can judge the original quantity/intensity

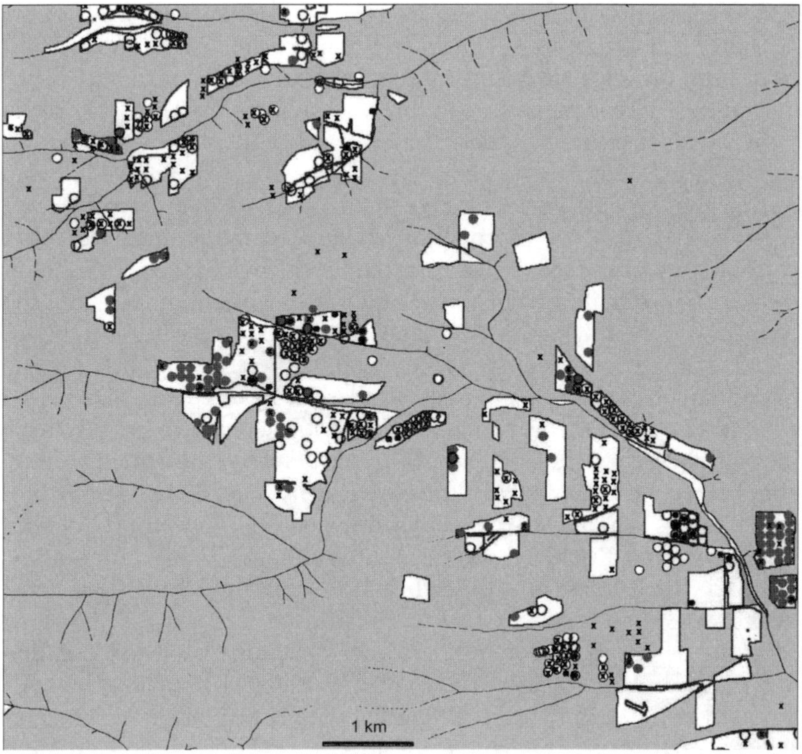

Figure 7.2 The Loděnice project area – an Iron Age industrial zone (distr. Rakovník and Kladno). Open circles: La Tène pottery. Gray circles: pieces of pelosiderite (concretions used as raw material in iron smelting). Black crosses: pieces of iron slag. White polygons: area surveyed (after Venclová 2001).

of the past production. In contrast, the surviving quantities of items of other materials, such as prehistoric pottery, have been considerably reduced over time.

Multivariate analysis (principal component analysis) was used to study combinations of artefacts and ecofacts and the significance (factor scores) of individual fieldwalking sectors to identify factors of variability. From the observed patterning of these factors, the organisation of activity areas (components) was determined and their mutual relationships were interpreted.

Iron production could have been actively undertaken within the residential areas, but it was more commonly located on their periphery and in areas with less agricultural value. This model evolved through time as production areas moved further away from residential ones as the supply of fuel (wood) dwindled. The generally quite high density of settlement components in this microregion indicates the intensity of the past production activities. However, not all of the activity areas could have been contemporary since the industrial activity was carried out during the La Tène period for a span of approximately 110 to 130 years.

In the northern part of the area under study (the Bakov stream valley), there are outcrops of sapropelite, which may have attracted people during the La Tène period. However, a comparison of their location with the spread of the sapropelite *débitage* accumulations shows that workshops were probably not directly dependent on the raw material sources. Sapropelite was quarried and partly worked on here, but the products were finished within the residential areas further away. There is only sporadic evidence for metalworking being found in the Bakov stream valley. In contrast, in the Loděnice stream valley (the next to the south) iron production appears to have been the main industrial activity (but not in all of the identified settlement areas), along with the finishing of sapropelite arm rings, usually from prepared pieces. There were also areas devoted only to iron production. This diverse and richly complex nature of the La Tène communities' industrial activities is still difficult to explain (Neustupný and Venclová 1996).

From the interpretation of the surface data it has been also possible to propose a hypothesis concerning the extent of prehistoric woodland cover in the landscape. Polygons with no surface occurrence of the pelosiderite were interpreted as an anciently deforested landscape (with all of the pelosiderite concretions collected as the raw material for the production of iron). Judging from this data, the landscape had to have been already widely deforested in the Iron Age.

According to E. Neustupný (2003) the example of the Bakov stream, which is characteristic of land with poor agriculture value, shows convincingly that the practical properties of landscapes did not impose any fatal limitations on prehistoric communities. People did overcome poorer environmental conditions if they found any practical function or meaning in the landscape (which supposedly happened in the case of sapropelite arm ring production). Seen from this angle, the human behaviour is far from being determined by nature environment – it is, therefore, better to speak of an adaptation of nature by man than of an adaptation of man to nature (Neustupný in prep.).

Burial Mounds in the Landscape

There are several reasons why the location analysis of funerary sites may appear to be (and in many aspects really is) more complicated than a similar analysis of settlement sites. Firstly, funerary components on the level of regions are usually less complete and/or less easily detectable by the available field methods than are settlement sites. Secondly, the relationships of funerary areas to the attributes of the natural environment are more variable and less comprehensible to a modern observer. While the location of settlement sites must have paid some attention to the practical aspects of the natural environment, the location of funerary areas was probably even less confined by these factors, and instead responded predominantly to individual, culture-specific norms.

There are, however, several areas in Bohemia where the evidence of prehistoric funerary activities is extremely rich. Such cases may be connected to regions with convenient taphonomy, namely those that have not been deforested since the Middle Ages. One such region lies in South Bohemia (Czech Republic), north of the regional centre České Budějovice. This hilly region was once part of the Schwarzenberg estates, and forest has survived here because of the conversion from the 16th century of the entire landscape into several deer parks that still exist today. Most of the prehistoric tumulus cemeteries in this area were excavated between the 1880s and the First World War; their results provide an indication of the dates of the cemeteries (mostly from the Middle Bronze Age and Hallstatt periods, with the Early and Late to Final Bronze Age periods also being represented; altogether ca. 2000–400 cal. BC), although the quality of the documentation is usually too poor to meet modern research needs (for the method of modern survey and the catalogue of sites see Beneš, Michálek and Zavřel 1999; Šimana 1999). The study area of 320 square kilometres consists for the most part of hilly relief, and about 45% of it is covered by forest. It is here that the deer parks were situated, preserving evidence of 96 prehistoric tumulus cemeteries with over 1,000 prehistoric burial mounds in total.

The study had two principal aims (Kuna 2006): to learn whether there were some general principles of locating mound cemeteries in the landscape, and to interpret the rules of burial location in terms of a more general settlement pattern. We considered whether there were settlement sites corresponding to each of the mound cemeteries, or whether there were large 'burial zones' being shared by more than one community living at a greater distance from the funerary areas. The latter solution seemed more probable at the beginning (there are clusters of cemeteries without any known settlement sites), but the crucial role of the different visibility of tumuli vs. settlement sites in forest is obvious. Our approach to the problem started at the location analysis of mound cemeteries by GIS, and stemmed from the idea that specialised 'burial zones' should have had different environmental characteristics from the cemeteries situated close to the residential areas.

The GIS analysis concerned several simple characteristics of the landscape, like the distance to the nearest water source, slope gradient and slope aspect, but it also included some more complicated procedures which we programmed ourselves. The latter set of procedures included an identification of hilltops, ridges, edges and upper and lower parts of the slopes and valley floors. For each of these geomorphological elements their area was calculated as well as the number of cemeteries occurring within their extent. Using simple statistical tests it was then easy to see if some of the relief elements (features) were preferred or avoided.

Surprisingly, even these simple analyses produced quite patterned results. For example, the distance from water source showed the increasing significance of higher distance categories for the cemetery location, the clear peak of the values being in the 300–400 metre class. From what we know about settlement sites (from other regions, of course: there are almost no settlement sites known in the area under study), most of them could be expected to occur less than 300 metres from the streams. Hence, cemeteries seem to have been in slightly more distant locations than the settlement sites, but they still seem to be connected to the settlement zone along the water streams. Another interesting point appeared from the analysis of slope aspect. The pattern observed in our data (Table 7.2) shows a clear preference for southern slopes. For the location of a cemetery such a preference would not make much sense if the immediate vicinity of a settlement site could not be assumed.

Table 7.2 Burial mounds in the landscape, the Schwarzenberg deer parks area at Hluboká nad Vltavou, in the district České Budějovice. Distribution of prehistoric tumulus cemeteries according to the slope aspect. Significant values marked bold

Slope aspect	*Spatial extent of the category in the landscape*		*Number of cemeteries*		*Index observed / expected*
	km²	*%*	*observed*	*expected*	
flat	2.6	0.8	0	0.8	0.0
N	37.2	11.6	5	11.2	0.4
NE	41.7	13.0	8	12.5	0.6
E	35.3	11.0	7	10.6	0.6
SE	37.6	11.8	15	11.3	**1.3**
S	50.7	15.8	20	15.2	**1.3**
SW	47.2	14.7	21	14.2	**1.4**
W	35.9	11.2	13	10.8	**1.2**
NW	31.9	10.0	7	9.6	0.7
total	320.1	100.0	96	96.0	1.0

Table 7.3 Burial mounds in the landscape, the Schwarzenberg deer parks area at Hluboká nad Vltavou, in the České Budějovice. Distribution of prehistoric tumulus cemeteries according to the relief features. Hilltops are defined as the highest points within the radii of 100, 250, 500 and 1000 metres; hence, for categories of hills could be distinguished. Significant values marked bold

Relief features	*Spatial extent of the category in the landscape*		*Number of cemeteries*		*Index observed / expected*
	km²	*%*	*observed*	*expected*	
hilltops100	28.3	8.8	11	8.5	**1.3**
hilltops250	17.8	5.6	12	5.3	**2.2**
hilltops500	6.6	2.0	2	2.0	1.0
hilltops1000	4.0	1.3	5	1.2	**4.1**
buffers 25 m	18.6	5.8	7	5.6	**1.3**
ridges	60.4	18.9	31	18.1	**1.7**
upper edges	1.3	0.4	0	0.4	0.0
above (upper) edges	3.8	1.2	2	1.1	**1.8**
lower edges	2.7	0.8	0	0.8	0.0
flood plain	7.1	2.2	0	2.1	0.0
lower plain	4.8	1.5	0	1.4	0.0
upper plain (plateau)	1.3	0.4	0	0.4	0.0
lower slope (< 10°)	82.6	25.8	16	24.8	0.6
upper slope (< 10°)	57.1	17.8	9	17.1	0.5
steep slope (> 10°)	23.7	7.4	1	7.1	0.1
total	320.1	100.0	96	96.0	1.0

Non-random preferences for the location of tumulus cemeteries are also illustrated by the analysis of relief landscape features. Table 7.3 shows the obtained values of significance for several relief elements, given by the ratio between the observed and expected numbers of cases. These values show that hilltops and ridges were clearly the most popular types of relief for the location of tumuli. Burial mounds often occur at dominant locations from which large parts of the landscape may be visually 'controlled' (Figures 7.3 and 7.4) (cf. Kuna et al. 2004: 253). This might argue for the special meaning of burial mounds, seeing them, for example, among many other things, as territorial markers or monumental symbols of community.

During a more careful analysis of the data, two additional aspects were observed. The tumulus cemeteries obviously tend to appear in dominant places but not necessarily in the highest or the most prominent (i.e. least accessible) of all those available. If a prominent area is too far from the stream (i.e. from a potential settlement location) it is usually not used; a location on a slope (above a potential settlement site) seems, then, to have been sufficient. Similar results were obtained from a model of the 'suitable' (not steeper than 10 degrees, within the

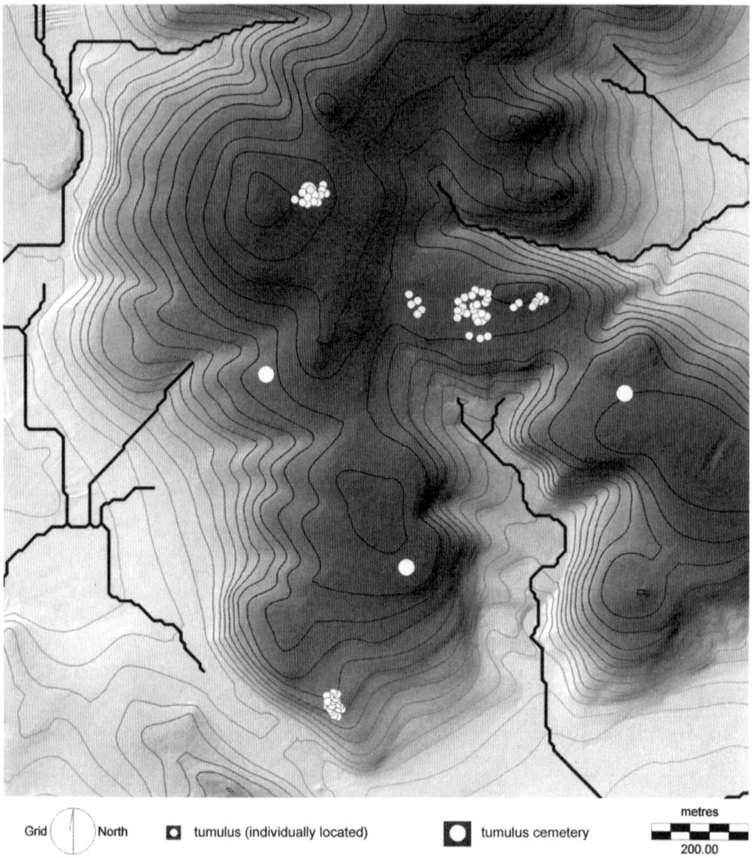

Figure 7.3 Burial mounds in the landscape, the Schwarzenberg deer parks area at Hluboká nad Vltavou, distr. České Budějovice. The position close to the edges of elevated plateaux, hilltops and the upper parts of slopes is typical. Small circles: burial mounds individually measured. Large circles: groups of burial mounds, not measured individually.

300-metre distance from the nearest water stream) and 'unsuitable' land. As seen in Figure 7.5, the tumulus cemeteries are not always situated on the 'suitable' land (i.e. within the area of the supposed settlement sites), but they are usually quite close to its peripheries and appear to avoid larger islands of the 'unsuitable' landscape.

All these observations have a common denominator: a close relation between the cemetery and a settlement site. Dominant places may have been preferred for the location of cemeteries, but not always; we must pay attention to other (and probably the decisive) aspects of the landscape, which in this case may have been the location of the settlement

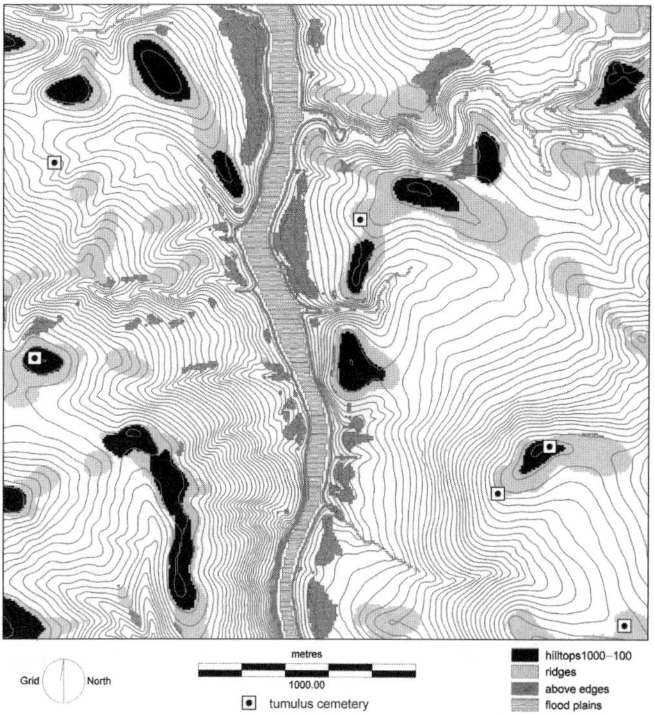

Figure 7.4 A GIS model of the landscape distinguishing the main relief elements (hilltops, ridges, edges and areas above them, floodplains) and the position of prehistoric tumulus cemeteries in the area of the Schwarzenberg deer parks (detail). The four categories of hilltops (see Table 7.2) have not been differentiated in this figure.

site. This, however, changes our opinion expressed at the beginning: the model which now seems more probable works with small settlement sites, each of which is closely accompanied by a cemetery. Such a model corresponds to what has been recently published on the Bronze Age settlement pattern in Germany and Denmark (Willroth 2001).

There are more hints to support this model. Firstly, in several sites in different areas of Bohemia some residual pottery fragments appear in the bodies of burial mounds. These have been interpreted as artefacts from settlement layers in the immediate vicinity of the cemetery brought during the mound construction (Čtrnáct 1954). From these cases we can also learn that if a sequence of a settlement and a cemetery is found, the cemetery is always later. This may mean that funerary areas were respected for a long time after they were created. In principle, the model of a close relationship between a settlement site and a tumulus cemetery could be verified by further survey, although any survey

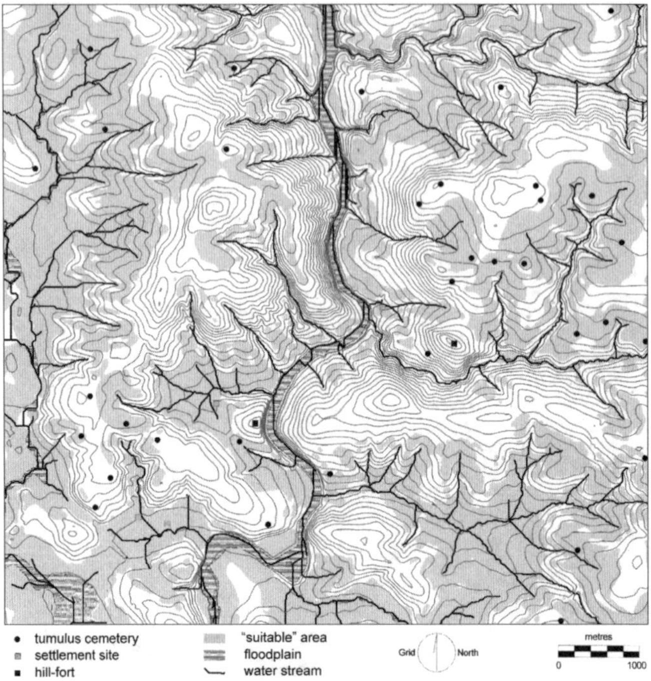

Figure 7.5 A predictive model of 'suitable' versus 'unsuitable' land for prehistoric settlement sites within the Schwarzenberg deer parks area (detail). 'Suitable' land (marked grey) is within 300 metres of water sources and on slopes below 10 degrees. The position of the tumulus cemeteries, two prehistoric hill forts and a few (generally not quite clear) settlement sites is marked.

looking for subsurface features (settlement sites) in a forested landscape is very difficult and demands special field methods.

This study of burial mounds shows that GIS as a methodological tool includes much more than simple procedures for measuring distances and slope gradients – it is a tool which can, if used creatively, help us to understand how people in the past perceived their landscapes and employed them in their social strategies. Studying the relation between residential and funerary areas may contribute to our knowledge of the past social structure, particularly if more field research is done in the future.

CONCLUSIONS

The development of landscape archaeology in central Europe has, as anywhere else in the world, been strongly influenced by the character of the region's archaeological resources. Compared with Atlantic Europe and Scandinavia, in central Europe there is a certain lack of what is normally understood to be the archaeological landscape – visible landscape features such as field boundaries or monuments. Rather, the central European archaeological record yields a lot of 'on-site' evidence of a high quality, mostly in the form of rich residential and funerary sites. These components can usually be structured with a high accuracy, providing the base for detailed chronologies, settlement histories and settlement patterns. No wonder that under such circumstances 'landscape archaeology' as a specific approach was mostly either neglected or – as in the case of contemporary German archaeology (*Landschaftsarchäologie*) – used to mean something else than in the English-speaking countries (a sophisticated version of modern settlement archaeology in the given example). The Czech theory of the community areas, too, derives from settlement archaeology; the term 'landscape' is used here in a marginal context and with a meaning that is incompatible with that of other archaeological schools. The different understanding of the term 'landscape' is, anyway, deeply rooted in the history of the discipline and certainly will not change in the near future.

Within the conceptual framework of archaeology in central Europe it is particularly difficult to identify a difference between the concept of 'landscape' on the one hand, and the concepts of 'environment' and 'settlement pattern' on the other. 'Landscape' is often used as a synonym for one of the other terms. To make our point of view clear, we would argue that 'landscape' as a concept should be related to human perception. Unlike 'environment' or 'settlement pattern', 'landscape' is a set of natural or man-made features on the Earth's surface, which are (or can be) perceived by people and given meaning by them. Such a definition, however, would hardly be agreed upon by most archaeologists in central Europe, mainly because of the very fact that human perception is understood as a sphere inaccessible to archaeology and, thus, inappropriate for the debate.

Since the 1960s, landscape archaeology has played an important role in broader shifts in central European archaeology, away from typologies and chronologies toward questions of the economy, demography, social structure and symbolic systems of the past cultures. As our two case studies show, while they may be different from the landscape archaeologies common in other parts of Europe or elsewhere in the world, archaeologists in central Europe are continuing to develop sophisticated approaches to landscape in archaeology.

REFERENCES

Barford, P. 2001. Space, Environment, and Cultural Landscape in Polish archaeology. In T. Darvill and M. Gojda (eds) *One Land, Many Landscapes*. Oxford: British Archaeological Reports (S987), pp. 19–32.

Barford, P., W. Brzezinski and Z. Kobylinski 2000. The Past, Present and Future of the Polish Archaeological Record project. In J. Bintliff, M. Kuna and N. Venclová (eds) *The Future of Surface Artefact Survey in Europe*. Sheffield: Sheffield University Press, pp. 73–92.

Beneš, A., J. Michálek and P. Zavřel 1999. *Archeologické nemovité památky okresu České Budějovice. Díl I–II – Archäologische Geländedenkmäler im Kreis Cecské Budějovice. Band I–II*, Praha (ARÚ AVČR – OÚ České Budějovice – Muzeum Strakonice).

Böhm, J. 1937. *Základy halštatské periody v Čechách – Die Grundlagen der Hallstattperiode in Böhmen*. Praha.

Böhm, J. 1941. *Kronika objeveného věku*. Praha: Družstevní práce.

Černý, E. 1973. Metodika průzkumu zaniklých středověkých plužin na Drahanské vrchovině. *Zprávy Československé společnosti archeologické při ČSAV* 15 (Supplement 6): 50–119.

Černý, E. 1979. *Zaniklé středověké osady a jejich plužiny, Metodika historickogeografického průzkumu v oblasti Drahanské vrchoviny*. Praha: Academia.

Chang, K.C. 1968. *Settlement Archaeology*. Palo Alto, CA: National Press Books.

Clarke, D. (ed) 1977. *Spatial Archaeology*. London: Academic Press.

Coones, P. 1985. One Landscape or Many? A Geographical Perspective. *Landscape History* 7: 5–12.

Čtrnáct, V. 1954. Mohylová chata a sídelní mohylové objekty v Plzeňsku – Die Hütte und Siedlungsobjekte der Hügelgräberkultur im Gebiete von Pilsen. *Památky archeologické* 45: 335–354.

Deeben, J. and D. P. Hallewas 2003. Predictive Maps and Archaeological Heritage Management in the Netherlands. In J. Kunow and J. Müller (eds) *Landschaftsarchäologie und geographische Informationssysteme*. Wünsdorf: Forschungen zur Archäologie im Land Brandenburg 8, pp. 107–118.

Dreslerová, D. 1995. A Socio-economic Model of a Prehistoric Micro-region. In M. Kuna and N. Venclová (eds) *Whither Archaeology? Papers in Honour of E. Neustupný*. Praha: Archeologický ústav AVČR, pp. 145–160.

Dreslerová, D. 1996. Modelování přírodních podmínek mikroregionu na základě archeologických dat – Modelling environmental conditions of microregion based on archaeological data. *Archaeologické rozhledy* 48: 605–614, 709–712.

Dreslerová, D. 1998. The Říčany Area: Field Walking and GIS in the Study of Settlement History. In E. Neustupný (ed) *Space in Prehistoric Bohemia*. Praha: Archeologický ústav AVČR, pp. 116–128, 170–176.

Dreslerová, D. 2002. Ein Modell zur Umweltrekonstruktion anhand archäologischer Daten. In C. Dobiat, S. Sievers and T. Stöllner (eds) *Dürrnberg und Manching – Wirtschaftsarchäologie im Ost-keltischen Raum*. Kolloquien zur Vor-und Frühgeschichte 7. Frankfurt a. M.: RGK, pp. 55–61.

Ebersbach, R. and C. Schade 2005. Modelle zur Intensität der bandkeramischen Landnutzung am Beispiel der Altsiedellandschaft Mörlener Bucht/Wetterau. In J. Lüning, CH. Frirdich and A. Zimmermann (eds) *Die Bandkeramik im 21. Jahrhundert*.Rahden/Westf.: Verlag Marie Leidorf GmbH (Internationale Archäologie 7), pp. 259–273.

Eggers, H.J. 1950. Das Problem der ethnishcen Deutung in the Frühgeschichte. In H. Kirschner (ed) *Festschrift für E. Wahle*. Heidelburg: Carl Winter-Universitätsverlag, pp. 49–59.

Filip, J. 1956. *Keltové ve střední Evropě*. Praha: Nakladatelství Československé akademie věd.

Gramsch, A. 2003. Landschaftsarchäologie – ein fachgeschichtlicher Überblick und ein theoretisches Konzept. In J. Kunow and J. Müller (eds) *Landschaftsarchäologie und geographische Informationssysteme. The Archeology of Landscape and Geographic Information Systems*. Wünsdorf: Forschungen zur Archäologie im Land Brandenburg 8, pp. 35–56.

Jankuhn, H. 1955. Methoden und Probleme siedlungsarchäologischer Forschung. *Archaeologia Geographica* 4: 73–84.

Jankuhn, H. 1973. Archaeologische Landesaufnahme. *Reallexicon der germanischen Altertumskunde* 1: 391–394.

Jankuhn, H. 1977. *Einführung in die Siedlungsarchäologie*. Berlin: Walter de Gruyter.

Knipper, C., S. Harris, L. Fisher, R. Schreg, J. Giesler and E. Nocerino 2005. The Neolithic Settlement Landscape of the Southeastern Swabian Alb (Germany), 1–33. http://www.jungsteinSITE.de (Consulted 15 March 2006).

Kossina, G. 1911. *Die Herkunft Germanen zur Methode der Siedlungsarchaeologie*. Würzburg Kabitzsch: Mannus-Bibliothek 6.

Kostrzewski, J. 1949. Post War Prehistory in Poland. *American Journal of Archaeology* 53(4): 355–356.

Kuna, M. 1991. The Structuring of Prehistoric Landscape. *Antiquity* 65: 332–347.

Kuna, M. 1994. *Archeologický výzkum povrchovými sběry – Archaeological survey by surface collection*. Praha: Zprávy ČAS – Supplément 23.

Kuna, M. 1997. Geografický informační systém a výzkum pravěké sídelní struktury. In J. Macháček (ed) *Počítačová podpora v archeologii*. Brno: Filosofická Fakulta Masarykovy University, pp. 173–194.

Kuna, M. 1998. Keramika, povrchový sběr a kontinuita pravěké krajiny. *Archeologické rozhledy* 50: 192–223.

Kuna, M. 2000. Surface Artefact Studies in the Czech Republic. In J. Bintliff, M. Kuna and N.Venclová (eds) *The Future of Surface Artefact Survey*. Sheffield: Sheffield University Press, pp. 29–44.

Kuna, M. 2001. Povrchový sběr a intenzita využití krajiny v pravěku. In J.K. Kozłowski and E. Neustupný (eds) *Archeologia przestrzeni. Metody i wyniki badań osadniczych w dorzeczach górnej Łaby i Wisły*. Kraków: Polska Akademia Nauk, pp. 27–54.

Kuna, M. 2002. Intruze jako doklad, nenalezených fází osídlení. In E. Neustupný and Z. Smetánka (eds) *Archeologie nenalézaného*. Dobrá Voda: Vydavatelství a nakladatelství Aleš Čeněk, pp. 119–132.

Kuna, M. 2006. Burial Mounds in the Landscape. In L. Šmejda (ed) *Archaeology of Burial Mounds*. Plzeň: University of West Bohemia, pp. 83–97.

Kuna, M. and D. Adelsbergerová 1995. Prehistoric Location Preferences: An Application of GIS to the Vinořský potok Project, Bohemia. In G. Lock and Z. Stančič (eds) *Archaeology and Geographical Information Systems: A European Perspective*. London: Taylor and Francis, pp. 117–131.

Kuna, M., J. Beneš and M. Dobes 2004. *Nedestruktivní archeologie: teorie, metody a cíle*. Praha: Academia.

Lüning, J. 1997. Landschaftsarchäologie in Deutschland, ein Programm. *Archäologisches Nachrichtenblatt* 2: 277–285.

Mähling, W. 1944. *Böhmen und Mähren gehören zum Reich*. Berlin: SS-Hauptamt.

Mazurowski, R. 1980. *Metodyka archeologicznych badań powierzchniowych*. Warszawa and Poznań: Państwowe Wydawnictwo Naukowe.

Meduna, P. and E. Černá 1991. Settlement Structure of the Early Middle Ages in Northwest Bohemia: Investigations of the Pětipsy Basin Area. *Antiquity* 65: 388–395.

Neustupný, E. 1986. Sídelní areály pravěkých zemědělců. *Památky archeologické* 77: 226–234.

Neustupný, E. 1991. Community Areas of Prehistoric Farmers in Bohemia. *Antiquity* 65: 326–331.

Neustupný, E. 1996. Polygons in Archaeology. *Památky archeologické* 87: 112–136.

Neustupný, E. 1998a. Structures and Events: The Theoretical Basis of Spatial Archaeology. In E. Neustupný (ed) *Space in Prehistoric Bohemia*. Praha: Archeologický ústav AVČR, pp. 9–44.

Neustupný, E. 1998b. Otherness in Prehistoric Times. In L. Larsson and B. Stjernquist (eds) *The World View of Prehistoric Man*. Stockholm: Vitterhets Historie och Antikvitets Akademie (Konferenser 40), pp. 65–71.

Neustupný, E. 2001. Hlavní problémy prostorové archeologie. In J.K. Kozłowski and E. Neustupný (eds) *Archeologia przestrzeni. Metody i wyniki badań osadniczych w dorzeczach górnej Łaby i Wisł,*. Kraków: Polska Akademia Nauk, pp. 7–26.

Neustupný, E. 2003. The Non-Practical Dimensions of Prehistoric Landscapes. In J. Kunow and J. Müller (eds) *Landschaftsarchäologie und geographische Informationssysteme*. Wünsdorf: Forschungen zur Archäologie im Land Brandenburg 8, pp. 291–295.

Neustupný, E. in prep. Vymezení archeologie. In L. Jiráň (ed) *Nejstarší dějiny Čech*.

Neustupný, E. and N.Venclová 1996. Využití prostoru v laténu: region Loděnice – Gebrauch des Raumes in der Latènezeit: die Region Loděnice. *Archeologické rozhledy* 48: 615–642, 713–724.

Schwellnuß, W. 1987. Braunkohlenarchäologie und Siedlungsgeschichte im Rheinland. In E. Černá (ed) *Archaeologische Rettungstaetigkeit in den Braunkohlengebieten und die Problematik der Siedlungs – geschichtlichen Forschung*. Prag: Archeologický ústav AVČR, pp. 169–183.

Shennan, S. 1985. *Experiments in the Collection and Analysis of Archaeological Survey Data. The East Hampshire Survey*. Sheffield: Department of Archaeology and Prehistory, University of Sheffield.

Sherratt, A.G. 1981. Plough and Pastoralism: Aspects of the Secondary Products Revolution. In I. Hodder, G. Issac and N. Hammond (eds) *The Pattern of the Past: Studies in Honour of David Clarke*. Cambridge: Cambridge University Press, pp. 261–305.

Šimana, M. 1999. Geodetická dokumentace mohylových pohřebišt – Geodätische Dokumentation der Hügelgräberfelder. In A. Beneš, J. Michálek and P. Zavřel (eds) *Archeologické nemovité památky okresu České Budějovice. Díl I–II – Archäologische Geländedenkmäler im Kreis České Budějovice. Band I–II*. Praha: Archeologický ústav AVČR, pp. 184–192.

Smejtek, L. 1994. Změny přírodního prostředí a vývoj mladobronzové sídelní struktury v mikroregionu Hříměždického potoka (jihozápad centrálních Čech). In J. Beneš and V. Brůna (eds) *Archeologie a krajinná ekologie*. Most: Nadace Projekt Sever, pp. 94–111.

Smrž, Z. 1981. Halštatské hradiště ve Stradonicích u Loun. Knovízské a halštatské osídlení mikroregionu Deberského potoka. *Archeologické rozhledy* 33: 487–503.

Smrž, Z. 1987. Entwicklung und Struktur der Besiedlung in der Mikroregion des Baches Lužický potok. In E. Černá (ed) *Archaeologische Rettungstaetigkeit in den Braunkohlengebieten und die Problematik der siedlungs-geschichtlichen Forschung*. Prag: Archeologický ústav ČSAV, pp. 17–30.

Smrž, Z. 1994. Výsledky studia pravěkého přírodního prostředí v mikroregionu Lužického potoka na Kadaňsku. In J. Beneš and V. Brůna (eds) *Archeologie a krajinná ekologie*. Most: Nadace Projekt Sever, pp. 84–93.

Soudský, B. 1966. *Bylany, osada nejstarších zemědělců z mladší doby kamenné*. Praha: Academia.

Storch, D. 2002. Krajina není. In P.Hájek (ed) *Krajina zevnitř*. Praha: Malá Skála, pp. 9–13.

Velímský, T. 1987. Die Rettungsgrabungen im nordböhmischen Braunkohlengebiet. In E. Černá (ed) *Archaeologische Rettungstaetigkeit in den Braunkohlengebieten und die Problematik der siedlungs-geschichtlichen Forschung*. Prag: Archeologický ústav ČSAV, pp. 11–15.

Venclová, N. 2001. *Výroba a sídla v době laténské. Projekt Loděnice*. Praha: Archeologický ústav AVČR.

Wahle, E. 1941. *Zur ethnischen Deutung frühgeschichtlicher Kulturprovinzen: Grenzen der frühgeschichtlicher Erkenntnis*. Heidelberg: Carl Winter (Sitzunsberichte der Heidelberger Akademie der Wissenschaften).

Waldhauser, J. 2001. Sakrální zóna Keltů v Českém ráji. Archeologický mýtus nebo realita? *Historická Olomouc* 12: 209–231.

Wetzel, G. 1987. Stand und Aufgaben der archäologischer Forschungen in den Tagebauvorfeldern des bezirkes Cottbus. In E. Černá (ed) *Archaeologische Rettungstaetigkeit in den Braunkohlengebieten und die Problematik der siedlungsgeschichtlichen Forschung*. Prag: Archeologický ústav ČSAV, pp. 49–60.

Willey, G.R. 1953. *Prehistoric Settlement Patterns in the Virú Valley, Perú*. Washington, DC: Smithsonian Institution (Bureau of American Ethnology Bulletin 155).

Willroth, K.-H. 2001. Haus, Acker und Grabhügel. Variable Konstanten im Siedlungsgefüge der älteren nordischen Bronzezeit. In M. Meyer (ed) ... *trans Albim fluvium. Forschungen zum vorrömischen, kaiserzeitlichen und mittelalterlichen Archäologie*, Festschrift für Aachim Leube zum 65. Geburtstag. Rahden/Westf.: Verlag Marie Leidorf GmbH, pp. 113–124.

Zimmermann, A. 2002. Landschaftsarchäologie I. Die Bandkeramik auf der Aldenhovener Platte. *Bericht der RGK* 83: 17–38.

Zimmermann, A., J. Richter, T. Frank and K.P. Wendt 2004. Landschaftsarchäologie II. – Űberlegungen zu Prinzipien einer Landschaftsarchäologie. *Bericht der RGK* 85: 37–97.

Zvelebil, M. 1994. Koncept krajiny, šance archeologie. In J. Beneš and V. Brůna(eds) *Archeologie a krajinná ekologie*. Most: Nadace Projekt Sever, pp. 20–36.

Zvelebil, M, J. Beneš and L. Kuna 1993. Ancient Landscape Reconstruction in North Bohemia, Landscape and Settlement Programme – Projekty Rekonstrukce krajiny v severních Čechách, Krajina a sídla. *Památky archeologické* 84: 93–95.

CHAPTER 8

Historical Archaeologies of Landscape in Atlantic Africa

Kenneth G. Kelly and Neil Norman

INTRODUCTION

During the past decade, archaeologies of landscape have made an increasingly important contribution to the anthropological investigation of societies of the distant past, as well as those of more recent periods (Bender 1993; Delle 1998; McIntosh and McIntosh 2003; see reviews by Anschuetz et al. 2001; Ashmore 2004; Knapp and Ashmore 1999; Layton and Ucko 1999). This recent emphasis upon landscape by no means represents a totally new way of looking at archaeological data. For example, in North America since the 1950s – especially since Gordon Willey's examination of prehistoric 'settlement patterns' in the Viru Valley, Peru (Willey 1953) – archaeologists have sought to develop perspectives that can acknowledge wider spatial contexts. Since then, archaeological uses of a landscape approach have diversified: whether inspired by the analysis of the spatial arrangement of archaeological data (e.g. Hodder and Orton 1976) or by how the spaces between sites provide 'connective tissue' between communities (Deetz 1990: 1).

This growth of the use of landscape approaches in North American archaeology has, however, proceeded in different ways in different contexts. In this chapter we, as archaeologists based at North American institutions, seek to explore archaeologies of landscape and space in the historical archaeology of Atlantic Africa in a reflexive manner: highlighting their development and potential in the study of African Diaspora historical archaeology in global perspective. Discussing a range of West African landscapes (Figure 8.1), we see archaeological approaches to landscape as encompassing efforts to understand material culture by linking it to the social geography that surrounded the items and framed the interactions of the archaeological actors (Conkey 1982, 1999). As such, we argue that the landscape is not simply the physical backdrop upon which archaeological sites are arrayed, nor is it the physical environment against which archaeological sites are

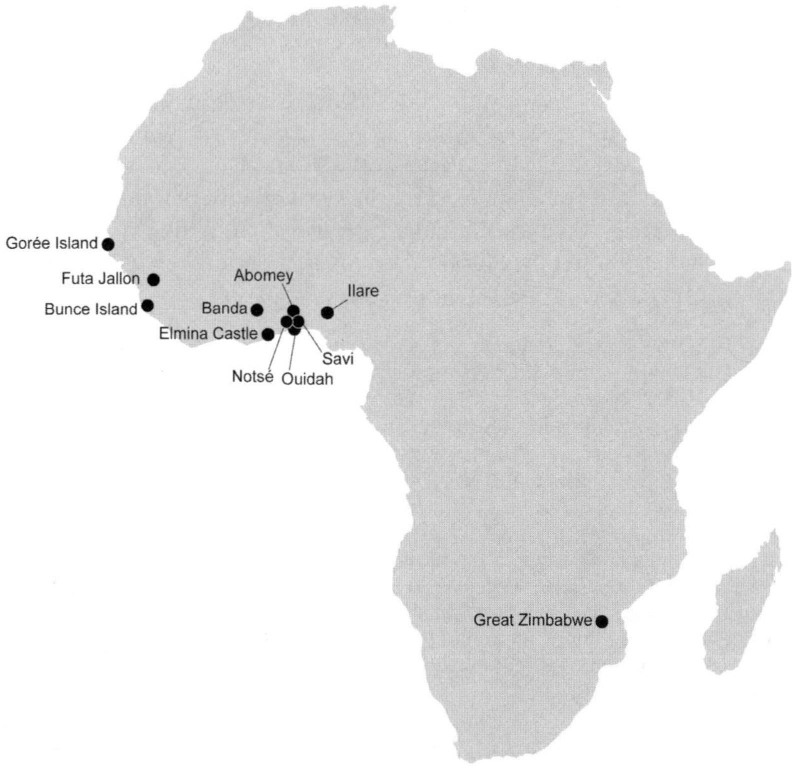

Figure 8.1 Map of African sites mentioned in chapter.

investigated. Landscape archaeology in Atlantic Africa addresses the social creation of the landscape within the complexities of the localities that were nested within global systems of distribution and exchange; and while it often focuses on the influence of factors external to the continent, it must not neglect local traditions, cosmologies and histories (Norman and Kelly 2004). We see the landscape approach as considering the contested and imagined locales that are in a constant state of cultural construction, deconstruction and reconstruction. Central to this perspective is the notion that there is not one single landscape, but rather that landscape varies though time and, importantly, as it is created through the interaction of human actors in space, that landscape can and does vary in the experience of different groups or individuals. Furthermore, the way that people read the materialisation of this interaction is internalised, and can be re-created, or re-experienced, in new or different settings.

APPROACHES TO LANDSCAPE IN AFRICAN ARCHAEOLOGY

The potential of focusing upon landscape has frequently been overlooked in historical archaeologies of the early modern world, despite its demonstrated efficacy by a variety of scholars. The distinction drawn by Wendy Ashmore between the empirical manner in which landscape approaches have traditionally been pursued in the United States and more recent attempts at more humanist, 'post-processual' or interpretive approaches represents a useful point of entry:

> In the United States, positivist archaeology remains strong, with widening exploration of postpositivist stances. In landscape research, most invoke theory from economic geography, ecology, and anthropology, for examining social and economic dimensions of land use. Some focus more closely on the physical landscape in itself, with theory drawn at least as often from the physical and natural sciences as from the social. In all of the foregoing, location and distribution of material resources figure importantly, with growing attention to monuments and rock art or other symbolic markings, and to landscapes materializing ideology or meaning. Those who explore the latter draw more explicitly on social theory with a humanistic cast. Historical archaeologists in the United States likewise tend to be more inclined to humanistic stances, by writing landscapes primarily (but not exclusively) in terms of colonial gardens. (Ashmore 2004: 258–259)

Since the 1970s, the focus in the New Archaeology, developed in the United States and the United Kingdom, upon landscape and space has greatly informed the study of African archaeology. A significant indication of this developing process was seen at the 1982 meeting of the Society for Africanist Archaeologists in America (now the Society of Africanist Archaeologists) in Berkeley, California, which addressed 'Landscape Archaeology, Settlement Patterns and Spatial Analysis'. The proceedings were dominated by processual approaches to prehistoric material ranging from regional settlement patterns to paleolithic artefact distributions and approaches that viewed landscape as broadly synonymous with environment (Nyame Akuma 1982). Since then, empirical studies drawing upon economic geography and ecology have made significant contributions to our understanding of prehistoric patterns of trade, movement and interaction in West Africa through the mapping of archaeological sites to resource bases (e.g. McIntosh 2005). From trans-Saharan salt caravans to networks that linked inland sources of iron and copper to coastal sources of oil palm nuts, salt and dried fish, these studies have sampled from the rich diversity of physical settings in West Africa. The region itself spans desert, sahelian grasslands, mixed woodlands, rain forests, mountains and marsh-like floodplains.

In human terms, the diverse environments of West Africa have been inhabited by an equally diverse range of ethnic and political groups, ranging from food producers to food gatherers, and from small-scale societies to highly complex social and political organisations.

Meanwhile, approaches closer to the 'humanistic' studies described by Ashmore above have developed. In prehistoric archaeology, Susan and Roderick McIntosh and their colleagues and students have combined processual and more humanistic perspectives to explore several thousand years of Niger and Senegal Valley prehistory (MacDonald 1994; McIntosh and McIntosh 2003; Togola 1996). These studies use regional approaches that question city-centric foci and expand the notion of settlement systems to outlying towns and villages that surrounded major centres. Furthermore, they problematise hierarchical models of power for urban landscapes by highlighting sites that exhibited diffuse heterarchical organisations and associations. But it is in Atlantic African historical archaeology that such humanistic perspectives have developed most rapidly, drawing especially upon traditions of historical archaeology developed in the United States (Deetz 1996; Leone 1984) and South Africa (Hall 2000).

The long history of European incursion beginning in the mid-to-late 15th century and colonialism in Atlantic Africa created reams of documentary sources such as travel logs, ship's manifests and colonial accounts from a wide range of European merchants, explorers, administrators and others. Recent research has sought to work across the tension between these accounts and complementary sources such as oral histories of local residents (Stahl 2004) to provide more detail and texture in accounts of economy, trade and migration (Quarcoopome 1993; Soumonni 1997; Greene 2002; Law 2004).

Historical archaeologists (e.g. DeCorse 2001a; Kelly 2002; Ogundiran 2002; Monroe 2003; Norman and Kelly 2004; Stahl 2004; Richard 2005) working in Atlantic Africa have often drawn upon the work of sociocultural anthropologists, social historians and architectural historians who have sought to broaden the definition of landscape to include the shifting and evanescent quality of experiencing landscape (e.g. Upton 1990) or the disruptive qualities of landscape, such that:

> the meanings attached to a particular landscape can both reinforce and disrupt our individual and social identities and cultural values while also generating tension and contestations about meaning, often among unequally empowered groups and individuals. (Greene 2002: 9)

Except for some early hyperdiffusionist attempts to demonstrate external origins for complex African societies, from the inland Niger delta to the Swahili coast and Great Zimbabwe (see examples in

Robertshaw 1990), current archaeologists recognise the Indigenous creation of African landscapes and do not seek their origin elsewhere. Indeed, when considering the majority of Diaspora-era historical archaeology in Africa, it is clear that a sophisticated landscape perspective – recognizing the multivalency of landscape meanings, the political and social power that landscapes become imbued with in the process of their construction and use, and the ways that the same place may have very different meanings to different individuals or groups experiencing it – is integral to current scholarship.

Archaeologists of Diaspora-era Africa have never viewed the landscapes in which they live, work and study as having been pristine or virgin. Just as with the archaeological study of landscapes in other colonial situations, such as the United States (Ashmore 2004: 257), those African landscapes affected by European colonisation cannot be conceived of as having been vacant prior to European arrival. These African landscapes were subject to dramatic anthropogenic transformations both before and after European colonisation. The recognition that all activities in the recent African past occurred in a populated, settled and frequently multiethnic setting has focused the direction of historical archaeological research in different ways. Thus, the majority of African historical archaeology has focused not on transplanted Europeans as it has elsewhere (e.g. North America, Australia and South Africa) but on Indigenous peoples and their negotiated relationships with European arrivals (see papers in DeCorse 2001b and Reid and Lane 2004 for examples). Furthermore, the continued importance of historically constituted landscapes upon which present-day activity is performed means that archaeologists are never far from, and must consider continually, the existing relationships of present(s) to past(s).

The challenge for historical archaeologists, therefore, lies in recognising both the continued influence of prehistoric landscapes as well as historical African influences upon landscape change. To meet this challenge, a shift from approaching human occupation of the landscape in terms of subsistence and access to material needs to an emphasis on the complex and often multiethnic dimensions of landscapes is required. One way in which such different perspectives are being explored by African historical archaeologists is by drawing upon oral historical, ethnohistorical and ethnographic data in the study of historical landscapes, especially by working with the descendants of those who lived in the area in the early modern period. In such work, the nature of archaeologists' and present-day inhabitants' relationships with the remains of the past is highly complex, requiring an approach that takes into account the dynamic process of creating historic accounts, one which often silences other interpretations (Lane 2004; Stahl 2004). Truly, the political and ethical challenges for researchers are great, as their work can give credence to a given group's claim of ownership to an area, or question the age and association of sacred sites.

In studying landscapes, such historical archaeologies have focused upon casting new light upon the social relations of the recent past. Thus, Kurt Anschuetz, Richard Wilshusen and Cherie Scheick have suggested that cultural landscapes shape peoples' interactions with the natural environment, provide an arena for communities' activities and develop layers of meaning over generations (Anschuetz et al. 2001: 160–161). In contrast, within studies of the relationships between humans and their physical environment, such approaches have much in common with interpretive approaches to the role of landscapes as social environments in which human actors are constantly creating senses of place (Preucel and Meskell 2004: 215) that constitute a humanised landscape (e.g. Bender 1998; Ashmore 2004; Preucel and Meskell 2004). Such 'social' approaches, developed in the United States and Europe, encourage us to move beyond accounts of single landscapes, exploring instead how landscapes emerge from the interaction of human actors in space, and how the experience of landscapes varies greatly according to the experience of different groups or individuals. As Julian Thomas has argued, 'it is not simply that [landscapes] are perceived differently: the same location may effectively be a different place for two different persons' (Thomas 2001: 176). In writing social archaeologies of landscape, therefore, while some archaeologists have emphasised how at times certain landscapes were designed and created by groups to express their political power or authority and minimise deviant behaviour (e.g. Leone and Hurry 1998), it is important to remember that understandings of landscapes through individualised experience may defy the intentions of the landscape creator.

HISTORICAL ARCHAEOLOGY AND LANDSCAPE IN WEST AFRICA

From the mid-1980s, archaeologists, especially those based in universities in the United States, built upon previous work on prehistoric sites by investigating historical sites associated with the African slave trade itself, such as Elmina Castle in Ghana (DeCorse 2001a), Bunce Island in Sierra Leone (DeCorse 1994) and Savi and Ouidah in Bénin (Kelly 2002; Norman and Kelly 2004). More recently, such approaches have broadened to focus upon European plantations in West Africa (Bredwa-Mensah 2004; DeCorse 1993), African settlements further inland from the coast (DeCorse et al. 2000; Stahl 2004), regional surveys to investigate long-term changes in social landscape in Senegal (Richard 2005; Thiaw 1999) and palaces associated with slave-trading states (Monroe 2003). While such work has rarely been couched in 'landscape' terms, all utilised, and in some cases presaged, the social archaeologies of landscape outlined above, especially in their emphasis on landscape design, the multivalency of landscapes and the relationships between landscapes, identity and population movements as is demonstrated in the discussion below.

Much of this focus upon sites associated with the trans-Atlantic trade has been conducted by Merrick Posnansky and his students and associates, who have conducted archaeological research in 14 African countries over the last 40 years. In the early 1970s the West Africa Trade Project, funded by the Leverhulme Trust, pioneered ethnoarchaeological field practice (MacEachern 1996: 253; Posnansky 1973, 1979, 1981, 1985, 2004) – an approach that Posnansky developed in relation to landscape at the site of Notsé, in Togo, in the late 1970s and early 1980s, drawing from oral histories in the area to interpret the ditch and wall system that encircled the site (Posnansky 1981). Posnansky argued that the ditches served to define social boundaries and orient social space in and around the site. Further research at Notsé outlines political manoeuvrings by modern groups vying for the rights to maintain ditches and walls associated with the settlement (Greene 2002). These landscape features are described as politicised zones drawn into challenges over association with the past glory of the site's inhabitants: a recurring theme in the archaeology of West Africa (Effah-Gyamfi 1985). Similarly, Posnansky's 'longitudinal' study at the site of Begho, Ghana, which aimed to examine change over extended periods of time, engaged with landscape by tracing the changing 17th and 18th century settlement boundaries as well as delimiting house structures contained within the central residential area (Posnansky 1979, 1981, 2004). While excavation at Begho defined spatial elements, the associated ethnohistorical research regarding the orientation and location of housing attempted to use contextual information from present-day communities in the area that trace their histories to Begho. The work of Posnansky and his students has always included a significant awareness of the importance of oral histories, ethnohistory and the utilisation of a diverse array of sources, in an 'interdisciplinary eclecticism' (MacEachern 1996: 258) that has aimed to develop interpretive perspectives that recognise the multiple meanings that may be assigned to archaeological sites and landscapes.

Christopher DeCorse's (2001a) study of the sites surrounding the Portuguese, and later Dutch, castle of Elmina in Ghana has addressed the phenomenon of colliding worldviews and cultures in relation to the built environment (Figure 8.2). DeCorse illustrates the marked continuity among the African communities located immediately around Elmina Castle from the 16th to the 19th centuries where – despite the European incursions that introduced manufactured items,

Figure 8.2 Ocean view of Elmina Castle. *From Bosman 1704, facing page 44.*

architectural styles and technologies into African life – these new expressions of material culture were incorporated in ways that resonated with earlier practice. The landscape outlined by DeCorse is one where European goods and styles were incorporated into African social networks or modified to meet with African aesthetics. DeCorse and his students (DeCorse et al. 2000) have also attempted the difficult work of establishing shifting settlement patterns through time for sites near Elmina through archaeological survey and ethnohistoric research. These studies seek to address the interplay of systems of exchange at the local level, and in doing so provide a window into the broader implications of such exchange for restructuring local societies in the face of trans-Atlantic trade.

Archaeological research into the landscapes of 18th and 19th century Danish plantations on the Gold Coast by DeCorse (1993) and Bredwa-Mensah (2004) has identified attempts to create an Europeanised landscape, or at least a landscape of European referents, as seen through the eyes of the plantation system in West Africa. Through the establishment of settled agricultural enterprises Danish planters were moving toward a permanent settler presence distinct from that of the transient trading post personnel typical in West Africa. The plantation organisation was itself a way of conceiving of space particular to early capitalist Europeans (Delle 1998), and the Danish Gold Coast examples were modelled on those of the West Indies, which in turn had been modelled on those established by England in Ireland. Their use of the principles of surveillance, the construction of worker housing in imposed, non-African ways and the considered use of tree plantings all embodied a novel, European-derived way of structuring space to the advantage of economy and power, as well as attempting to domesticate the natural environment to make it more acceptable to European sensibilities with shaded alleys, vistas and well-ventilated exposures.

In contrast, the landscapes of West African trade entrepôts of the pre-colonial period (16th to mid-19th century), such as Gorée, Elmina and Savi (Figure 8.1), demonstrate little attempt to domesticate, or make familiar, the landscapes into which they were thrust. Rather, in these early European colonial ventures in Africa, the built environment communicated ideologies of power, economics and social relations. These large-scale European trade castles of the Gold Coast were built with an explicit expectation of the message(s) they communicated both to African and European alike (see Lawrence 1963 for images of these impressive structures).

Such studies are not restricted to European-constructed landscapes. Patrick Darling (1984) focused on the region surrounding the large-scale urban centres in southwestern Nigeria dating to c. AD 1100. He attempts to recast the vast 16,000-kilometre network of ditches and banks that encircle these centres in southwestern Nigeria in terms of the ethnohistorical and historical data available from the region. He suggests that the ditches represent the abilities of kings in the area to mobilise compulsory, or enslaved, labour as well as attempts to construct monumental architecture. Beyond physical spaces, he surveys ritual specialists and modern keepers of oral history in the area to build a case for the association of these spaces with local cosmologies.

As well as highlighting the range of created landscapes in the processes of contact, trade and colonialism, historical archaeologists have also explored in increasingly sophisticated ways the contemporary understandings of, and debates over, historical landscapes. Away from the coast, Ann Stahl has carried out a long-term ethnohistorical and archaeological study of the Banda Chieftaincy in central western Ghana: a hinterland that was beyond the spheres of control for the well-documented inland polities of central Ghana (Stahl 2004). Stahl presents landscapes that are the sites of contested histories, comparing alternative present-day interpretations in order to highlight silences and selectiveness: the tension associated with the creation, presentation and negotiation of histories. In Nigeria, Akin Ogundiran (2002: 26) has also explored the use of archaeological sites to create meanings in the present:

> The physical ruins of ancient settlements and the locations of significant events on the landscape (usually marked by natural features) serve as the reference points in Ilare's oral traditions, a sort of mnemonic device for recognizing the milestones in Ilare's history ... historical consciousness in Ilare is anchored in the abandoned settlements and historical landmarks, 'real' places where traditions, language, and values are believed to have developed and carried to the present. (Ogundiran 2002: 26)

HISTORICAL LANDSCAPE ARCHAEOLOGY IN WEST AFRICA: THE CASE OF SAVI AND OUIDAH

The potential of a landscape focus in historical archaeological research in West Africa can be demonstrated by a consideration of a number of sites including Savi, Ouidah and Cana in Bénin and the Rio Pongo in Guinea. The first to be examined are the important trade entrepôts of Savi and its successor, Ouidah, in coastal Bénin. Research undertaken by Kenneth Kelly between 1991 and 1999 focused on the site of Savi, capital of the slave-trading Hueda polity from c. AD 1670 to 1727 (Kelly 1997a, 1997b, 2001, 2002; Monroe 2003; Norman and Kelly 2004). Savi was the site of a series of interactions unique to the 'Slave Coast' of the Bight of Bénin: through astute observation and control of the trading arena, the Hueda political elite were able to manipulate European traders from a variety of nations to the advantage of the Hueda. The Hueda elite recognised that European traders represented an important and powerful commodity in themselves, because in large part Hueda independence was due to their usurping the middleman role in the trade in slaves and other commodities from Allada (an inland polity) in the 1670s. The European trade was at the same time the means by which Hueda independence had been achieved, through controlling the trade in slaves and other commodities, yet the Hueda also recognised that the European presence brought with it a series of risks, including the establishment of semi-autonomous European trading enclaves.

Through a range of strategies described in numerous 18th century European accounts, the Hueda elite aimed to shape the form and perceptions of the urban landscape of Savi in order to place European traders at a continued disadvantage, and so to minimise their risk and maximise their gain. Strategies included restricting European traders to unfortified trading lodges in the centre of Savi (over 10 kilometres from the sea, and away from the support of European warships), prohibiting the construction of fortified trading forts, prohibiting the monopolisation of trade by a single European nation and incorporating Europeans in key ceremonies emphasising Hueda authority (see accounts in Barbot 1732; Jones 1985; Labat 1731; Law 1991a, 1991b; Norris 1968 [1789]; Phillips 1732; Smith 1967 [1744]; Snelgrave 1734). To demonstrate that the Europeans lived in Savi at the pleasure of the Hueda king, a Dutch letter of 4 February 1715 is relevant: 'When it was pointed out that if the King of Fida [Hueda] decided to exercise his right to revoke his permission to any nation to trade in his country or to search other ships, the RAC [Royal African Company] could not prevent him from doing so, as the English live, just as we do, in his land only by his tolerance' (quoted in Van Dantzig 1978: 180). At the same time, the close physical proximity of the traders and the Hueda elite also signified to other Africans the close relationships enjoyed between Hueda and European traders (Kelly 1997b, 2002).

Of particular interest in this historical urban landscape is the way that European traders in late-17th and early-18th century Savi were provided with earthen lodges constructed according to local standards within the palace complex of the Hueda elite and separated from the rest of Savi by a substantial ditch system – at places up to eight metres in depth (Figure 8.3). The controlled landscape of these lodges, clustered adjacent to the palace, embodied a series of strategies in which different messages were communicated to different occupants of the same town. By prohibiting trade from being monopolised by a single European nation – as happened in most other places along the coast, such as Elmina and the Gold Coast trading posts, Gorée and elsewhere – the Hueda were able to play one European trader against another. This served both to overturn the usual dynamic where a single European trading nation was accustomed to exclusive 'possession' of their trading turf, and to demonstrate Hueda power as they were able to play trader against trader. The provision of lodges in compounds adjacent to the Hueda palace accommodated European traders' desire for high-status locations at which to conduct trade, but conversely held them in a degree of captivity, kept within the confines of the elite district demarcated by the ditch system. While European traders understood their location as providing preferential access to the elite, Hueda commoners and other Africans recognised Hueda control over the trading arena. By prohibiting the construction of European-style fortified lodges, the Hueda elite again demonstrated their physical control of the situation, and also prohibited the Europeans from displaying their difference through architecture. As Barbot (Hair et al. 1992: 644) wrote, '[I]n the Kings town, the company have a factory house, a place of very considerable trade: but it is a wretched place, as all other European settlements, to live in. ...' The European traders, their lodging and their activities became a part of a landscape communicating elite power in ways that would have been

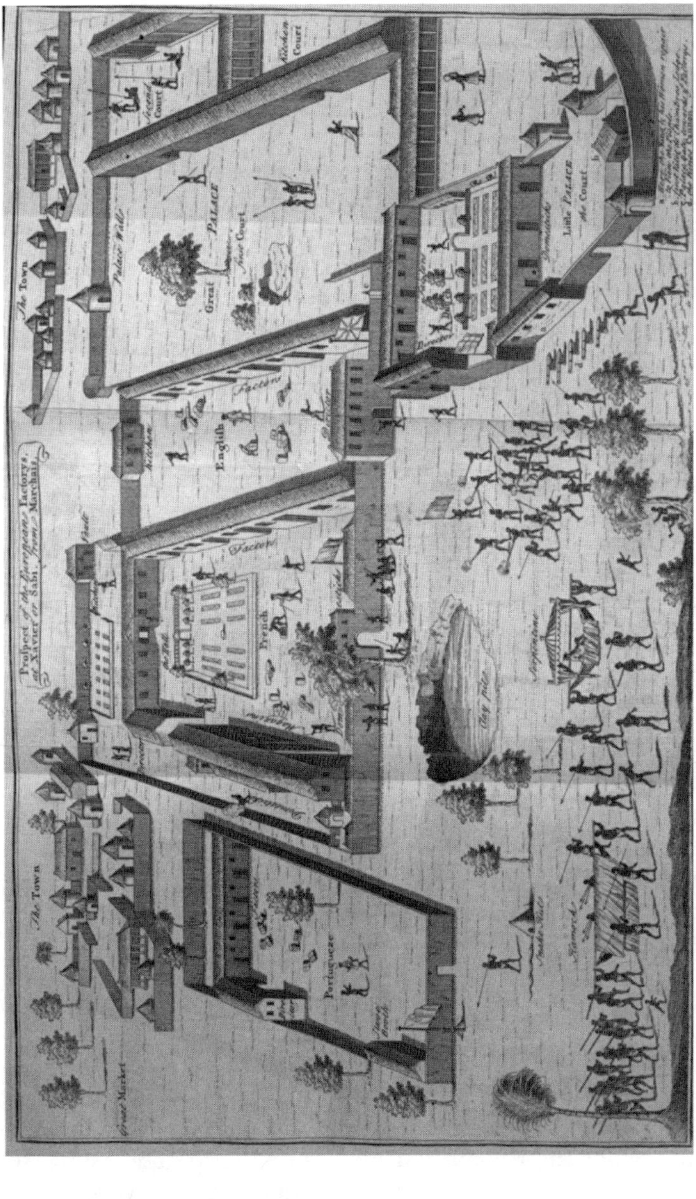

Figure 8.3 Early 18th century view of the Savi Palace Complex. *From Labat 1747, plate 9, facing page 64.*

recognisable to African observers, but would have probably remained obscure to the Europeans being so used.

After the conquest in AD 1727 of Hueda by Dahomey, the neighbouring state to the north, aspects of the social landscape continued to be employed to demonstrate power and control over the trading arena. While fortified lodges now developed, apparently more against inter-European conflict than against African aggression (Law 1992: 108; Verger 1976: 122), the African landscape strategies of the 17th century persisted during the Dahomey period, as monopoly of trade by any single European power was prohibited by keeping all the European trading establishments confined to the town of Ouidah, still 3 kilometres from the sea. However, Dahomey employed landscape in some novel ways to communicate relations of power and status. One of the clearest manifestations of control concerned the location of the Dahomean military garrison and the Dahomean governor of the trade and traders, an office known as the Yovogan, both established during the 1730s, shortly after the conquest of Savi. The Yovogan's residence was established immediately north of the British and French trading forts, and adjacent to the shrine to Dangbe, the python deity that had been one of the most powerful shrines of the Hueda. The military garrison was also located north of Ouidah, straddling the principal road to the Dahomean capital of Abomey, which lay 100 kilometres inland. It physically blocked access to the interior by Europeans, except for the one time each year during the so-called 'Annual Customs' when they were permitted to visit the capital. Thus, Dahomey created a landscape at Ouidah that communicated many things, including the appropriation of Hueda territory by placing the Yovogan adjacent to the Dangbe shrine, and the control over Europeans by concentrating their forts together at some distance from the sea, and also far from the seat of power, and blocking that access physically with the Yovogan and the garrison (Kelly 2002).

An unusual example of Europeans attempting to communicate some of their ideas about landscape is hinted at in a map of Ouidah drawn in 1776 that shows a large, apparently European-style garden adjacent to the French fort (Berbain 1942). Given the extensive historical archaeological literature on the roles played by formal gardens in 18th century North America (see contributions to Yamin and Metheny 1996), it is fascinating to consider just what message or messages were being communicated – or received – through this French garden in an African landscape. Was it aiming to create a place of familiarity for French traders far from home? Was it attempting to demonstrate the rational power of Europeans over nature, and therefore over society and the world, as has been suggested for elite Georgian gardens in North America (Leone 1984, Kryder-Reid 1994 and others)? Would that have been apparent to African observers? In another interesting use of landscape, in the late 19th century when a Catholic church was established at Ouidah, it was placed between the Dangbe shrine and the Yovogan's compound, perhaps severing the tie that Dahomey had tried to materialise following their conquest of Hueda, and certainly strongly expressing ideological distance between Christianity and local practices (Kelly 2002).

Since 1999, Neil Norman has continued research at Savi, investigating the areas outside the central elite zone in the capital. This research has explored the association of ditches in the Hueda kingdom with the python deity Dangbe, worshipped from at least the 17th century to the present (Norman and Kelly 2004). These landscape features were employed both to divide social space and to evoke cosmological elements for the protection and fortune of the kingdom. Undoubtedly, certain members of the Hueda populace garnered prestige from the close proximity of their residences to such ditches. Recent field seasons (2003–2006) conducted by Norman in the Savi area have revealed a complex mosaic of 17th through 18th century settlement types surrounding the palace. Several of the larger sites are surrounded by ditch segments, some that even rival the length and depth of those at the palace centre. Quite probably, these settlements surrounding Savi relate to accounts in the historical and oral record that tell where regional governors and other elite individuals attempted to usurp political control from the various kings of Hueda. If, as has been argued above, the ditches represent a materialisation of political authority, and provided certain residents with symbolic and physical protection, it is reasonable to assume that these political dissidents would create new ditches in order to make claims of authority and challenges to the king.

Again, this situation illustrates how the different perceptions of West African landscapes can address the same spaces. In the 17th and 18th centuries, European travellers rarely mentioned the ditches, but when they did, they suggested that the ditches were of European design and construction. European visitors to the palace site in the 19th century picked up on these accounts and perpetuated them, although modern oral histories and archaeological evidence suggest that they were of Huedan design and construction (Kelly 1997a; Norman and Kelly 2004).

HISTORICAL LANDSCAPE ARCHAEOLOGY IN WEST AFRICA: THE CASE OF THE ABOMEY PLATEAU

Cameron Monroe (2003, 2005) has considered the role of palace centres on the Abomey plateau, constructed between the 17th and 19th centuries, as politicised hubs that were closely related to the emerging bureaucratic order of the Dahomean kingdom. Monroe weaves together ethnohistorical and archaeological research to demonstrate the use of these palace complexes within an ethos of conquest and consolidation: the kings of Dahomey expanded the boundaries of the kingdom and at the same time built palaces and secondary residences to support these efforts. As political systems within the emerging Dahomey kingdom became increasingly integrated, the palace structures developed in architectural complexity in order to accommodate the burgeoning administrative class. Likewise, architecture associated with religious pageantry was augmented to accommodate increasingly politicised religious rituals and public displays that frequently incorporated Europeans.

HISTORICAL LANDSCAPE ARCHAEOLOGY IN WEST AFRICA: THE CASE OF THE RIO PONGO

A recent preliminary archaeological survey conducted in 2006 by Kelly at a series of sites along the upper Rio Pongo in Guinea provides further insight into the multiple constructions of landscape that were present in the recent West African past, and some of the ways those landscapes continue to be inhabited today. The Rio Pongo region witnessed some of the earliest European trade contacts dating to the late 15th and 16th centuries, and yet was also one of the last regions of West Africa to continue to play a role in the slave trade, with so-called 'illegal trade' continuing after the abolition of the slave trade by Denmark in 1803, and by Britain and the United States in 1807. Indeed, the trade in captive Africans continued on the Rio Pongo until at least the 1860s, and possibly later. The upper Rio Pongo is particularly interesting from a landscape perspective, in that it is a region that was conceived in a variety of ways: to the American and British slave traders who married into elite families along the river the region had meanings distinct from those of the local Indigenous inhabitants, who in turn, viewed areas such as the Mullataria, where descendants of the slave traders lived, as usurped. The traders from the Futa Jallon, in the interior, who brought captives to trade, experienced a different landscape. The British anti-slave-trade squadrons who were unable to enter the river due to sandbars, but knew of the trade, experienced yet another different landscape. Others perceived the landscape in yet different ways, such as the Afro-Caribbean missionaries who began to establish churches in the area from the mid-19th century, while it was still a significant slave trading destination. Furthermore, the slave traders sent their children to the mission schools so that they could master the languages and skills that would make them better traders. And in the early 20th century, Lebanese and Syrians moved into the region to begin wholesale trading in cola, cotton and other products.

LANDSCAPE ARCHAEOLOGY IN SOUTHERN AFRICA

The studies of historical landscapes in West Africa are complemented by the traditions of pre-colonial and colonial landscape studies in southern Africa, a region more characterised by a colonial population motivated by settlement, unlike the predominantly trade motivation of West Africa (Brink 1993). Jill Kinahan's (2000) archaeological research in Namibia investigated the changing interactions between Khoekhoen pastoralists and the European whalers, guano diggers and traders they encountered. This study is unusual in that it investigates the different ways the trading arena was negotiated by a pastoral group, unlike the more common interaction of European traders with settled agricultural societies. Whilst primarily focusing on the material culture associated with the trading relationship, Kinahan clearly identifies different cultural ways for making sense of intercultural interaction. She sees that Khoekhoen pastoralists, who have stereotypically been seen as being engaged in unrestrained and irrational trade for trinkets, were instead careful consumers of trade items that made sense in the internal logic of Khoekhoen culture. Although not principally a landscape study, it is through the analysis of trade items found at a variety of sites occupied from the 18th to the 20th century that Kinahan is able to track the changes in the way the Khoekhoen cultural system worked, as the presence of European traders increasingly destabilised Khoekhoen life.

The most influential studies of historical landscapes in southern Africa have, however, developed from the early efforts led by American historical archaeologist James Deetz, who aimed to explore landscape as 'that part of the terrain which is modified according to a set of cultural plans' (Deetz 1990: 2). Under Deetz's influence, South African archaeologists have described attempts by 17th and 18th century English colonists to (re)create an idealised 'memory' of rural England in South Africa (Scott and Deetz 1990). In a similar vein, in their study of the Vergelegen, a farmstead in the Western Cape of South Africa constructed in the first years of the 18th century by the governor of the region, Ann Markell, Martin Hall and Carmel Schrire (1995) have explored attempts by Dutch colonists to impose the strict symmetry and rationalised architecture on the landscape. These studies have underlined how Europeans have constructed landscape in attempts to adjust to alien settings: that is, by physically changing part of the setting and re-establishing a sense of place that evoked familiar elements of European landscape and architecture.

Such studies of European colonial landscapes contrast with Gavin Lucas' more nuanced study of the diversity of South African colonial and postcolonial identities in the Dwars Valley, which examines the role of material culture in the construction of these identities from the Dutch East India Company to free settlers and enslaved Africans (Lucas 2004).

Examining the 'anatomy' of a late-17th and early-18th century settler farm, Lucas alerts us to the importance of the wider global contexts of the movements of material culture and people in the Atlantic world and the Indian ocean (Lucas 2004: 86, ff.). Several authors have pointed to the importance of explicitly acknowledging the complexities of localities that were nested within global systems of distribution and exchange (e.g. Hall 2000); such regional, comparative or multi-site perspectives remain relatively unexplored in the archaeology of historical landscapes in West Africa. Conversely, the growth of archaeologies of the African Diaspora is witnessing the development of new examinations of wider, global 'landscapes' (cf. Hauser and Hicks this volume).

CONCLUSIONS

As we noted above, historical archaeologies of landscape in Africa have tended to focus upon the influence of external, European forces, while neglecting the complexities of African traditions, cosmologies and histories in relation to landscape (Norman and Kelly 2004). Too often, archaeologists have followed the well-worn path to key nodes on global systems of exchange where Europeans traded and interacted, leaving scholarship on the African coast that moves beyond the forts, lodges, palaces and castles underdeveloped. Having described the potential significance of a landscape approach in acknowledging these complexities in West Africa, we argue that part of the significance of such work lies in its potential to contribute to historical archaeologies of the African Diaspora, through comparison with contemporary landscapes in the New World (Kelly 2004).

As Atlantic African archaeologies of landscape develop, the recognition of the role of enslaved Africans in the construction of landscapes in the Diasporic communities of the Americas is a crucial area for further research. As well as the labour of enslaved Africans in creating and maintaining New World plantation and urban landscapes, these new landscapes were understood by Africans in a variety of ways. Moreover, historical archaeology has an important role to play in identifying African landscapes in the New World – landscapes which were either entirely obscure to European observers, or which were subversive environments created by Africans and their descendants to make sense of the alien social, cultural and natural environments in which they were held in bondage. Archaeological research in the Caribbean and North America has demonstrated that plantation settings, although most obviously characterised by imposed landscapes of order, discipline and surveillance (Delle 1998), were also settings in which the same landscape was experienced in different ways by the enslaved Africans (Armstrong and Kelly 2000; Benoît 2000; Brown 2002;

Pulsipher 1994; Pulsipher and Goodwin 2001). For example, landscapes that were constructed by planters to limit autonomy could be subverted by enslaved Africans by using alternative pathways to and from their tasks, re-aligning their houses to create family-focused compounds, creating garden spaces that allowed expression of individuality and incorporating bush or uncultivated lands into the realm of regular activities. Furthermore, diminutive structures (both chosen by and imposed on African descendant communities) necessitated that many of the activities of daily life took place outside the house proper. Thus, the spaces between houses were often central locales for creating and maintaining communities, and evoked similar settlement strategies and aesthetic choices recorded in West African settings. Such African ideas of landscape as a way of organising culture travelled across the Atlantic in the minds of enslaved Africans, just as European ideas of landscape were brought to Africa by both traders and settlers.

As landscape approaches have demonstrated their validity in a variety of settings – and as it has been demonstrated that one does not necessarily need a regional survey or inventory to develop an understanding of the ways landscape played into the cultures of particular regions and times – the landscape perspective will increasingly become a standard part of the analytical toolkit brought to bear on all manner of questions. In virtually all of the African cases referred to in this chapter, a sophisticated understanding of the importance of what we may call landscape perspectives is integral to the interpretations of the sites. While few of the studies make explicit reference to 'landscape', it is clear that they are all addressing questions surrounding the creation and experience of the multiple meanings a particular landscape or place may invoke. The theoretical perspectives of landscape archaeology help illuminate the ways African environments were created by people to express specific messages of control and power, ties to the land and domination in ways that could be read by others in some cases, and in other cases worked without being understood by all the participants. Landscape perspectives in Diaspora-period African archaeology have permitted much more sophisticated understandings of the interactions between Europeans and Africans, and between different African societies.

REFERENCES

Anschuetz, K.F., R.H. Wilshusen and C.L. Scheick 2001. An Archaeology of Landscapes: Perspectives and Directions. *Journal of Archaeological Research* 9(2): 157–211.

Armstrong, D.V and K.G. Kelly 2000. Settlement Patterns and the Origin of African Jamaican Society: Seville Plantation, St. Ann's Bay, Jamaica. *Ethnohistory* 47(2): 369–397.

Ashmore, W. 2004. Social Archaeologies of Landscape. In L.Meskell and R.W. Preucel (eds) *A Companion to Social Archaeology*. Malden, MA: Blackwell Publishers, pp. 255–271.

Barbot, J. 1732. *A Description of the Coasts of North and South Guinea*. London.
Bender, B. 1993. *Landscape: Politics and Perspectives*. Oxford: Berg Publishers.
Bender, B. 1998. *Stonehenge: Making Space*. Oxford: Berg Publishers.
Benoît, C. 2000. *Corps, jardins, mémoires – Anthropologie du corps et de l'espace à la Guadeloupe*. Paris: Editions de la Maison des sciences de l'homme/Editions du CNRS.
Berbain, S. 1942. Le Comptoir français de Juda (Ouidah) au XVIIIe siècle. *Mémoires de l'Institut Francaise d'Afrique Noire* (no. 3).
Bredwa-Mensah, Y. 2004. Global Encounters: Slavery and Slave Lifeways on Nineteenth Century Danish Plantations on the Gold Coast, Ghana. *Journal of African Archaeology* 2(2): 203–227.
Brink, Y. 1993. Figuring the Landscape: Land, Identity and Material Culture at the Cape in the 18th Century. *South African Archaeology Review* 52: 105–112.
Brown, Ras M. 2002. Walk in the Feenda: West-Central Africans and the Forest in the South Carolina-Georgia Lowcountry. In L. Heywood (ed) *Central Africans and Cultural Transformations in the American Diaspora*. Cambridge: Cambridge University Press, pp. 289–317.
Conkey, M. 1982. Boundedness in Art and Society. In I. Hodder (ed) *Symbolic and Structural Archaeology*. Cambridge: Cambridge University Press, pp. 115–128.
Conkey, M. 1999. Inside Grotte Chauvet: Encountering the First Cave Paintings. *California Wild* 54(3).
Darling, P. 1984. *Archaeology and History in Southern Nigeria: The Ancient Linear Earthworks of Bénin and Ishan*. 2 vols. Oxford: British Archaeological Reports (International Series) 216.
DeCorse, C.R. 1993. The Danes on the Gold Coast: Culture Change and the European Presence. *African Archaeological Review* 11: 149–173.
DeCorse, C.R. 1994. An Archaeological Reconnaissance of Bunce Island, Sierra Leone. *Nyame Akuma* 42: 36.
DeCorse, C.R. 2001a. *An Archaeology of Elmina: Africans and Europeans on the Gold Coast, 1400–1900*. Washington, DC: Smithsonian Institution Press.
DeCorse, C.R. (ed) 2001b. *West Africa During the Atlantic Slave Trade: Archaeological Perspectives*. Leicester: Leicester University Press.
DeCorse, C.R., E. Carr, G. Chouin, G. Cook and S. Spiers 2000. Central Region Project, Coastal Ghana. *Nyame Akuma* 53: 6–11.
Deetz, J.F. 1990. Prologue: Landscapes as Cultural Statements. In W. Kelso and R. Most (eds) *Earth Patterns: Essays in Landscape Archaeology*. Charlottesville: University of Virginia Press, pp. 1–4.
Deetz, J. 1996. *In Small Things Forgotten: An Archaeology of Early American Life*. New York: Anchor Books.
Delle, J.A. 1998. *An Archaeology of Social Space: Analyzing Coffee Plantations in Jamaica's Blue Mountains*. New York: Plenum Press.
Effah-Gyamfi, K. 1985. *Bono Manso: An Archaeological Investigation Into Early Akan Urbanism*. African Occasional Occasional Papers No. 2. Calgary: Department of Archaeology, University of Calgary Press.
Greene, S. 2002. *Sacred Sites and the Colonial Encounter: A History of Meaning and Memory in Ghana*. Blomington: Indiana University Press.
Hair, P., A. Jones and R. Law (eds) 1992. *Barbot on Guinea: The Writings of Jean Barbot on West Africa 1678–1712*. London: Hakluyt Society.
Hall, M. 2000. *Archaeology and the Modern World: Colonial Transcripts in South Africa and the Chesapeake*. London: Routledge.
Hodder, I. and C. Orton 1976. *Spatial Analysis in Archaeology*. Cambridge: Cambridge University Press.
Jones, A. 1985. *Brandenburg Sources for West African History; 1680–1700*. Stuttgart: Franz Steiner Verlag Wiesbaden GMBH.

Kelly, K.G. 1997a. The Archaeology of African-European interaction: Investigating the Social Roles of Trade, Traders, and the Use of Space in the Seventeenth and Eighteenth Century Hueda Kingdom, Republic of Bénin. *World Archaeology* 28(3): 77–95.

Kelly, K.G. 1997b. Using Historically Informed Archaeology: Seventeenth and Eighteenth Century Hueda/European Interaction on the Coast of Bénin. *Journal of Archaeological Method and Theory* 4(3/4): 353–366.

Kelly, K.G. 2001. Change and Continuity in Coastal Bénin. In C.R. DeCorse (ed) *West Africa During the Atlantic Slave Trade: Archaeological Perspectives*. Leicester: Leicester University Press, pp. 81–100.

Kelly, K.G. 2002. Indigenous Responses to Colonial Encounters on the West African Coast: Hueda and Dahomey from the 17th through 19th Centuries. In C.L. Lyons and J. Papadopoulos (eds) *The Archaeology of Colonialism*. Los Angeles: Getty Research Institute, pp. 96–120.

Kelly, K.G. 2004. The African Diaspora Starts Here: Historical Archaeology in Coastal West Africa. In P. Lane and A. Reid (eds) *African Historical Archaeologies*. New York: Kluwer Academic/Plenum Press, pp. 219–241.

Kinahan, J. 2000. *Cattle for Beads: The Archaeology of Historical Contact and Trade on the Namib Coast*. Windhoek: Namibia Archaeological Trust.

Knapp, A.B. and W. Ashmore 1999. Archaeological Landscapes: Constructed, Conceptualized, Ideational. In W. Ashmore and A.B. Knapp (eds) *Archaeologies of Landscape: Contemporary Perspectives*. Malden, MA: Blackwell Publishers, pp. 1–30.

Kryder-Reid, E. 1994. 'As Is the Gardener, So Is the Garden': The Archaeology of Landscape as Myth. In P.A. Shackel and B.J. Little (eds) *Historical Archaeology of the Chesapeake*. Washington, DC: Smithsonian Institution Press, pp. 131–148.

Labat, J.B. 1731. *Voyage du Chevalier des Marchais en Guinée, isles voisines et à Cayenne, fait en 1725, 1726 et 1727*. Amsterdam.

Lane, P. 2004. Reconstructing Tswana Townscapes: Toward a Critical Historical Archaeology. In A. Reid and P. Lane (eds) *African Historical Archaeologies*. New York: Springer, pp. 269–299.

Law, R. 1991a. *The Slave Coast of West Africa, 1550–1750: The Impact of the Atlantic Slave Trade on an African Society*. Oxford: Clarendon Press.

Law, R. 1991b. *Correspondence of the Royal African Company's Chief Merchants at Cabo Corso Castle with William's Fort, Whydah, and the Little Popo Factory, 1727–1728: An Annotated Transcription of Ms. Francklin 1055/1 in the Bedfordshire County Record Office*. Madison: University of Wisconsin Press.

Law, R. 1992. Warfare on the West African Coast, 1650–1850. In R. B. Ferguson, and N. L. Whitehead (eds) *War in the Tribal Zone: Expanding States and Indigenous Warfare*. Santa Fe, NM: School of American Research Press, pp. 103–126.

Law, R. 2004. *Ouidah: The Social History of a West African Slaving 'Port,' 1727–1982*. Athens: Ohio University Press.

Lawrence, A.W. 1963. *Trade Castles and Forts of West Africa*. London: Jonathan Cape.

Layton, R. and P. Ucko 1999. Introduction: Gazing on the Landscape and Encountering the Environment. In P. Ucko and R. Layton (eds) *The Archaeology and Anthropology of Landscape: Shaping Your Landscape*. New York: Routledge, pp. 1–19.

Leone, M.P. 1984. Interpreting Ideology in Historical Archaeology: Using the Rules of Perspective in the William Paca Garden in Annapolis, Maryland. In D. Miller and C. Tilley (eds) *Ideology, Power and Prehistory*. Cambridge: Cambridge University Press, pp. 25–35.

Leone, M.P. and S.D. Hurry 1998. Seeing: The Power of Town Planning in the Chesapeake. *Historical Archaeology* 32(4): 34–62.

Lucas, G. 2004. *An Archaeology of Colonial Identity: Power and Material Culture in the Dwars Valley, South Africa*. New York: Springer.

MacDonald, K.C. 1994. *Socio-Economic Diversity and the Origin of Cultural Complexity along the Middle Niger (2000 BC to AD 300)*. Unpublished Ph.D. dissertation, University of Cambridge.

MacEachern, S. 1996. Foreign Countries: The Development of Ethnoarchaeology in Sub-Saharan Africa. *Journal of World Prehistory* 10(3): 243–304.

McIntosh, R.J. 2005. *Ancient Middle Niger: Urbanism and the Self-Organizing Landscape*. Cambridge: Cambridge University Press.

McIntosh, R.J. and S. McIntosh 2003. Early Urban Configuration on the Middle Niger. In M.L. Smith (ed) *The Social Construction of Ancient Cities*. Washington, DC: Smithsonian Institution Press, pp. 103–120.

Markell, A., M. Hall and C. Schrire 1995. The Historical Archaeology of Vergelegen, an Early Farmstead at the Cape of Good Hope. *Historical Archaeology* 29(1): 10–34.

Monroe, J.C. 2003. The Dynamics of State Formation: The Archaeology and Ethnohistory of Pre-Colonial Dahomey. Unpublished Ph.D. dissertation, University of California, Los Angeles.

Monroe, J.C. 2005. American Archaeology in the Republic of Bénin: Recent Achievements and Future Prospects. *Antiquity* 79 (305). http://antiquity.ac.uk/projgall/monroe/ (Consulted 1 March 2006).

Norman, N.L. and K.G. Kelly 2004. Landscape Politics: The Serpent Ditch and the Rainbow in West Africa. *American Anthropologist* 106(1): 98–110.

Norris, R. 1968 [1789]. *Memoirs of the Reign of Bossa Ahadee, King of Dahomy, an Inland Country of Guiney*. London: Frank Cass and Co.

Nyame Akuma 1982. Society for Africanist Archaeologists in America, Berkeley Conference. *Nyame Akuma* 20: 4.

Ogundiran, A. 2002. *Archaeology and History in the Ilare District: Central Yorubaland Nigeria 1200–1900 AD*. BAR International Series Cambridge Monographs in African Archaeology 55.

Phillips, T. 1732. A Journal of a Voyage Made in the Hannibal of London, Ann. 1693, 1694. In A. Churchill and J. Churchill (eds) *Collection of Voyages and Travels* (vol. 6). London, pp. 173–239.

Posnansky, M. 1973. Aspects of Early West African trade. *World Archaeology* 5(2): 149–162.

Posnansky, M. 1979. Excavations at Begho, Ghana, 1979. *Nyame Akuma* 16: 23–27.

Posnansky, M. 1981. Notsé Town Wall Survey. *Nyame Akuma* 18: 56–57.

Posnansky, M. 1985. Togo and New Directions in the Study of West Africa's Past. *Actes du Séminaire UCLA – UB sur les Sciences Sociales. Volume 1: Arche?logie, Histoire, Géographie*. Lomé, Université du Bénin.

Posnansky, M. 2004. Process and Change: A Longitudinal Ethnoarchaeological Study of a Ghanaian Village, Hani 1970–1998. *African Archaeological Review* 21(1): 31–47.

Preucel, R.W. and L. Meskell 2004. Places. In L. Meskell and R.W. Preucel (eds) *A Companion to Social Archaeology*. Malden, MA: Blackwell Publishers, pp. 215–229.

Pulsipher, L.M. 1994. The Landscapes and Ideational Roles of Caribbean Slave Gardens. In N. F. Miller and K.L. Gleason (eds) *The Archaeology of Garden and Field*. Philadelphia: University of Pennsylvania Press, pp. 202–221.

Pulsipher, L.M. and C. Goodwin 2001. 'Getting the Essence of It': Galways Plantation, Montserrat, West Indies. In P. Farnsworth (ed) *Island Lives: Historical Archaeologies of the Caribbean*. Tuscaloosa: University of Alabama Press, pp. 165–203.

Quarcoopome, N.O. 1993. Notsé's Ancient Kingship: Some Archaeological and Art-Historical Considerations. *African Archaeological Review* 11: 109–128.

Reid, A. and P. Lane (eds) 2004. *African Historical Archaeologies*. New York: Springer.

Richard, F.G. 2005. The Siin Landscape Archaeological Project. *Nyame Akuma* 63: 5–14.

Robertshaw, P. (ed) 1990. *A History of African Archaeology*. London: James Currey Ltd.

Scott, P. and J.F. Deetz 1990. Buildings, Furnishings, and Social Change in Early Victorian Grahamstown. *Social Dynamics* 16: 76–89.

Smith, W. 1967 [1744]. *A New Voyage to Guinea*. London: Frank Cass & Co.

Snelgrave, W. 1734. *A New Account of Some Parts of Guinea and the Slave Trade*. London: J. & J. Knapton.

Soumonni, E. 1997. The Neglected Local Source Material for Studying the Slave Trade in Dahomey. In R. Law (ed) *Source Material for Studying the Slave Trade and the African Diaspora: Papers from a Conference of the Centre of Commonwealth Studies, University of Stirling, April 1996*. Stirling: University of Stirling, pp. 79–89.

Stahl, A. 2004. *Making History in Banda: Anthropological Visions of Africa's Past.* Cambridge: Cambridge University Press.

Thiaw, I. 1999 Archaeological Investigations of Long-Term Culture Change in the Lower Falemme (Upper Senegal Region), AD 500–1900. Unpublished Ph.D. dissertation, Rice University.

Thomas, J. 2001. Archaeologies of Place and Landscape. In I. Hodder (ed) *Archaeological Theory Today*. Cambridge: Polity Press, pp. 164–186.

Togola, T. 1996. Iron Age Occupation in the Mema Region, Mali. *African Archaeological Review* 13: 91–110.

Upton, D. 1990. Imagining the Early Virginia Landscape. In W. Kelso and R. Most (eds) *Earth Patterns: Essays in Landscape Archaeology*. Charlottesville: University of Virginia Press, pp. 71–86.

Van Dantzig, A. 1978. *The Dutch and the Guinea Coast, 1674–1742: A Collection of Documents from the General State Archive at the Hague*. Accra, Ghana Academy of Arts and Sciences.

Verger, P. 1976. *Trade Relations Between the Bight of Bénin and Bahia from the 17th to 19th Century*. Ibadan, Nigeria: Ibadan University Press.

Willey, G. 1953. *Prehistoric Settlement Patterns in the Viru Valley, Peru*. Washington, DC: Smithsonian Institution.

Yamin, R. and K.B. Metheny (eds) 1996. *Landscape Archaeology: Reading and Interpreting the American Landscape*. Knoxville: University of Tennessee Press.

CHAPTER 9

Landscape, Time, Topology: an Archaeological Account of the Southern Argolid, Greece

Christopher L. Witmore

INTRODUCTION

Throughout the Mediterranean, the last 35 years have witnessed considerable progress in the archaeology of landscape. In a region once dominated by the grand traditions associated with Greco-Roman antiquity – southern Italy, Greece and western Turkey in particular (Morris 1994; Shanks 1996) – archaeological practices have been powerfully transformed, modified and refined. This progress is due in large part to advances made in intensive and systematic survey. A locus of experimentation and innovation, survey archaeology has benefited from an explosion of activity since the early 1970s, resulting in an increased awareness of the broader regional context of archaeological sites, and of a broader range of activities (e.g. manuring, woodland management, companion species relations, etc.) associated with an increased variety of historic periods (e.g. Venetian, Ottoman and modern) (for discussions of this 'explosion' of survey activity refer to Alcock and Cherry 2004: 1–4; Athanassopoulos and Wandsnider 2004; Cherry 2003).

This emphasis upon landscape among Mediterranean archaeologists has several implications. In Greece, for example, archaeologists now engage with a wider range of sites, features, contexts and things. It should be noted that thirty-five years ago, everything from contemporary terracing to random ceramic scatters to non-diagnostic surface materials was simply not considered (at least in publication). At the same time, we have seen an increased emphasis upon the diachronic study of long-term change: compare, for example, the Bronze Age focus of McDonald and Rapp (1972) to the multi-period work of the contributors in Cavanagh et al. (1996, 2002). Most recently, issues of macro-scale comparability (Alcock and Cherry 2004) and reiterative practice (i.e. the reworking of survey materials through multi-stage and multi-scalar endeavours) have received renewed attention (cf. Cherry 1983; e.g. Cavanagh et al. 2005; Davis 1998; Davis et al. 1997; Zangger et al. 1997). New techniques of remote sensing, including magnetometry or resistivity, and new sampling procedures such as phosphate analysis have also contributed to such long-term and wide-ranging studies (e.g. Cavanagh et al. 2005).

In building upon the achievements of this increasingly rich, broad and dynamic field of archaeological research – as so clearly demonstrated by the recent five-volume publication of *The Archaeology of Mediterranean Landscapes* (Barker and Mattingly 1999–2000; see also Athanassopoulos and Wandsnider 2004; Cherry 2003) – this chapter delves into two aspects of landscape that remain little explored in the region: the material complexities of the land (cf. Witmore 2004b) and the character of both human and nonhuman relations with various pasts – in short, the polychronic nature of landscape. In addressing these two aspects, the chapter questions how we archaeologists understand time itself. Therefore, in what follows, I attend to three questions related to landscape and time. First, how do regional archaeologies produce temporalities from what is encountered on the ground in a given region? Second, how might we better understand the polychronic, even poly-*kairotic*,[1] nature of landscape? And, third, given a different understanding of time and landscape, what other archaeological syntheses of the land might we produce?

In addressing these questions I move from a discussion of chronology (clock time) to entropic time to the notion of 'percolating' time in relation to survey archaeology in the region. In so doing, I argue that it is not time which produces the ensemble of landscape or the relations between people, things and companion species; rather, it is the ensemble – the relations themselves – which produce time. While the aim is to excavate beneath the past/present divide, this is not to proclaim that 'that pasts are produced in the present' (Shanks and Tilley 1992: 7), though they are collectively produced. Rather, it is to say that pasts are thoroughly blended into the present; that pasts push back and have an impact within contemporary relations in a multiplicity of ways (Olsen 2003; Witmore 2004a); and that these relations, these simultaneous transactions, as things have a stake, are what beget time. Moreover – and unlike Geoff Bailey (1983, 1987), who characterises entropic time as archaeological time – I hold this dynamic, percolating time to be fundamentally archaeological (Witmore 2006a).

Still, this understanding of time requires a very different account of landscape from those deployed by recent archaeological surveys in the Mediterranean. While linear time is conventionally oriented by unidirectional historical narrative, 'percolating' time may best be synthesised topologically. Michel Serres defines topology as 'the science of proximities and ongoing or interrupted transformations' (Serres with Latour 1995: 105). Rather than delineating stable and well-defined temporal distances, a topology plots the relations, the nodes of interaction and the passages between various pasts (Serres with Latour 1995: 60). A topology, I suggest, provides an archaeological synthesis that is more faithful to the ontologies, the material presences, the raw complexities of landscape. To suggest this (since to demonstrate it would require many more pages than are available in this volume), the second half of this chapter presents an abbreviated topology of the southern Argolid in Greece. In conclusion, I define nine aspects of percolating time, which might be considered in Mediterranean landscape archaeologies.

ARCHAEOLOGY, LANDSCAPE AND MOTIVE

A few more words are necessary with regard to archaeology, landscape and motive. While archaeology is a diverse discipline, it is united by the study of material pasts. Archaeology, if we dig to its etymological roots, is the 'discipline of things,' quite literally *old things – 'ta archaia'* (cf. Olsen 2003: 89). Ontologically, archaeologists begin with relations between people and things. If we are to take these matters seriously then we must understand time in terms of the relations between various entities, whether people, materials, things or companion species; we must understand time as an entangled aspect of the land and not solely as an external parameter along which to order, delineate and measure its development (Serres with Latour 1995; Bergson 1998; Assad 1999). Equally, while 'landscape' is an ideologically charged concept which arose out of a very particular European and American relationship between people and the natural world, and which was based upon a predominately visual aesthetic (Andrews 1999: 2–22), it has come to be associated with a more complex array of extra-urban activities, practices and features in the context of archaeology. In this chapter I regard landscapes as aggregate mixes of multiple material pasts (be they caves, ruins, derelict terraces, olive trees or water wells), which are not necessarily linear in association. Whether fragmented and dispersed or accreted, whether torn or folded and pleated, multiple pasts constitute the polychronic ensemble of landscape and have action in peoples' lives now. However, I deny landscape the status of simply being 'out there' because the products of transactions with the land circulate as a multiplicity of materials (worked flint, pine resin, olive oil, broken millstones and so on) and media (stone inscriptions, text, maps, photographs, video, etc.). It is in regard to this multiplicity that we must seek understandings, not oriented around a linear and unitary temporality, but rather around a percolating time characterised by nonlinearity, instability and fluctuation.

The motive of this chapter is not to critique or supplant conventional chronological treatments of landscape change. We as archaeologists need our measures (Lucas 2005, 114): without them we would have less understanding of time's complex and chaotic nature. Rather, the motive is to help shift the notion of time in archaeology from the solely successive to that which is also simultaneous. The motive is to suggest possible modes for enriching our syntheses of material pasts and understanding how these pasts have action for the contemporary world as it is lived in ways which undercut divides between present and pasts, humans and things, humanities and sciences (Webmoor and Witmore 2005). For unless we wish to succumb to the problems associated with such bifurcations (e.g. the social and the material, or culture and nature) which have marked our various disciplinary turns (Hodder 1999; Olsen 2003; Shanks and Tilley 1992; Witmore 2004a), we must rework our conventional understandings of our archaeological fields – not only landscapes, but also features and things (including their circulating forms) – in ways that cut across the sciences and the humanities. There is no way to sufficiently address the trans-disciplinary nature of our practices without 'reshuffling' some of the most basic ingredients in archaeologies of landscape – space, materials and time – locally. In this regard, a mixed vocabulary that builds upon both the scientific and humanistic character of archaeology is deployed.

THE ARCHAEOLOGICAL MATERIALS OF LANDSCAPE: THE MEASURE OF TIME

All the way from Hatchery West in central Illinois, USA (Binford 1972), to east of the Apennines, Italy (Barker 1995), to northern Keos, Greece (Cherry et al. 1991), to southern Turkey (Wandsnider 2004a and 2004b), 'probabilistic survey' has come to play a key role in landscape archaeology as it has developed in the Mediterranean. A probabilistic survey is one which involves the sampling of a subset of a larger area in order to make generalisations about the whole, whether the whole be a 'site' or an entire landscape. At Hatchery West, having ploughed the surface of a portion of the site, Lewis Binford divided the area up into 416 6-by-6-metre squares. Items in each square were collected and studied in order to 'gain some impression' of the character of the site (Binford 1972: 164). Variations of this method in the form of sampling along transects and within material clusters are at the heart of systematic survey practices in Greece and the wider Mediterranean.

It is with probabilistic survey that we as archaeologists begin to search for pasts in places we would not have visited otherwise (Redman 1974).[2] We begin to deal with whole classes of material that were previously regarded as random and irrelevant 'background noise' – 'the junk you find on the surface' (Flannery 1976: 51; also refer to Cherry et al. 1991: 47–52). Archaeology, to be sure, has a long history of demonstrating how raw complexity holds productive possibility. Here the messy and ambiguous nature of previously random materials is 'domesticated' through an ongoing, yet circulating, process of classification, standardisation and measurement (for an excellent discussion of classification in science and more generally, refer to Bowker and Star 2000). This process provides us with compatible and comparable bases. And it is this passage from seeming chaos to practical order that not only defines survey practices but also establishes how contemporary surveys would come to make 'time' (Witmore 2004a). But along the path from central Illinois to contemporary intensive survey in the Mediterranean – in fact, in the very same year that Binford republished his work at Hatchery West in *An Archaeological Perspective* – other texts added to the mix necessary for intensive and systematic regional survey.

The year 1972 saw the appearance of *The Minnesota Messenia Expedition* (McDonald and Rapp 1972) and the English translation of the second revised edition of Ferdinand Braudel's *The Mediterranean and the Mediterranean World in the Age of Phillip II.*[3] In the wake of these works a 'new wave' of intensive survey washed the ceramic-littered shores of the Mediterranean (Cherry 2003: 141). Other factors, of course, contributed to this deluge. The development of landscape survey projects in Greece was facilitated by the relative simplicity of obtaining survey permits before 1988 (at least, compared to the excavation permit limitations in place since 1932), and the relatively low costs of survey when compared to excavation (Cherry 1983, 2003; Bintliff and Snodgrass 1985; Witmore 2005). Still, by 1972 the ingredients crucial for intensive regional surface survey were in place.[4] There were predecessors, to be sure (examples of regional study are to be found in the topographical tradition from William Martin Leake [1777–1862] to W. Kendrick Pritchett or in the early collaborative endeavours of the French *Expédition Scientifique de Morée* [1829–1831]), but after 1972 matters of multi-disciplinarity (McDonald 1972) would blend with the ordered materials produced by probabilistic survey in detailing the relations between people and the land on focused regional scales within a long-term diachronic perspective.

The alternative perspectives of Binford, Braudel, and McDonald and Rapp were first blended together on a tiny volcanic island in the midst of the Aegean. On Melos intensive probabilistic survey practices were pioneered in systematic pedestrian survey under the direction of John Cherry between 1976 and 1977 (Cherry 1982: 16; Renfrew and Wagstaff 1982). Pedestrian survey or 'fieldwalking' is a technique of ground reconnaissance where individuals walk across landscapes in regular intervals collecting exposed artefacts from the ground surface.

By plotting these find spots, patterns of past activity and occupation and land use can be identified. Intensity, measured in the distance between find spots, often varies depending on the presence or absence of surface materials. On Melos, the distances between fieldwalkers ranged from 15 to 25 metres (though subsequent surveys would reduce these distances), and the areas randomly selected for pedestrian survey were staggered by transect lines 1,000 metres in width, oriented north/south. The transects were separated by even distances (4 kilometres) across an island of 151 square kilometres (Figure 9.1; Cherry 1982: 18). As a site-based survey, success was measured in terms of the increase in known sites on the island – from 47 to 130. Here we should note that later surveys have focused on contiguous areas of land, covering much smaller areas of a region – 20 square kilometres with the Keos Survey (Cherry et al. 1991) and 44 square kilometres with the Argolid Exploration Project (hereafter AEP) (Jameson et al. 1994) – and yet they have located many more archaeological sites per square kilometre – 7.5 per square kilometres with the AEP (for a critique of such increasing intensities in survey see Blanton 2001).

Few materials were collected during the fieldwalking on Melos. Instead, diagnostic materials were identified, photographed and, with the exception of 'special finds' such as ceramic fragments for specialist consultation, figurines, loomweights or obsidian points, were left behind (Cherry 1982: 19). Nevertheless, the identification of

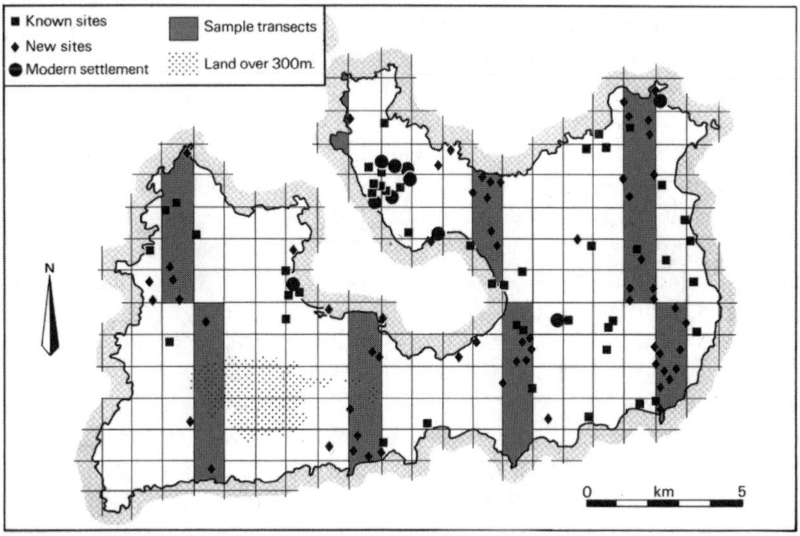

Figure 9.1 Map of Melos, Greece, with the sites (both previously known and newly encountered) and the eight transects. *After Cherry 1982: 18.*

these surviving materials was central to the project's characterisation of the archaeological landscape. In interpreting these items, the project made use of material chronologies established earlier in the course of excavation, such as those at Phylakopi, where typological analyses and sequencing resulted in an external linear temporality composed of organised, chronologically specific referents (Renfrew 1982: 35–36). The use of such refined temporal management schemes as organisational structures allowed survey personnel to sort materials into stacked chronometric boxes on the basis of diagnostic comparanda. While none of this was new, the extension of the spatial scope of classification to a wider and expanding array of sites, features and materials across the whole of the landscape was new (Cherry 1982). In this way the Melos field team translated things identified on the ground into a combination of coordinate (location on the ground), date (period of use) and function (what it probably was while in 'use') in their documentation. In this process, 'linear time' is as much about information storage and management as it is about relations in various pasts (cf. Bowker 2005). Moreover, sorting regional materials into stacked temporal boxes does not produce time; at best it generates a particular temporality.

MULTI-TEMPORAL PROCESSES: ENTROPIC TIME

Various pitfalls in the interpretation of materials identified during fieldwalking have been highlighted, especially the problem of purely functional interpretation, as when we might infer that a cluster of roof tiles combined with domestic ceramics and with other 'habitational' features are representative of a 'farmstead' (see Cherry et al. 1991: 337; Pettegrew 2001). No one has addressed the temporal ramifications of this interpretive issue in the context of Mediterranean survey as thoroughly as LuAnn Wandsnider (2004a, 2004b). She has argued (*after* Bailey 1983) that to render 'formational entities as functional units, whether empirical or conceptual, is to make them temporally flat – i.e. as attributable to narrow spans of time, and thus, to deny their formational heritage' (Wansnider 2004a: 53).

By classifying materials solely in terms of use, or occupational or settlement relations, then, archaeologists sieve away processes of accretion and entropy. Privileging functionality – 'synchronic' relations – fails to take into account post-depositional processes (cf. Binford 1981; Schiffer 1987). At a landscape scale, such a concern with event, functionality and spatial relations in the past even fails to acknowledge the presence of, and potential transactions with, those sites, features and materials (albeit transformed) in subsequent periods – their status as material pasts within other pasts (Alcock 2002; Bradley 2002; Lucas

2005: 37–42; Olivier 2003; van Dyke and Alcock 2003). Wandsnider argues that we need to treat survey materials as temporal traces irreducible to aspects of function. In the study of landscape, such an approach must also take into account transformative processes.

Still, in the wake of the Melos survey, understanding long-term change on a regional scale became a major concern. Survey practitioners sought a diachronic perspective on landscape change, and it was in this respect that initial affinities (in terms of timescales and chronological resolution) were articulated with *Annales* history, and particularly the work of Braudel. Braudel famously presented a tripartite scheme, consisting of long-term durations (*la longue durée*) which were predominately geographical, environmental or even technological; medium-term socio-economic fluctuations (*conjoncture*); and short-term sociopolitical histories of the event or the 'individual' (*événements*) (cf. Bintliff 1991: 6–9; Knapp 1992: 6; Lucas 2005: 15–18). The full impact of Braudelian history upon survey archaeology was of course not immediate; the wider ramifications of Braudel's work and the *Annales* perspective were not felt in archaeology at large until the late 1980s (e.g. Bintliff 1991; Hodder 1987; Knapp 1992). What is significant here is the particular character of the *Annales* scheme in complicating temporal perspectives.

The attraction of Braudel's writings for survey archaeologists lay in its multi-temporal alternative to chronology – a dimension of landscape that has often been highlighted (e.g. Ingold 1993). Much emphasis, at least initially, was placed upon the long duration, as survey materials could, it was argued, point to issues of long-term change (Cherry 1983: 388) and the 'slow but perceptible rhythms' in the landscape (Cherry et al. 1991: 10). This concern with the long term frequently amounted to an equal emphasis upon all periods of human occupation. Such emphasis revealed patterns of ebb and flow, of fluctuation in the intensity of land use and density of population (Cherry et al. 1991; Jameson et al. 1994). Because geomorphology, pedology, paleoecology and other sciences came together in regional survey, these ebbs and flows could take into account even the seemingly most durable and stubborn transformative aspects of landscape (Jameson et al. 1994: 149–213). But rarely has there been any substantive engagement with the broader multi-temporal ramifications of Braudel among survey archaeologists in the final publications from regional surveys (however, see Barker 1995; Bintliff 1991). However, this situation is changing.

The temporal resolution of diagnostic materials recovered during landscape survey is much coarser than are the 'real-time' (what Bailey and Binford referred to as 'ethnographic time') experiences of everyday life (Cherry et al. 1991: 458); though the perception of quotidian temporalities too is highly variable (Bergson 1999). Here, Wandsnider's identification of a 'functional metaphysic' versus a 'formational metaphysic' in Mediterranean survey is of interest (Wandsnider 2004a: 50–56; Wandsnider 2004b: 74–75). The former refers to synchronic sets of relations during occupation and use, while the latter takes a more nuanced analytical approach to the taphonomy of survey materials. Building upon Bailey's time perspectivism (Bailey 1981, 1983, 1987), Wandsnider draws attention to the ontological nature of archaeological deposits as they are encountered in the raw. Here materials on the ground are regarded as rich aggregates of actions and processes that accrete over hundreds or thousands of years. With 'time perspectivism', through which processes of change are regarded as occurring at different temporal rates, these multi-temporal transformations are described as an 'archaeological time' which operates and needs to be understood differently from processes occurring in an event-based or 'ethnographic time', as the latter unfolds over the course of a year or a decade (Wandsnider 2004a: 54).

One way forward is to classify various processes of entropy. The challenge is to temporally map post-depositional processes associated with archaeological features, materials and things. In accepting this challenge, Wandsnider and other project members at Rough Cilicia, a landscape survey based in south coastal Turkey from 1998, treat landscape as a 'formational phenomenon' (Dillon et al. 2001; Wandsnider 2004a: 56–57). Instead of concerning itself with the 'completeness of settlement pattern', the Rough Cilicia approach takes up the definition of various temporal processes at work across a polychronic mosaic of surfaces, features or things (Wandsnider 2004a: 56–57). Of course, these 'formational' processes are by no means consistent; indeed, they are highly variable.

Consider, for instance, an example pertaining to the dynamics of geomorphologic processes from the region of Laconia, in the south-central Peloponnesus, Greece. In identifying a subset of 20 small rural sites from the original Laconia Survey (Cavanagh et al. 1996, 2002) with conditions favourable for the conservation of archaeological remains, the Laconia Rural Sites Project (LRSP) (Cavanagh et al. 2005) factored out those areas where post-depositional processes amplified the rate of material transformation. Here, factors such as angle of slope, the depth of plough, the maintenance (or lack thereof) of terracing, pedology, soil creep, fluvial erosion, the presence of woodland and so forth all contributed to variable rates of change. So, while some features would have decayed or dispersed due to their location on unstable slopes many centuries ago, others might persist today due to burial at the base of such slopes. While these factors all played a part in the choice of sites suitable for the LRSP (Cavanagh et al. 2005: 281–282), they are also exemplary of disparate viscosities in 'entropic flow' and, as such, highlight the complexity and variability of time's 'passage'.

In creatively attempting to break up an 'event-based' notion of the past as it once was, the Rough Cilicia Project represented a contemporary intervention into what is left of different pasts. But despite acknowledging multi-temporal processes of entropy, in this work the arrow of time remains unidirectional. Landscape archaeologists have yet to account for reverse movements – the eddies and counter-currents of time (see Prigogine and Stengers 1984; Serres with Latour 1995). The latter are attributes of what we may term, following Michel Serres, 'negentropic'[5] (negative entropy) processes (Serres 1982: 71–83). The very presence of well-preserved archaeological features from Classical Laconia is indicative of such temporal 'eddies', but to get at the ontology of 'old things' we have to begin with our own practices.

Archaeology does not involve the 'discovery' of the past 'as it was', but explores relations between a diversity of pasts in entangled networks of relation in the present (Shanks 2004; Witmore 2004a, 2006a, 2006b). Archaeological landscapes are made up of what is left of various pasts, of dynamic 'processes of matter' (Shanks 2005). Temporalities of use, re-use and transformation are therefore more profitably understood in terms of transactions between people and things or other entities. Landscape archaeologies can explore the shifting roles of various entities (humans, things, companion species) in these processes, drawing on an archaeological 'record' that is always dynamic (Lucas 2005: 53). Archaeological landscapes are continually caught up in processes of transformation, but these are neither wholly linear, nor necessarily cumulative, nor totally progressive.

LANDSCAPE: THE PERCOLATION OF TIME

In *The Archaeology of Time* Gavin Lucas concludes with a paradox: is it possible, he asks, 'to re-think the nature of time, the nature of temporalisation that archaeology creates and sustains, or is archaeology, in fact, defined by this temporality as much as it defines it?' (Lucas 2005: 136). This section addresses this question. As Lucas quite rightly points out, much of archaeology regards time solely by its measurement (Lucas 2005: 2–15; cf. Serres with Latour 1995: 60–61; Witmore 2006a). Drawing on the work of Michel Serres, I explore how time is not simply an external parameter, and suggest that while archaeology has conventionally operated under a 'modernist' (Witmore 2006b) conception of time, it is still possible to rethink the nature of time in archaeology.

The Newtonian conception of time as orderly, consistent and predictable, much like a clock, has been radically transformed by the sciences of chaos (e.g. meteorology, thermodynamics, non-linear dynamics, etc.)

in which time is more appropriately understood as turbulent, unpredictable and highly variable (Hayles 1991; Prigogine and Stengers 1984). This notion of 'dynamic time,' of time characterised by instability and chaotic fluctuation, has been argued to hold profound possibility for the discipline by archaeologists such as James McGlade and Sander van der Leeuw (McGlade 1999; McGlade and van der Leeuw 1997). Taking direction from the sciences of chaos, these archaeologists have sought to break up the progressive image of linear change in the evolution of 'societal' structures by enrolling features of nonlinear, dynamic process – nonlinear causality, aleatoric variation, chaotic fluctuations which are mathematically modelled, etc. – in dealing with change in human/environment relations. For example, social systems such as larger urban agglomerations are regarded as open, dissipative and inherently unstable over the long term (McGlade and van der Leeuw 1997; see summary by Lucas 2005: 17). Change occurs when instability is amplified above certain thresholds, an effect which can either 'enhance the robustness of the system' or induce 'catastrophic decline' (McGlade and van der Leeuw 1997: 338). But these processes have different temporalities – different rates of change – from those processes operating on environmental scales or those associated with micro-level phenomena (McGlade 1999: 156). Here, landscape is shaped at the confluence of these various processes.

Using chaos theory, McGlade and van der Leeuw seek to model the complexity and variability of change, but while I follow them in exploring the complexity of our understandings of time and change I do not wish to begin with these propositions at a distance in the past. It is often the case that archaeology begins with the presupposition that the past is separate, demarcated, distant and distinct, rather than with our own engagements with the material past, our own relations with things. Our understanding of the nature of time and landscape cannot begin with assumptions of the bifurcation of past and present, or of people and things. These false separations, as Bruno Latour (1993) has argued in the context of modernist thought, lead to the presentation of time as a linear temporality (Witmore 2006a, 2006b).

For instance, in her important discussion of functional versus formational metaphysics in Mediterranean landscape archaeology, Wandsnider characterises these alternative understandings as shifting 'paradigms'. Here, Wandsnider invokes a model of radical succession in which previous approaches to the past are relegated to disciplinary history. By repeating the gesture of the Copernican (Kantian) revolution, the scientific paradigm (Kuhn 1970; also refer to Meltzer 1979) – like its counterpart the humanistic episteme (Foucault 1972) – appears to break with a past considered to be old-fashioned, outmoded, outdated (Latour 1993: 67–72; Serres with Latour 1995: 57). The very presence and persistence of that which is now rhetorically claimed to be a previous way of interpreting archaeological features is for Wandsnider to be considered a 'disorder' (2004a: 50 and 59). However, such simultaneities are only disorderly when filtered through the particular modernist historicism which presents a progressive model of transformation in which a new era breaks definitively with the past. My point is that our iconoclastic and in some cases maniacal approaches to previous ideas, practices and our former ways of understanding also produce a particular temporality that we know to be oversimplified.

Furthermore, and as Lucas points out, the nonlinear approach of McGlade and van der Leeuw posits 'a conception of time as the tension between continuity and change' within a particular social system – a conception which they share with the *Annales* history (Lucas 2005: 17). But what phenomena are continuing and changing? Our understanding of time also depends upon what we understand to be assembled under the banner of the 'social' (Latour 2005a; Olsen 2003; Witmore 2006b). So long as we understand society to also be made up of things, and so long as things – whether fragmented, worn or pristine, whether ancient, modern or futuristic – are present, a different conception of time and of our relations to the past is possible.

In landscape archaeology, the past is right there under the feet of the fieldwalker, present in the hand of the surveyor. So too it is stacked in the dusty museum storeroom, or translated and inscribed as books on library shelves. In other words, these many material pasts, whether anachronistic, archaic or seemingly abolished, are not *past*, but are also here, simultaneous and present. Because something of them remains, something of them can return, recirculate and enter into renewed sets of relations (Latour 1993: 69). Time doesn't simply pass away. As Michel Serres has written, time flows in astonishingly complex and turbulent ways (Serres 1982: 71–83; Serres 1995a; Serres with Latour 1995: 44–70). Time is marked by calms, thunderous accelerations, counter-currents and eddies: time percolates (Prigogine and Stengers 1984; Serres with Latour 1995; Serres 1995a; Witmore 2006a). What might be very distant in linear time can be quite proximate and near in percolating time. To illustrate this point, Serres offers the image of an ironed and perfectly flattened handkerchief:

> You can see in it certain fixed distances and proximities. If you sketch a circle in one area, you can mark out nearby points and measure far-off distances. Then take the same handkerchief and crumple it, by putting it in your pocket. Two distant points suddenly are close, even superimposed. If, further, you tear it in certain places, two points that were close can become very distant. (Serres with Latour 1995: 60)

This crumpled and torn manifestation is analogous to the landscapes that archaeologists work with, and contribute to, in their practice.

In the course of walking, collecting and documenting, practitioners engage with a rich multi-temporal accretion of various pasts. While some pasts are distant or buried, others are proximate and have action in the course of people's lives today. It is this complex arrangement of a multiplicity of material pasts which comprises landscape. Moreover, landscape is as much about how it circulates as it is about how it transforms. As such, landscape is not solely constituted by the material presence of slopes, valleys, terrace lines or cereal fields out there: so many aspects of landscape circulate in multiplicities of materials, other media, other voices – ceramic fragments, building stone, pine resin, travelogues, archaeological reports, oral testimonies, memories, digital video and so on – beyond the spacio-temporal range of the Mediterranean countryside (also Witmore 2004a). By working *with* (as distinguished from *on*) these diverse material pasts through practices such as excavation, survey, typological analyses, documentary presentation and so on, archaeologists enter into renewed sets of relations with these material pasts.

AN ABBREVIATED TOPOLOGY OF THE SOUTHERN ARGOLID

Having all too quickly summarised the ways in which Mediterranean landscape surveys since 1972 have produced temporalities from survey materials; having briefly addressed some of the ways in which time is coming to be understood in archaeology at large; and having offered an alternative understanding of time, now i turn to the question: what other archaeological syntheses of landscape we might produce? Drawing upon the work of Michel Serres (1995a), I offer a 'topology' of an archaeological landscape as an example of this synthesis. For Serres, topology, as the 'science of nearness and rifts', is a means of mapping which is distinct from metrical geometry, the 'science of stable and well-defined distances' that dominates the geographical ordering of space and time (Serres with Latour 1995: 60). Rather than follow a linear orientation, the movement and associations are nonlinear and vectoral.

From topography to topology to toponymy, a glance at any English dictionary will reveal the Greek root *topos*. *Topos* or *topoi* are place or places, respectively. *Topos* can also refer to a room in a house, a part of the body, a position in the zodiac, a place of burial, a passage in a text or even a topic or theme within rhetoric. In regarding landscape as a mixed ensemble of *topoi*, whether on the ground, through the glass (e.g. with theodolites or photography) or as a graph, I am purposely denying some of landscape's classic associations and distinctions – from the surrounding environs of human habitations to a perceived version of the natural world set in opposition to the city (Andrews 1999: 1–22). The *topoi* which comprise this topology are teased out through empirical vignettes illustrative of nine key themes of percolating time: 1) actuality; 2) sociotechnical genealogy; 3) palimpsest and chiasmus; 4) entropy and negentropy; 5) katachresis; 6) distributed memory; 7) active pasts; 8) life-trajectories; 9) the very long term. Definitions of these themes will be offered as a conclusion.

In moving through these nine themes, the following topology focuses on the landscapes of the southern Argolid in Greece (Figure 9.2) and builds upon the work of the AEP (see van Andel and Runnels 1987; Jameson et al. 1994; Runnels et al. 1995; Sutton 2000). The AEP grew out of the earlier topographical work of archaeologist M.H. Jameson. During the 1950s, Jameson (along with his wife Virginia during the 1950 season) undertook a series of topographical surveys in the area (Jameson et al. 1994: 7–10). This fieldwork developed into excavations during the 1960s at the sites of Halieis (Jameson 1969) and the Franchthi Cave (Jacobsen 1969, 1981). The AEP began in

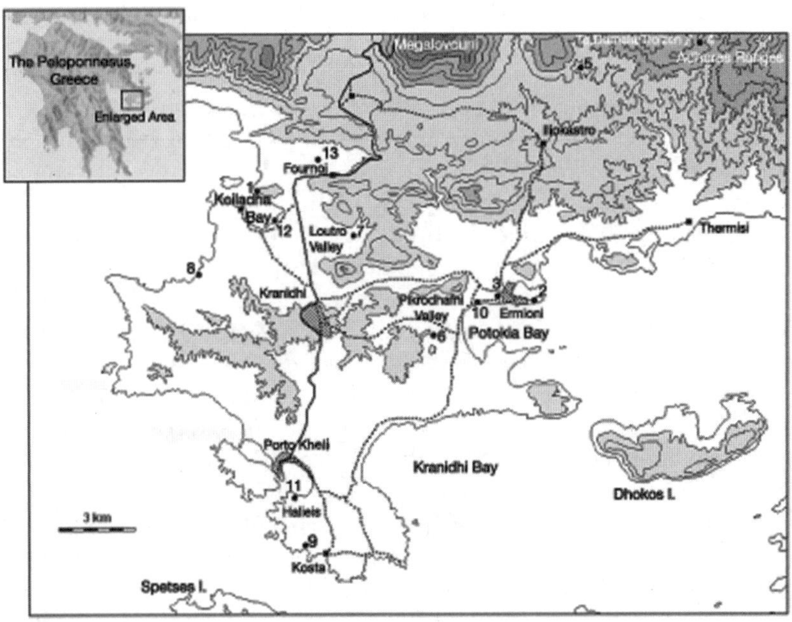

Figure 9.2 Map of locations discussed in this chapter: (1) Franchthi cave and paralia; (2) the *Bisti* of Ermioni; (3) the *Pron* of Ermioni; (4) Soros, 'G10'; (5) *Kastro* (*Eileoi*), 'G2'; (6) the Monastery of Ayioi Anaryiroi; (7) Loutro well, 'B20'; (8) Panayitsa, 'B4'; (9) Bay of Lorenzo across from Spetses; (10) Ermioni Magoula; (11) the lower town of Halieis; (12) brick kilns and spit on edge of Koiladha Bay; (13) Philanoreia 'F60'. *Note: map digitised from AEP base map of southern Argolid (Figure 3.2, Jameson et al. 1994: 152). Contour levels set every 100 metres; shading begins at 100 m and alternates every 300m thereafter.*

1972 as an archaeological survey designed to address wider landscape questions associated with these excavations. After a pause, an intensive and systematic (probabilistic) surface survey was subsequently undertaken between 1979 and 1981 (Jameson et al.1994; also Witmore 2004a, 2005). The following topology complements the long-term history presented in the final project publication (Jameson et al. 1994).

20 July 1958 and c. 48,000 BC

Returning on foot from the annual festival of the Prophet Elijah (Profitis Ilias), Michael H. Jameson enters a small valley known as Riniza on the eastern slopes of Megalovouni in the southern Argolid, Greece. Here, while walking through a field, Jameson encounters a single worked flake on the surface. The flake is made of a dense, dull black flint. It has two faces: a multi-faceted striking platform and a bulbar face with a prominent bulb of percussion (Bialor and Jameson 1962: 181). Its dimensions are 53 millimetres × 36 millimetres × 13 millimetres. The qualities of the flake, the characteristics of its design and the aspects of its material all place its origin somewhere in the Paleolithic c. 50,000 years ago. But 'somewhere in the Paleolithic' is neither when nor where we should begin in order to understand its temporal presence.

Our entry point for understanding this lithic tool begins on the ground nearly half a century ago in a chance encounter between an archaeologist and this single worked flake. So much comes together in the relations between practitioner and artefact in 1958 – an archaeologist's unique mode of engagement with downcast eyes acutely observant of details on the surface (an acumen analogous to that of a keen detective at the scene of a crime [Shanks 1992: 53–54]), combined with years of disciplinary training and experience, combined with positive relations with local communities, and so on. Still, the moments of knapping, manipulation and the process of reduction are all inscribed into the body of the flint flake, and they too are present. Here we encounter a pleat, a fold between two presents separated by a vast distance in linear time – that of the encounter between Jameson and the flake, and between that of Jameson's knowledge and of the flake's making. This simultaneity is actuality.

Indeed this folding of two presents spawns a series of investigations in a number of nearby caves, which in the end turn up little more than a few pottery sherds, a couple of flint fragments and a number of fossilised bits of Pleistocene fauna (Bialor and Jameson 1962). For Bialor and Jameson at least, the caves in the Riniza valley could be crossed off the list of potential areas as candidates for the presence of Paleolithic materials. But something of the early prehistory of Greece would soon re-emerge.

'Origins'

The story of Franchthi Cave is a story of transition – from hunting and gathering to sedentism and agriculture. Its stratified deposits span 20,000 years across three critical periods – the Upper Paleolithic, the Mesolithic and the Early Neolithic. Franchthi, it is widely held, contains a story of the 'origins' of agriculture (cf. Jacobsen 1981). Here beginnings lay buried (Perlès 2001: 46–49).

Discussions of agricultural origins tend to focus on how human beings can adopt a radically new way of life. On one side of a perceived temporal threshold are Mesolithic hunter/gatherers; on the other, settling down in the Neolithic, are agriculturalists and pastoralists. Following Catherine Perlès, four possible scenarios for the transition into the Neolithic are posited for Greece: 1) autochthonous or local development of agriculture; 2) cultural diffusion or acquisition of ideas from others; 3) demic diffusion or the importation of ideas by newcomers; and 4) a more complex interchange between existing and incoming peoples (Perlès 2001: 38–51). While scholars debate the exact nature of the transition, all agree that a radical shift occurred – there was an origin to agriculture (cf. Price 2000). Yet there are problems with the presupposition of a revolutionary transition.

The idea of a Mesolithic/Neolithic transition rests upon a misconceived and modernist notion of historicity regarding what it is to be human and how human beings in turn relate to the world. Innovation, for Perlès and others, is about discovery. This implies radical shifts in how human beings live with the world.

However, human beings, as distributed mixtures and collectives with our material worlds, are situated within a network of association and understanding which absorbs and allows for changes which we 'moderns' in hindsight regard as radical. While others also may well have thought in such terms in the past, this is not to be assumed. Certainly, new implements, new 'ideas,' new entities entering the scene can cause shifts in another entity or other entities' paths of relation to the world (a process exemplified through Jameson's encounter with the Paleolithic flake – the flake's 'path of relation' would have been to remain on the surface).

The transformations occurring around 7000 BC, however, are not solely about how new things, new understandings or new members are enrolled within a community – within a collective; rather, they are about how the role of members already present or the relations of those already present change. In other words, instead of leaving some *Lens nigricans* or *Lens ervoïdes* behind to germinate, human beings now help *Lens orienntalis*, which appears in the Franchthi deposits around 7000 BC, along the way to maturation. A new member has entered the collective. And though the roles of lithics may change, people can still fish for tuna, forage for barley and hunt for game just as others did before them, all in the context of modified sets of relations with things.

Radical revolutions are not the only explanation for the emergence of new collectives, new hybrids as suggested by the things, which circulate from the cave floor deposits of Franchthi. More subtle genealogical shifts are also to be traced.

'Time's Two Arrows'

But consider that, while it makes perfect sense to deploy linear sequencing in the Franchthi Cave and trace the processes of accretion, deposition and stratification, by habitually seeking more ephemeral activities we tend to ignore the seemingly obvious fact that the most durable, unchanging and stubborn feature of the locale is the rock which surrounds and encases the deposits (Serres 1987: 302). The material presence of the cave cross-cuts and connects all of the activities which have taken place over the millennia at Franchthi. The cave provided shelter for Mesolithic hunters, formed part of an enclosure for companion species in the recent past and acted as a convenient shaded canopy for the archaeological excavations (Jacobsen 1981).

Indeed, we must remember that an eddy in the flows of time is spawned through the archaeological excavation itself. Entities once distant through burial under several millennia of sedimentation are made proximate through our actions as archaeologists. In this way time does not simply pass in even flows of accreted layers. Time percolates at the turbulent confluence of entropic and negentropic processes.

Entropy and 'Negentropy'

Ten kilometres east of the Franchthi headland, along the eastern coastline of the southern Argolic peninsula (anciently Akte) lies a small peninsula, 1.2 kilometres long and 0.3 kilometres wide, known as the *Bisti* (Albanian for 'tail'). Roughly divided in two, the western half is occupied by the town Ermioni (known anciently as Hermion and in the recent past as Kastri), which continues to rise up the *Pron* (the 'foreland'). The eastern half, which begins at the remains of a medieval crosswall 0.6 kilometres from the tip, was planted with pines in the early 20th century and is now an archaeological park.

The position of ancient Hermion on this peninsula bound by two harbors meant that the site was utilised as a convenient 'quarry' for building stone, which was removed by boat. William Martin Leake (1777–1860), in his discussion of the site, comments: '[I]ts situation near the sea, and not far from some islands of recent populousness, has been very unfavourable to the preservation of its remains of antiquity' (Leake 1830: 461). Leake was referring to the common practice of removing ready-cut stones for the construction of buildings in rapidly expanding towns such as Hydra (several stones left in mid-transportation are present near the eastern end on the northern side of the *Bisti* today). Antiquarians quarried stone for rather different purposes.

In 1729 the notorious (as his findings were often fraudulent) Abbé Fourmont claimed to have dismantled a medieval crosswall, which fortified the eastern end of the town of Kastri, while searching for stone inscriptions. In his displacement of such a durable material past, Fourmont may have located upwards of three dozen inscriptions (Omont 1902).

Inscriptions are capable of being transformed into many things – lintels, thresholds, wall fabric, etc. – and many of the inscriptions of ancient Hermion met with such fates in the various transformations of the *Bisti*. Though many of these stone inscriptions are now lost, Fourmont's harvest was also transformed into entries for *Inscriptiones Graecae IV* (Jameson et al. 1994: 587).

Whether for mason, antiquary or archaeologist, the land is laden with the simultaneous processes of entropy and negentropy (negative entropy). Media, whether stone- or paper-based inscriptions, though reductions of an original situation, amplify and circulate something of another time at a spatio-temporal distance. Their ability to manifest other times, moments and situations makes them negentropic entities. Media add to the turbulence of time.

The Path from Damala

At some point between 1801 and 1806, Sir William Gell (1777–1836), an English diplomat, classical scholar and antiquarian, crossed the Adheres range by way of a path from Damala, near the ruins of Troizen. Along the way Gell recorded distance between features worthy of observation with to-the-minute precision. Locales observed from afar were often situated with a compass bearing along the line of sight from a specific place along the path. Gell carried a pocket compass and a sextant, which he claims were often taken to be instruments of magic rather than measure by locals (Gell 1823: 352). Gell also took time to sketch views along the way with pen and ink, for him sometimes a more suggestive and 'accurate' means of articulating 'the face of a country' (Gell 1810: xiv).

In his published *Itinerary of Greece* (1810), Gell juxtaposed select translations of Pausanias or Strabo with his own descriptions of a particular route. On horseback, with the *Periegesis* (Pausanias' *Description of Greece*) in hand, with a number of attendants and with a diversity of instruments, accoutrements and conveniences, Gell reiterated

Pausanias' path from Troizen. Thus a text from the 2nd century AD mediates an engagement with the landscape during the early 19th century AD. Here we encounter another pleat in the fabric of time – aspects of Gell's travels are proximate with those of Pausanias.

The Road from Troizen

Pausanias' *Description of Greece*, written in the 2nd century AD, is notoriously sparing with details of his exact route (Alcock 1993: 28–29; Habicht 1985: 104; Snodgrass 1987: 86). He tells us:

> [T]here is a road from Troizen to Hermion by the rock which was formerly called the altar of Strong Zeus, but which the moderns have named the rock of Theseus ever since Theseus picked up tokens here. Following the mountain road which runs by this rock we pass a temple of Apollo, surnamed Apollo of the Plane-tree Grove. (*Description of Greece* 2.34.6; reproduced in Frazer 1898)

Given the nature of the topography in this mountainous terrain, it is extremely likely that Pausanias passed, if not nearby, at least within view of Soros, a large, more or less circular stone cairn that was much later denoted as 'G10' by the AEP (Jameson et al. 1994: 521–522). At over 8 metres in height and 40 metres in diameter at its base the cairn lies exposed on the ridgeline at an elevation of just over 680 metres. It commands a view north to the upper heights of the Troizen acropolis and southwest to the area of Kastro (anciently Eileoi) and beyond. Given its monumentality it can be seen from a great distance.

Though of earlier date, Soros is indicative of other boundary markers on the borders of the region of ancient Hermion. In passing by this cairn, a lone sentinel upon the ridgeline, Pausanias entered into the drainages above the plateau of Eileoi and into the region demarcated by the AEP as focus of an intensive pedestrian survey.

Polychronic Ensemble: AEP Transects

Between 26 July and 31 July of 1981, members of the AEP 'Red Team' walked transects of the Pikrodhafni Valley just south of the dirt road from Kranidhi to Ermioni. On 26 July the team began near a fork where a second dirt road leads off to the south toward the Monastery of Ayioi Anaryiroi. They moved east. Pockets of thick vegetation, along with the steep scarp of the streambed, forced the team to deviate from an area between the roads.

Rarely are survey transects perfectly straight. Transects are oriented and directed by features in the landscape. Plots of land, whose patterning fluctuates over several generations, demarcate where one series of transects ends and another begins. In this fashion material pasts exert force upon our paths of engagement. They have as much a stake in survey practice as irate and intractable landowners. The narrative of the field logs captures something of these interactions.

By the end of the day on 31 July the team had encountered, delimited and mapped several concentrations of surface materials. They had also mentioned various features which caught their attention in the body of the field notebook labelled 'Red Team 1981'. These were also marked on a series of 1:5000 trace maps. On the ground the things and features worthy of mention were encountered in the following order:

- 'E40' (the E for commune Ermioni), an obvious scatter of diverse materials which were subsequently identified with various periods – the Late Geometric, Archaic, Classical, Hellenistic and Medieval periods (Jameson et al. 1994: 493).
- A scatter of roof tiles and sherds (identified as Medieval to early Modern) eroding from the slopes to the north (later designated as 'E50' by the 'Blue Team').
- An 'old' cistern of mortar, limestone rubble/slabs and thick roof tiles.
- Some limestone blocks arranged in an 'E' exposed in the surface of the dirt road from Kranidhi to Ermioni. This road was subsequently paved.
- A 'moderately dense scatter of roof tiles' (site 'E41'), which was later identified as Late Roman in date.
- A 'significant scatter' located on the southern slopes of a conical hill ('E42'), which was medieval (1000–1500 AD) in date.
- East of 'E42', a lime kiln. This feature was subsequently labeled 'MS12' by the Modern Sites Survey (Murray and Kardulias 2000).
- Nearby the kiln, an abandoned 'house', which was not marked on the map.
- A lime kiln, adjacent to the branch road leading toward the Monastery ('MS15').
- A small but very dense scatter of sherds and roof tiles; this site 'E43' was located in a deep-plowed field.
- A *mandria*, or brush sheep fold ('MS13').
- A scatter of sherds and tiles significant enough to be labeled as a 'site' located on the third terrace up from the bottom of the conical hill on the northeastern side ('E44').
- An ashlar block, 0.56 metres × 0.70 metres × 0.30 metres, with two cuttings for metal clamps.
- 'E46', the remains of a church, approximately 6 metres × 4 metres. Rubble and tile construction. Medieval in date.

We should note that the complexity of this polychronic ensemble was realised through a circulating process of transformation (Witmore 2004a). In order to further elucidate the temporal nature of this process, let us briefly return to a rough sketch of survey practices from 1981.

Between 26 and 31 July 1981, some objects, such as ceramic sherds scattered about the patches of field and maquis, enter into an already ongoing process of definition. These ceramic sherds are both produced through their translation and amplification in various modes of documentation (Latour 1999; Witmore 2004a, 2004b) and themselves productive through their material qualities – i.e. durability, fabric, firing signatures and so on – and even through their very presence on the surface. Other pasts have action in orienting and facilitating the survey – the layout of plots of land, the orientation of terraces, the paths of the dirt road. Some are still relegated to the status of 'background noise' and have yet to be realised. Others point to more distant pasts, such as a three-metre in diameter olive tree in the middle of a cereal field. Some relate to just moments before, such as the freshly harrowed plot surrounding 'E43'. And though some things were caught up in a process of gaining new status through the iterative processes of archaeology, all were simultaneously present in the course of a series of archaeological transects on the ground in late July 1981. Moreover, the various media, which testify to the transects themselves, allow us to juxtapose 'Red Team Log 1981' with the final published texts cited above.

The wonderful ontology of such simultaneous 'anachronisms' is precisely the point. This random sorting of multiple pasts has *katachretic* effects (Shanks 2004: 152). This sorting also leads to surprising blends, mixtures and associations; it creates points of transaction between various entities otherwise separated by linear temporal distances, which are often vast. At these points of contact and transaction we encounter pleats in the fabric of time. Such folds are (often) observable in contexts where more stubborn pasts persist.

Active Pasts: Hermion

Demeter was the chief deity of the ancient city of Hermion. In his work *On the Nature of Animals*, the 2nd to 3rd century AD author Aelian (Claudius Aelianus) relates how 'the Hermionians worship Demeter and sacrifice to her magnificently and grandly; and they call the festival Chthonia' (*On the Nature of Animals* 11.4). Pausanias describes the sanctuary where the altar of Demeter was located as the most remarkable of all the sanctuaries at Hermion. He states that it is on the *Pron*.

Along the eastern hillside of the *Pron*, portions of a trapezoidal wall, which may have formed a continuous stretch for up to 100 metres, provide foundations for several houses. Comprised of hard, grey limestone blocks with rough, quarried faces, this wall is as high as three metres in places. It has been identified as potentially that of the *temenos* or precinct wall for the Demeter sanctuary (Jameson et al. 1994: 593). A more durable and stubborn element of material past, this wall orients, delineates, provides foundation for and plays a role in the layout of a series of houses at some point in the more recent past, most likely in the later 18th century.

Durable Pasts

In the Loutro Valley, at the low junction of a series of olive and fallow cereal fields with spreads of Archaic, Classical and Hellenistic materials and framed by terraces containing worked blocks at the intersection of contemporary farming roads accessing these fields, is a well of potentially ancient date (labelled by the AEP as 'B20'). Interactions between people, digging implements and a low water table near the centre of the upper Loutro Valley watershed resulted in the construction of a stone-lined well at some point between the 4th and 6th centuries BC.

In 2003 a bluish plastic container, 0.25 metres x 0.25 metres x 0.20 metres, attached with a metal rod to a worn and knotted hemp rope, sat on the edge of the wellhead (Figure 9.3). This makeshift bucket mediates relations between a farmer, a shepherd or a random passerby and the well and the water it contains, just as a ceramic water jug would have done over two millennia ago. An Archaic, Classical or Hellenistic articulation is folded into the contemporary.

Repetitive engagements between people, water containers and a well occurred here. The splash of the water pot or pail; the sloshing water in the hollow stone-lined shaft; the clank of a laden vessel on the stone lining; the drip of excess water as the rope is pulled skyward – all constitute transient yet recurrent background noises for people who gathered at this well across the centuries. Providing that we hear the clamour of similar things (the clanking of a plastic bucket differing of

Figure 9.3 Photograph of 'B20' wellhead with plastic bucket.

course from that of a ceramic *hydra*), the past may, through such sounds, still be heard in the Loutro Valley (Witmore 2006a).

Noise Diversified

For Michel Serres, the archaic French term for noise, *noiseuse*, is appropriate to a discussion of the multiplicity, sensorial complexity and presence of the material world. *Noiseuse* carries connotations of ruckus, strife and commotion (Serres 1995a). It speaks to the chaos of landscape. These are more chaotic qualities which we too often ignore and filter out; and yet they are fundamental to human 'being'. Such 'noise' permeates the Greek countryside, saturating every cove and crevasse.

The noises of the past have been obfuscated or even replaced by others. A cacophony of lorries, cars, mopeds, tractors, boat engines, horns and construction equipment can be heard almost anywhere in the southern Argolid. The repetitive noises of the past recede. They are drowned out by oceans of ever more complex things. Sea noise has both receded in the wake of other noises and advanced inland far beyond the coastlines of 23,000 years ago. Yet despite these changes it has remained ever present across the millennia.

The Agitation of Proteus

Across the very long term, one that far exceeds the *longue durée* of Braudel, the accreted pasts of the southern Argolid have been formed largely by the sea. Proteus, the shape changer, continually transforms the outlines of the land in the southern Argolid. Since groups frequented a cave on the western end of the Franchthi promontory roughly 23,000 years ago, the sea has risen 120 metres. The pace of the transformation has slowed down significantly over the last 10,000 years and still people with boats and ships, oars and sails adjust to the rhythms of its caprice. From the coast travelled by hunters/gatherers in the Upper Palaeolithic to the shores measured in the portulan (a text listing sequences of territories, harbours and coastal landmarks with distances between them), attributed to Pseudo-Scylax of the 6th century BC (Jameson et al. 1994: 568–572), the shape and size of the southern Argolid coastline has transformed by upwards of 250 square kilometres.

Whether we consider the exposure of buried features and materials in a beach scarp at Panayitsa (AEP 'B4'); a roof tile kiln in the small bay of Lorenzo across from Spetses (part of which was excavated in 1967 [Jameson 1969: 341–342]); the great walls of the *Bisti*, structural features at Ermioni Magoula, the lower town of Halieis; a spit constructed from the rejects and waste of brick kilns on the eastern edge of Koiladha Bay; or the early Neolithic settlement area at Franchthi Paralia, we observe that, through a combination of both exposure and erosion, the ebbs and flows of the sea have made the presence of once-distant pasts proximate while eventually washing others away. Entropic processes often provide occasion for negentropic transactions.

The Life-Trajectories of Monuments and Things

The stones and inscriptions of the *Bisti* at Hermion were redeployed as building materials for a medieval crosswall, or sit on museum floors and steel shelves in Nafplion, or in display cases around Europe. Marble lion heads, perhaps from the *sima* (the crowning gutter) of an ornate roof, in what is probably ancient Philanoreia are transformed into waterspouts for the Fournoi village fountain (AEP 'F60'), constructed in the 20th century. Rough-faced blocks of blue/grey ashlar, probably once the wall stones of structures associated with the Classical settlement in the Loutro Valley, are now incorporated into the fabric of a linear terrace wall roughly 10 metres to the west of the 'B20' well.

Transformations of people and transformations of things occur simultaneously not only in space but also in time. Such change provides an impetus for the things to be forgotten and to decay, only in many circumstances to be later remobilised, redeployed, remembered. These transformations – entropic processes of corrosion and ruin, processes of fragmentation and displacement, of circulation and accretion, these negentropic processes of materialisation and (re)deployment – are the conduits through which often distant pasts percolate in other eras. It is in this way that we may begin to understand time not only as a series of successions, but also as a series of simultaneities (Latour 2005b: 39–40).

Archaeology too is a process of percolation. As a discipline it adds to the aggregate mixture of times present today. It too is part of the transformative processes present on the ground in the southern Argolid. It too must be factored into the stories of the land.

CONCLUSIONS: LANDSCAPES AND PERCOLATING TIME

The notion of percolation, as defined here, holds that 'it is the sorting that makes the times, not the times that make the sorting' (Latour 1993: 76). We may now summarise further the key features of percolating time, of dynamic sorting, as articulated through this quickly sketched topology. I have explored nine aspects of percolating time in relation to landscape:

1. *Actuality* is the non-arbitrary conjunction of presents (see Shanks 1997: 246): the past's present, which in the case of Jameson's Levallois-Mousterian flake is the moment of its knapping and the present of the encounter – either the moment of 'discovery' in 1958 or the protracted chain of subsequent engagements. Actuality is a quality of the relatedness

between various entities (people, things, companion species) across otherwise vast temporal distances. Actuality is an aspect of temporal folding.

2. *Genealogy – sociotechnical mixtures.* There are neither pure subjects nor pure objects. Rather we are deep entanglements and thorough mixtures of people and things. Because humanity cannot be simply boiled down to 'immaterial' ideas, subjectivity, we may define our linkages in a different way. Things are part of the collective that comprises humanity. *Things are us.* Instead of laminar and progressive chronologies or radical revolutions, genealogies plot successive relations more subtly. Through genealogy we identify the entry points of new entities – hybrids of ideas and things – into past collectives.
3. *Palimpsest* and *chiasmus.* The former metaphor connotes erasure and superimposition; the latter denotes crossing, intersection and intertwining. It indicates the points of contact, the pleats in the fabric of time. These metaphors fluctuate, but are always possible and simultaneous in sites, features and landscapes.
4. *Entropy* and *negentropy* are time's two arrows. Entropy is an irreversible flow toward aging, degradation, decay, ruination and death. But negentropy is the action/force/energy of information and life, which re-form through activities such as recycling, memory and *un*forgetting (activities exemplified through archaeological practices such as the documentation of matter otherwise caught up in a process of decay). These flows are coupled and coincident and yet turbulent (for more on these terms see note 5).
5. *Katachresis* is a juxtaposition of two seemingly disparate things, accounts or situations which can create aleatoric frictions, transactions and associations. As an empirical method deliberately employed by the contemporary archaeologist, katachresis is designed as a forcible juxtaposition, the effect of which may lead to the unexpected (Shanks 2004: 152). Whether by chance or design, understandings and confusions arise which would, perhaps, have not occurred otherwise. The random sorting of various pasts in landscape often has a katachretic effect.
6. *Distributed memory.* While memory can be defined as 'a sense of time attained through the senses' (Assad 1999: 103), it is both extended through its necessary prostheses (Witmore 2006a) and shaped through the mediating roles of things (Olivier 2003). Recall is constituted through a distributed set of relations with media (letters, field notebooks, photographs and so on – see Bowker 2005), things (mementoes of an event, features indicative of previous activities) or mnemonic traces (material manifestations of an event – e.g. plough marks through a Hellenistic floor surface). Distributed memory is both katachretic and vectoral. It can be about the relations between various moments effected at the speed of thought or, through archaeological memory practices, at the speed of the pin or processor or, operating at slower paces, it can be comprised of mnemonic traces indicative of activities which occurred two years, two decades, two centuries, or two millennia ago. Distributed memory is replete with multi-temporal pleats and yet it is full of voids, often devoid of what is forgotten.
7. *Active pasts.* As a corollary of the analytical levelling of people and things, nonhuman entities are understood to exert force. Things have action. Just as a Classical *temenos* wall can have action in the construction of a subsequent building, or an ancient well can still provide a point of confluence for daily practices in the Loutro Valley, so too can other material pasts have action in the present.

8. *Life-trajectories*. Things may be identified not according to their place in a life cycle, but rather in terms of their shifting relations with other entities. Does one identify a sherd as an aging, discarded fragment of a former whole, or in terms of its shifting role, as a base for an Athenian black-figure skyphos, a piece of rubbish, an ostrakon, a burnisher, a filler for a mudbrick or a catalogued diagnostic? Life-trajectories are tracings of this metamorphosis and these various relations that ensue.
9. *The very long term* is a temporal vastness, not couched in terms of continuity or its counterpart, change, but rather to be expressed as a vector of time predicated on connection, relevance and the percolation of pasts. While a c. 50,000-year-old Levallois-Mousterian flake can shape people's lives today and the course of their research (e.g. Bialor and Jameson 1962), the blade has been part of our human collective for dozens of millennia. Understanding the very long term is to 'think in accordance with the rhythms and scope' of things and of the land (Serres 1995b: 29).

* * *

As I remarked at the outset of this chapter, percolating time results from the sorting of various pasts caught up within mixed sets of both simultaneous and successive relations. The nine aspects of percolating time defined above, though not exhaustive, have been isolated in order to map and characterise these relations. Moreover, they were exemplified and synthesised by following a prescribed path through select empirical *topoi*. A topology was presented as a means of translating the crumpled, percolating nature of time and the landscapes of the southern Argolid. As a science or mathematics of connection and interruption, a topology does not proceed by free association. This vectoral movement, as manifest in our multifarious relations with multiple material pasts, is, I suggest, what distinguishes archaeological from historical notions of time. A topology is an archaeological synthesis of time – a time which has thereby been complicated (even) further.

To embrace such temporal complexity is to engage in the active reshuffling of the most basic ingredients in the archaeology of landscape – space, time and materials. In working through the example of intensive and systematic survey practices from Melos, I have argued that linear temporalities are associated with measurement and organisational structures. As external parameters, such notions of time have served as measures against which to orient our understandings of landscape change. To be sure, various archaeologists have challenged these understandings over the last three decades or so, yet even here a modernist historicism oriented toward progressive models of knowledge production has persisted. This is not to claim that progress does not exist, but rather to stand back and understand that what came before is relevant to the present as more than obsolete heritage or out-of-date history. We must, in following a more difficult and convoluted path, re-characterise our relations with the things of the past differently. Most crucially, we have yet to undercut the radical division between past and present. Here, we must understand not only our own practices as part of the stories of the landscapes we wish to document, but we must recast material pasts as having action, as having a stake, as being co-present, co-creative and co-constitutive in contemporary landscape processes.

The past is no longer past. We may recognise and acknowledge a multitude of relations with material pasts, which previous schemes sieved away into the debris pile – as some future ones will doubtless do as well. Understanding time as a turbulent, percolating multiplicity does not pose a threat to what we as archaeologists have always done. On the contrary, such an approach to time enriches our understanding by taking into account more entities, more complex relations, and a wider range of variables. It is this rich and percolating multiplicity that constitutes the material basis of archaeology, the discipline of things.

ACKNOWLEDGEMENTS

In the process of preparing this chapter many people have either contributed to various discussions or have provided direct feedback. I thank Ewa Domanska, Sebastian de Vivo, Alfredo Gonzalez-Ruibal, Jack Mitchell, Bjørnar Olsen, Michael Shanks, Alain Schnapp, Michel Serres and Timothy Webmoor. I am grateful to the editors for the invitation to contribute to this volume and I also thank them for their patience, tenacity and detailed comments. I extend my thanks to those practitioners who have fearlessly wrestled with the issue of time in archaeology. Responsibility for the idiosyncrasies of this chapter is mine alone.

NOTES

1. The Greek term *kairos* refers to both time and weather and while the distinction between *chronos* and *kairos* can and should be pushed further, there is no space for it here. For more on this distinction in archaeology see McGlade 1999: 144–146; Shanks 1992: 185.
2. This is usually couched in terms of intensive versus extensive survey. For example, archaeologists such as Richard Hope Simpson have investigated regions on more extensive scales through a combination of questioning local informants, aerial photography and surface reconnaissance focused on the more obvious archaeological features of a given area (cf. Cherry 2003).
3. The second volume of Braudel's tour de force was released in 1973. Still, the full impact of Braudelian history was not as immediate; there was some lag time before the full ramifications of Braudel's work and the *Annales* perspective would be felt in archaeology (Hodder 1987; Bintliff 1991; Knapp 1992).
4. The year 1972 also saw the publication of *The Emergence of Civilization* (Renfrew 1972), and *Models in Archaeology* (Clarke 1972). As John Cherry has pointed out, both texts had an impact upon regional studies in the Aegean (2003: 141).

5. The physicist Leon Brillouin (1889–1969) coined the term 'negentropy' in his book *Science and Information Theory* (1956) in referring to the relationship between information and entropy. Information, and the ability to retain it, for Brillouin, decreased the degradation of a system and thereby could be described as having a negative entropic effect: hence 'negentropy' (1956: 152–161). However, the scientific community soon shied away from the term, as the notion of entropy at the macroscopic scale, it was contended, encompassed what Brillouin had defined as negentropy (Wilson 1968). For Michel Serres, life, though susceptible to death, nevertheless persists, 'going up the entropic stream by means of phylogenetic invariances and the mutations of selection' (Serres 1982: 74). Matter *and* life must be (mixed) properties of the system or the system doesn't exist. While information, life – negentropy – as an eddy in the stream may be undone by entropic flow it may be re-formed elsewhere (Serres 1982: 75).

REFERENCES

Alcock, S.E. 1993. *Graecia Capta: The Landscapes of Roman Greece*. Cambridge: Cambridge University Press.

Alcock, S.E. 2002. *Archaeologies of the Greek Past: Landscape, Monuments, and Memories*. Cambridge: Cambridge University Press.

Alcock, S.E. and J.F. Cherry (eds) 2004. *Side-by-Side Survey: Comparative Regional Studies in the Mediterranean World*. Oxford: Oxbow Books.

Andrews, M. 1999. *Landscape and Western Art*. Oxford: Oxford University Press.

Assad, M.L. 1999. *Reading with Michel Serres: An Encounter with Time*. Albany: State University of New York Press.

Athanassopoulos, E.F. and L. Wandsnider (eds) 2004. *Mediterranean Archaeological Landscapes: Current Issues*. Philadelphia: University of Pennsylvania Museum of Archaeology and Anthropology.

Bailey, G. 1981. Concepts, Time Scales and Explanation in Economic Prehistory. In A. Sheridan and G.N. Bailey (eds) *Economic Archaeology*. Oxford: British Archaeological Reports, pp. 97–117.

Bailey, G.N. 1983. Concepts of Time in Quaternary Prehistory. *Annual Review of Anthropology* 12: 165–192.

Bailey, G.N. 1987. Breaking the Time Barrier. *Archaeological Review from Cambridge* 6: 5–20.

Barker, G. 1995. *A Mediterranean Valley: Landscape Archaeology and Annales History in the Biferno Valley*. London: Leicester University Press.

Barker, G. and D. Mattingly (series eds) 1999–2000. *The Archaeology of Mediterranean landscapes* (5 vols.). Oxford: Oxbow Books.

Bergson, H. 1998. *Creative Evolution*. New York: Dover.

Bergson, H. 1999. *Duration and Simultaneity: Bergson and the Einsteinian Universe*. Manchester: Clinamen Press.

Bialor, P.A. and M.H. Jameson 1962. Palaeolithic in the Argolid. *American Journal of Archaeology* 66(2): 181–182.

Binford, L. 1972. *An Archaeological Perspective*. New York: Seminar Press.

Binford, L. 1981. Behavioural Archaeology and the Pompeii Premise. *Journal of Anthropological Research* 37: 195–208.

Bintliff, J.L. 1991. *The Annales School and Archaeology*. Leicester: Leicester University Press.

Bintliff, J.L. and A.M. Snodgrass 1985. The Cambridge/Bradford Boeotian Expedition: The First Four Years. *Journal of Field Archaeology* 12: 123–161.

Blanton, R.E. 2001. Mediterranean Myopia. *Antiquity* 75: 627–629.

Bowker, G.C. 2005. *Memory Practices in the Sciences*. Cambridge, MA: MIT Press.

Bowker, G. and S.L. Star 2000. *Sorting Things Out: Classification and Its Consequences*. Cambridge, MA: MIT Press.

Bradley, R. 2002. *The Past in Prehistoric Societies*. London: Routledge.

Brillouin, L. 1956. *Science and Information Theory*. New York: Academic Press.

Cavanagh, W., J. Crouwel, R.W.V. Catling and G. Shipley 1996. *Continuity and Change in a Greek Rural Landscape: The Laconia Survey. Volume II, Archaeological Data*. London: British School at Athens.

Cavanagh, W., J. Crouwel, R.W.V. Catling and G. Shipley 2002. *Continuity and Change in a Greek Rural Landscape: The Laconia Survey. Volume I, Methodology and Interpretation*. London: British School at Athens.

Cavanagh, W., C. Mee and P. James, with N. Brodie and T. Carter 2005. *The Laconia Rural Sites project*. London: British School at Athens.

Cherry, J.F. 1982. A Preliminary Definition of Site Distribution on Melos. In C. Renfrew and M. Wagstaff (eds) *An Island Polity: The Archaeology of Exploitation in Melos*. Cambridge: Cambridge University Press, pp. 10–23.

Cherry, J.F. 1983. Frogs around the Pond: Perspectives on Current Archaeological Survey Projects in the Mediterranean Region. In R. Keller and D.W. Rupp (eds) *Archaeological Survey in the Mediterranean Area*. Oxford: British Archaeological Reports, pp. 375–416.

Cherry, J.F. 2003. Regional Survey and Its Future. In J.K. Papadopoulos and R.M. Leventhal (eds) *Theory and Practice in Mediterranean Archaeology: Old World and New World Perspectives*. Los Angeles: Cotsen Institute of Archaeology, University of California, pp. 137–159.

Cherry, J.F., J.L. Davis and E. Mantzourani 1991. *Landscape Archaeology as Long-Term History: Northern Keos in the Cycladic Islands*. Los Angeles: Institute of Archaeology, University of California.

Clarke, D.L. (ed) 1972. *Models in Archaeology*. London: Methuen.

Davis, J.L. 1998. *Sandy Pylos: An Archaeological History from Nestor to Navarino*. Austin: University of Texas Press.

Davis, J.L., S.E. Alcock, J. Bennet, Y.G. Lolos and C.W. Shelmerdine 1997. The Pylos Regional Archaeological Survey. Part I: Overview and the Archaeological Survey. *Hesperia* 66: 391–494.

Dillon, M., N. Rauh, L. Wandsnider, M. Hoff, R. Townsend and L. Theller 2001. Juliosebaste in Rough Cilicia: Discovery of a lost Roman City. http://pasture.ecn.purdue.edu/~rauhn/ (Consulted 15 July 2006).

Flannery, K.V. (ed) 1976. *The Early Mesoamerican Village*. London: Academic Press.

Foucault, M. 1972. *The Archaeology of Knowledge*. New York: Tavistock Publications.

Frazer, J.G. 1898. *Pausanias's 'Description of Greece'* (6 vols). London: Macmillan.

Gell, W. 1810. *The Itinerary of Greece with a commentary on Pausanias and Strabo and an account of the monuments of antiquity at present existing in that country compiled in the years MDCCCI:II:V:VI*. London: T. Payne.

Gell, W. 1823. *Narrative of a Journey in the Morea*. London: Longman, Rees, Orme and Brown.

Habicht, C. 1985. *Pausanias' Guide to Ancient Greece*. Berkeley: University of California Press.

Hayles, N.K. 1991. Complex Dynamics in Literature and Science. In N.K. Hayles (ed) *Chaos and Order: Complex Dynamics in Literature and Science*. Chicago: University of Chicago Press, pp. 1–33.

Hodder, I. 1987. *Archaeology as Long-Term History*. Cambridge: Cambridge University Press.

Hodder, I. 1999. *The Archaeological Process: An Introduction*. Oxford: Blackwell Publishers.

Ingold, T. 1993. The Temporality of Landscape. *World Archaeology* 25(2): 152–174.

Jacobsen, T.W. 1969. Excavations at Porto Cheli and Vicinity, Preliminary Report, II. The Franchthi Cave, 1967–1968. *Hesperia* 38(3): 343–381.

Jacobsen, T.W. 1981. Franchthi Cave and the Beginning of Settled Life in Greece. *Hesperia* 50(4): 303–319.

Jameson, M.H. 1969. Excavations at Porto Cheli and Vicinity, Preliminary Report, I. Halieis, 1962–1968. *Hesperia* 38(3): 311–342.

Jameson, M.H., C.N. Runnels and T.H. van Andel 1994. *A Greek Countryside: The Southern Argolid from Prehistory to the Present Day*. Stanford, CA: Stanford University Press.

Knapp, A.B. (ed) 1992. *Archaeology, Annales and Ethnohistory*. Cambridge: Cambridge University Press.

Kuhn, T.S. 1970. *The Structure of Scientific Revolutions*(2nd ed). Chicago: University of Chicago Press.

Latour, B. 1993. *We Have Never Been Modern* (trans. C. Porter). Cambridge, MA: Harvard University Press.

Latour, B. 1999. *Pandora's Hope: Essays on the Reality of Science Studies*. Cambridge, MA: Harvard University Press.

Latour, B. 2005a. *Reassembling the Social: An Introduction to Actor-Network-Theory*. Oxford: Oxford University Press.

Latour, B. 2005b. From *Realpolitik* to *Dingpolitik* – or How to Make Things Public. In B. Latour and P. Weibel (eds) *Making Things Public – Atmospheres of Democracy*, Cambridge, MA: MIT Press, pp. 14–41.

Leake, W.M. 1830. *Travels in the Morea*. Vol. II . London: J. Murray.

Lucas, G. 2005. *The Archaeology of Time*. London: Routledge.

McDonald, W.A. 1972. The Problems and the Program. In W.A. McDonald and G.R. Rapp (eds) *The Minnesota Messenia Expedition. Reconstructing a Bronze Age Regional Environment*. Minneapolis: University of Minnesota Press, pp. 3–17.

McDonald, W.A. and G.R. Rapp (eds) 1972. *The Minnesota Messenia Expedition. Reconstructing a Bronze Age Regional Environment*. Minneapolis: University of Minnesota Press.

McGlade, J. 1999. The Times of History: Archaeology, Narrative and Non-Linear Causality. In T. Murray (ed) *Time and Archaeology*, London: Routledge, pp. 139–163.

McGlade, J. and S.E. van der Leeuw (eds) 1997. *Time, Process and Structured Transformation in Archaeology*. London: Routledge.

Meltzer, D.J. 1979. Paradigms and the Nature of Change in American Archaeology. *American Antiquity* 44(4): 644–657.

Morris, I. 1994. Archaeologies of Greece. In I. Morris (ed) *Classical Greece: Ancient Histories and Modern Archaeologies*. Cambridge: Cambridge University Press, pp. 4–47.

Murray, P.M. and P.N. Kardulias 2000. The Present as Past: An Ethnoarchaeological Study of Modern Sites in the Pikrodhafni Valley. In S.B. Sutton (ed) *Contingent Countryside: Settlement, Economy and Land Use in the Southern Argolid since 1700*. Stanford, CA: Stanford University Press, pp. 141–168.

Olivier, L. 2003. The Past of the Present: Archaeological Memory and Time. *Archaeological Dialogues* 10(2): 204–213.

Olsen, B. 2003. Material Culture after Text: Re-membering Things. *Norwegian Archaeological Review* 36(2): 87–104.

Omont, H. 1902. *Missions archéologiques françaises en Orient aux XVIIe* et XVIIIe *siècles* (vol. I). Paris: Imprimerie nationale.

Perlès, C. 2001. *The Early Neolithic in Greece*. Cambridge: Cambridge University Press.

Pettegrew, D.K. 2001. Chasing the Classical Farmstead: Assessing the Formation and Signature of Rural Settlement in Greek Landscape Archaeology. *Journal of Mediterranean Archaeology* 14(2): 189–209.

Price, T.D. (ed) 2000. *Europe's First Farmers*. Cambridge: Cambridge University Press.

Prigogine, I. and I. Stengers 1984. *Order Out of Chaos: Man's New Dialogue with Nature*. Toronto: Bantam Books.

Redman, C. 1974. *Archaeological Sampling Strategies*. Reading, MA: Addison-Wesley.

Renfrew, C. 1972. *The Emergence of Civilization: The Cyclades and the Aegean in the Third Millennium B.C.* London: Methuen.

Renfrew, C. 1982. Bronze Age Melos. In C. Renfrew and M. Wagstaff (eds) *An Island Polity: The Archaeology of Exploitation in Melos*. Cambridge: Cambridge University Press, pp. 35–44 .

Renfrew, C. and M. Wagstaff (eds) 1982. *An Island Polity: The Archaeology of Exploitation in Melos*. Cambridge: Cambridge University Press.

Runnels, C.N., D.J. Pullen and S. Langdon (eds) 1995. *Artifact and Assemblage: The Finds from a Regional Survey of the Southern Argolid, Greece* (vol. I). Stanford, CA: Stanford University Press.

Schiffer, M. 1987. *Formation Processes of the Archaeological Record*. Albuquerque: University of New Mexico Press.

Serres, M. 1982. *Hermes: Literature, Science, Philosophy*. Baltimore: Johns Hopkins University Press.

Serres, M. 1987. *Statues: Second livre des foundations*. Paris: Editions Julliard.

Serres, M. 1995a. *Genesis* (trans. G. James and J. Nielson). Ann Arbor: University of Michigan Press.

Serres, M. 1995b. *The Natural Contract* (trans. W. MacArthur and W. Paulson). Ann Arbor: University of Michigan Press.

Serres, M. with B. Latour 1995. *Conversations on Science, Culture, and Time* (trans. R. Lapidus). Ann Arbor: University of Michigan Press.

Shanks, M. 1992. *Experiencing the Past: On the Character of Archaeology*. London: Routledge.

Shanks, M. 1996. *Classical Archaeology of Greece: Experiences of the Discipline*. London: Routledge.

Shanks, M. 1997. L'archéologie et le passé contemporain: Un paradigme. In A. Schnapp (ed) *Une Archéologie du Passé Récent?* Paris: Fondation Maison des Science de l'Homme, pp. 245–252.

Shanks, M. 2004. Three rooms: Archaeology and Performance. *Journal of Social Archaeology* 4(2): 147–180.

Shanks, M. 2005. Symmetrical Archaeology. http://traumwerk.stanford.edu:3455/MichaelShanks/1207 (Consulted 12 June 2006).

Shanks, M. and C. Tilley 1992. *Re-constructing Archaeology*. London: Routledge.

Snodgrass, A. 1987. *An Archaeology of Greece: The Present State and Future Scope of a Discipline*. Berkeley: University of California Press.

Sutton, S.B. (ed) 2000. *Contingent Countryside: Settlement, Economy and Land Use in the Southern Argolid since 1700*. Stanford, CA: Stanford University Press.

van Andel, T.H. and C. Runnels 1987. *Beyond the Acropolis: A Rural Greek Past*. Stanford, CA: Stanford University Press.

van Dyke, R. and S.E. Alcock (eds) 2003. *Archaeologies of Memory*. Oxford: Blackwell Publishers.

Wandsnider, L. 2004a. Solving the Puzzle of the Archaeological Labyrinth. In S.E. Alcock and J.F. Cherry (eds) *Side-by-Side Survey: Comparative Regional Studies in the Mediterranean World*. Oxford: Oxbow Books, pp. 49–62.

Wandsnider, L. 2004b. Artifact, Landscape, and Temporality in Eastern Mediterranean Archaeological Landscape Studies. In E.F. Athanassopoulos and L. Wandsnider (eds) *Mediterranean Archaeological Landscapes: Current Issues*. Philadelphia: University of Pennsylvania Museum of Archaeology and Anthropology, pp. 69–79.

Webmoor, T. and C. Witmore 2005. Symmetrical Archaeology. http://traumwerk.stanford.edu:3455/Symmetry/Home (Consulted 12 June 2006).

Wilson, J.A. 1968. Entropy, not negentropy. *Nature* 219: 535–536.

Witmore, C.L. 2004a. On Multiple Fields. Between the Material World and Media: Two Cases from the Peloponnesus, Greece. *Archaeological Dialogues* 11(2): 133–164.

Witmore, C.L. 2004b. Four Archaeological Engagements with Place: Mediating Bodily Experience through Peripatetic Video. *Visual Anthropology Review* 20(2): 157–172.

Witmore, C.L. 2005. Multiple Field Approaches in the Mediterranean: Revisiting the Argolid Exploration Project. Unpublished Ph.D. dissertation, Stanford University. http://traumwerk.stanford.edu:3455/multiplefields/Home (Consulted 12 June 2006).

Witmore, C.L. 2006a. Vision, Media, Noise and the Percolation of Time: Symmetrical Approaches to the Mediation of the Material World. *Journal of Material Culture* 11(3): 267–292.

Witmore, C.L. 2006b. Archaeology and Modernity, or Archaeology and a Modernist Amnesia? *Norwegian Archaeology Review* 39(1): 49–52.

Zangger, E., M.E. Timpson, S.B. Yazvenko, F. Kuhnke and J. Knauss 1997. The Pylos Regional Archaeological Project. Part II. Landscape Evolution and Site Preservation. *Hesperia* 66: 549–641.

CHAPTER 10

A Landscape of Ruins: Building Historic Annapolis

Christopher Matthews and Matthew Palus

INTRODUCTION: MODERN AND ANCIENT

After the American Revolution, the city of Annapolis, Maryland, became a landscape of American ruins. Among the wealthiest and most influential residents, the city's image came to be based on the notion that the city never progressed. Increasingly during the 19th and 20th centuries, this notion was materialised through the actions of the city's administrators and its elected governing council, if not those in the county and the state. The urban form was still inhabited, but among the houses, streets, gardens and shops the era of the American Revolution was continually evoked. The commemoration of the city's close association with the events of this period in the urban landscape is of interest to archaeologists not only because of the historical significance of American independence, but also because for many the Revolution was a moment that in some ways marked an 'end of history' in Annapolis. The landscape of Annapolis became an instant symbol for American history, and the city's antiquity became central to the experience of inhabitants and visitors alike.

These were not the forgotten remains of a lost culture. Rather, the surviving 18th century urban forms represented the exquisitely and actively remembered remains of the accomplishments of a golden generation. Annapolis' landscapes of ruin were central to the experience of modernity in the city, as a powerful though ambivalent relationship with the Annapolis past developed, especially for the social elite. The past became tradition, through the curation of the urban landscape as a cultural artefact. This chapter explores these ruins, considering how they were constructed during the pre-Revolutionary period, and how they were preserved in the lived urban landscapes of the 19th and 20th centuries.

INTRODUCING HISTORY: FOUNDATION MYTHS

Maryland was established as the fourth permanent English colony in North America in 1634, with its capital city chartered in 1667 at the first English settlement, St Mary's City, in southern Maryland (Miller 1988: 58–59). The first proprietor of the colony, Cecil Calvert, 'envisioned a society based upon a hierarchical model of social relations with the proprietor having princely powers over the colony and its development' (Leone and Hurry 1998: 35). In other words, the colony was to be administered as an extension of the English monarchy and derive a similar authority. However, in the late 17th century power and entitlement swung away from Maryland's proprietary family, the Calverts, due to their Roman Catholicism. As part of a shift that in other ways paralleled the earlier Protestant Reformation in England, Maryland saw the appointment of a new royal governor for the colony, Francis Nicholson. During the winter of 1694–1695 the provincial government was relocated from St Mary's City, its home since the 1630s, to a new capital city for the colony, which was called Annapolis (Miller 1988).

Nicholson superimposed a new city plan for Annapolis onto a small community that had been settled by Protestants in 1649, called Anne Arundel Town. As Annapolis grew, the 'Nicholson plan' for the city was largely fulfilled. Like the plan for St Mary's City, it was rooted in baroque principles, and though the two town plans differed in important ways they shared this basic derivation, using monumentality and landscape to underscore state authority by aggrandisement, ostentation and by controlling or directing what people – the Crown's subjects – saw.

> The French, English, and later Italian traditions of baroque design put buildings of authority at the end of vistas. This fixed the spectator's eyesight and kept it from flight. The point was to capture the attention of individuals repeatedly and to orient them to symbols of authority … St Mary's City and Annapolis were designed using the same principles, which constituted one major way European monarchs aggrandized themselves. (Leone and Hurry 1998: 41–42)

In St Mary's City, the 'symbols of authority' of the state were established during the 1660s and 1670s as a Jesuit chapel, a State House and a prison, all built in brick and visible from the public square at the centre of the settlement; four major thoroughfares radiating from the centre of St Mary's City terminated at these buildings, putting them at the focus of these vistas. In Annapolis, the equivalent key symbols were the State House and an Anglican church, each constructed within separate landscapes surrounded by paved circles, from which radiated a number of streets. In both colonial capitals, state and religious authority were juxtaposed; in Annapolis the State House, on higher ground and within a more substantial circle, clearly had primacy over the religious authority symbolised by the adjacent church building (Miller 1988; Leone and Hurry 1998).

The strength of survival of the historic fabric of Annapolis' colonial period is remarkable, and Nicholson's 1696 planned landscape can still be easily experienced. The modern 'historic district', included on the National Register of Historic Places since 1966, relates closely to the earliest known map of Annapolis, the 1718 Stoddert survey (Figure 10.1; Leone 2005: 84).

The most studied feature of the Nicholson plan in Annapolis is the circle that was designed to display the State House. This portion of the city's landscape illuminates a great deal about Nicholson's baroque plan, but archaeological analysis has offered alternative interpretations for the plan and its operation. For instance, John Reps represented the Nicholson plan as an appeal to continental fashion, but concluded that Nicholson himself was only partly fluent in baroque principles of urban design (Reps 1972; cf. Leone and Hurry 1998: 39). When the city's plan is viewed in two dimensions, as on a map, there is prominent asymmetry in the way streets enter and leave State Circle, taken by Reps as a failure in Nicholson's competence and contrasted against his more symmetrical 1699 plan for the urban form of Williamsburg, Virginia (Reps 1972).

In contrast, landscape archaeology has shown that the Nicholson design bears substantial nuance: the shape of State Circle was never a circle, but rather a masterfully crafted egg-shape that created the appearance of a harmonious landscape around the State House, while at the same time accommodating both an awkward existing topography and existing property claims when the plan was arranged in the late 17th century. Additionally, when the vicinity of State Circle is viewed from a different perspective, not as a map but as a three-dimensional landscape through which one moves, the design becomes much more elegant. The various paths converge not on the centre of the geometric area, but on the State House that was constructed on a site that lies off-centre. As archaeologist Mark Leone has observed, Nicholson's plan works better for the site on which Annapolis was to be built than a more symmetrical plan would have, thus attesting to the knowledge of the principles of baroque landscape design that Nicholson wielded. He knew the rules well enough to know where they could be bent or broken (Leone 2005: 84; Leone et al. 1998: 293).

The application on the ground of the Nicholson plan is well understood from landscape archaeological study of State Circle. First, the

Figure 10.1 The 1743 redraft of Stoddert's 1718 survey of Annapolis, Maryland. *Courtesy of the Maryland State Archives, Special Collections, MSA SC 1427-1-6.*

application of baroque principles created the appearance of uniformity and circularity out of State Circle, which actually fell across a challenging piece of terrain with a bluff or bench somewhat off-centre. Second, streets entering the sub-circular State Circle, with vistas that terminate at the off-centre Maryland State House, show that Nicholson 'knew he was dealing with optical principles in the service of authority' (Leone and Hurry 1998: 41). Third, Nicholson used converging and diverging lines of sight in a very systematic way, which reflects intentionality and deliberate use of such principles in landscape design. The longest five streets entering the circle widen out as they approach it, making the State House appear closer and 'shortening' the approach. In contrast, the two shortest streets narrow towards the State House, causing the perception that the streets are lengthened and seeming to 'prolong' the journey to the State House. In sum, these qualities of the Nicholson plan are not accidental or incidental; they are measurable from the Stoddert survey of Annapolis, and they are extant in the city today. The dimensions and volumes have remained consistent in that the *trompe l'oeil*-like illusions designed by Nicholson still operate.

BUILDING AGAINST HISTORY: THE GEORGIAN LANDSCAPE

Nicholson's plan for Annapolis almost completely replaced pre-existing structures and property, but it also created a more formalised system for the description of land ownership. In this respect, Nicholson's colonial city plan enabled the growth of Annapolis. The basis of this development, however, was guided by the actions of property-owners themselves, and the character of the landscape of modern Annapolis emerged from specific episodes of planning and building on the holdings of a small number of residents. A number of archaeological studies of the city have thrown light upon these stories by focusing on the development of the private elite landscapes of the 18th century (Kryder-Reid 1998; Leone 1984; Leone and Shackel 1990; Matthews 2002; Yentsch 1994). The common finding among these studies has been that there was an adherence by builders to an orderly neo-classical design. Projects were not chosen because of their familiarity or necessarily their beauty, but because they were correct. The consumption in Annapolis of the numerous architectural pattern books and builders' guides which described the basis of the 'orders' put forth by Palladio (Leone 1984; Matthews 1998b) led to the adoption of principles of design that drew upon classical forms as a universal, timeless and essentially anti-historical aesthetic (Placzek 1965; Wittkower 1971 [1962]). The neo-classical architecture and spatial forms developed in Annapolis were informed by a belief that that the principles of perfection lay not within the ideas of men but within the principles and rules of nature. Moreover, as the private elite landscapes built in Annapolis during the Revolutionary era materialised this perspective, those who were building them were also claiming that a new and just society following the same natural principles could also be built under their guidance. These constructions resolved a contradiction between the reflections of hierarchy and monarchical rule embedded in Nicholson's city plan and the new relationships between citizens and new rationales for the right to rule that would develop leading up to and following the Revolution.

The William Paca House and Garden (Figure 10.2) is the most well known of these landscapes both in the archaeological literature (Leone 1984; Leone 2005: 63–73) and physically within the city (as the centrepiece of the Historic Annapolis Foundation's properties). The Paca mansion was constructed between 1763 and 1765, and was renovated in the 1960s as a house museum (Wright 1999). The garden was rebuilt during the restoration process, as it had been buried by the construction of a hotel during the early 1900s. Archaeological research identified several major surviving features of the garden, such as the foundations of the brick garden wall, portions of the surface topography within the planned wilderness in the furthest part of the garden, a bridge over a pond, several lined drainage features and a garden outbuilding that might have been a spring house. These features guided the garden's 1960s renovation (Shellenhamer 2004).

Mark Leone's analysis of the Paca Garden in the 1980s focused on the use of the rules of perspective. As with other elite gardens built in the Chesapeake, the Paca Garden consisted of a terraced landscape comprised by five falling terraces containing *parterres* (level spaces occupied by ornamental flowerbeds). Paca's townhouse was located on the highest terrace, overlooking the other terraces descending towards a bridge that rose over a pond and a small structure known as the summer house. The rules of perspective were applied to this garden in a very strict manner. As the garden fell, each terrace was narrower than the one above it, except for the lowest area that comprised a curvilinear 'wilderness'. Furthermore, the angle and drop in elevation from terrace to terrace was not uniform but varied, with the upper terraces being higher above those below them and having steeper drops than the lower terraces. A viewer could not see these effects because upper terraces obscured parts of the terraces below them, and this created an illusion that the wilderness at the base of the garden was further from the main house than it really was. From the wilderness, the house appeared further away and grander. The garden employed specific proportional ratios between the terraces and the house that were so precise that *parterres* within the garden were the proportional equivalent to rooms inside the house. The use of these precise and harmoniously proportional measures lay behind the success of the garden's spatial illusions. Variance from this geometry would have thrown off the illusion and ruined the garden's effect of tying the whole landscape together (Leone 1984; Paca-Steele and Wright 1987).

In all, 14 properties were developed along these lines by the Maryland elite in Annapolis between 1763 and 1774 (Leone and Shackel 1990). Whilst each had its own individual character, they

Figure 10.2 View of the William Paca Garden from the lowest terrace. *Photograph by Jason Shellenhamer.*

followed the same basic geometric rules found at the Paca Garden to create perfect and integrated forms in architectural and garden space. These new constructions radically changed the look of Annapolis just before the beginning of the Revolutionary War. They created a new urban landscape within the Chesapeake region.

To understand these polite constructions of houses and gardens we need to consider the men who built them. William Paca, Charles Carroll of Carrollton and Samuel Chase were not residents of Annapolis prior to building their townhouses there. They were, however, leading figures of a dominant planter class that had ruled over Maryland society for the previous two generations. This generational depth is one of the key issues explaining interest in building at Annapolis in the 1760s. Class formation in the Maryland colony was incremental. Wealth was generated not only through the tobacco plantation economy, but also through the ability of a fortunate few to acquire sufficient holdings to pass on an inheritance. With an inheritance, the children of some early settlers became a 'native elite' (Jordan 1979). This status allowed those individuals to start where many of their neighbours hoped to end up: as owners of the land, buildings, tools and especially labour that made tobacco planting extremely profitable. Because few early settlers actually succeeded (a result of disease, short life, late marriage and debt relations tied to servitude and later class standing), the native elite easily secured positions of authority, and the most powerful of these were enlisted to act as local creditors.

Wealth, especially when equated with education, taste and leisure time, prompted the native elite to become active consumers of luxury goods imported from Britain. These goods were available from London merchants who took tobacco futures as payment. Poorer planters in Maryland were eager to gain access to such products as well, and the wealthier planters therefore became middlemen extending credit to their neighbours in exchange for their crops. The wealthy thus inserted themselves as lynchpins in the local economy, making their private market interests into the common interest of all of those indebted to them.

Another product of the planter-dominated system was that it limited urban growth (Kulikoff 1986). The presence of navigable rivers and a centripetal, almost private, economy meant that there was no need for market towns for transport or redistribution. Annapolis was a small port serving the colonial government with taverns and inns and very poorly financed minor industrial ventures such as tanneries and ropewalks. In the absence of the activities of the colonial government, by the mid-18th century the previous capital St Mary's City had been largely abandoned (Leone and Hurry 1998). This situation changed for many Maryland towns, however, with the arrival in the mid-18th century of Scottish merchants, who challenged the status quo by setting up independent stores in many of the Chesapeake's few towns. Their stores were well-stocked and offered generous credit to smaller planters. By the 1760s, these attractions turned places like Bladensburg, a crossroads settlement 25 miles west of Annapolis, into active market towns with several merchant stores, taverns and artisan and smith shops. The availability of generous credit spurred new ways of imagining class in the Chesapeake, and smaller planters in turn developed a more self-confident awareness of their place in these emerging economic situations. This shift was already visible in the widespread rejection of the planter-dominated Anglican church during the Great Awakening, in which the overbearing control of the vestries by the elite was often cited (Isaac 1973).

The emerging alternative consciousness of smaller planters was not readily tolerated by the colony's ruling class. The legislature in Annapolis had been attracting leading planters from the ruling class since the early 1700s, but during the mid-century the new threat to the gentry system led many leading planters to the city to secure their authority. The gardens and townhouses described above were essential components of their performance of authority, and its retention. The creation of these new landscapes developed alongside a new engagement with colonial politics for the sake of Maryland people: an engagement that culminated in the Revolution. Men like Paca, Carroll and Chase came to Annapolis as elected officials representing the rural districts their families had constructed. Yet, while their fervour in participating in colonial politics was tied to the need to perform as representatives, they were for the most part preserving their own interests.

This elite performance of authority engaged with discursive and material cultures. In print and speech, these men attacked the arbitrary nature of British colonial authority. James Otis' famous 'taxation without representation' stance against the Stamp Act of 1765 was part of this wider colonial resistance strategy (Morgan and Morgan 1995). In place of British domination, American politicians began to propose a political order that they claimed was based in natural law, and that would surpass the merely historical events of the everyday to discover the underlying principles that guided all life (Leone 1984). These principles drew on the ideology of the Enlightenment, which argued for the possibility of a just human society in place of the arbitrary or dogmatic rule of the few. This ideology was built into the new constructions in Annapolis by the rural elite. Their new landscapes demonstrated that men could indeed construct perfect forms, and contrast culture against wilderness. Thus, the guidebooks for building gardens like Paca's urged masters to guide their peers through the spaces, explaining the illusions of perspective and proportion. These performances in designed landscapes showed that the illusions were not meant to deceive but to reveal, below the surface of everyday life, principles that brought order to both the natural and the social world.

Those who understood these principles demonstrated their suitability to lead society, offering those who followed the security of knowing that the society being built was not based in the interest of the few, but on the common, natural associations that all persons had with each other. This ideological construction was supplemented by a solidarity-building claim by the ruling elite that they too suffered under the colonial system. They cited their position as debtors to large British merchant concerns and their equally difficult times in functioning under the limitations of the new Parliamentary taxes. This appeal to a common stance with their poorer neighbours recast class distinction as a common national interest. With the support of the common planter and artisan, and their demonstration that a new natural society could be built, these men led the American colonies into the Revolution.

The reward of this effort was for Paca immediate, as he 'inherited' the Governorship of the State of Maryland. He and his peers had wrested from the colonial authority any controls over their accumulation of wealth and power, while for the common planter, they established a system that appeared to have few barriers to personal independence. For Annapolis, however, the reward of independence was limited. The city was subject to the interests of those who subsequently used it as a backdrop for their independent progressive advancement, and the creation of a newly designed urban landscape was central to that process. As we argue in the next section, the significance of the Georgian landscape continued after 1776, into the 19th and 20th centuries. Its preservation was the agent of later change.

One important additional key to the Georgian landscape in Annapolis is the society it contained within itself. The Annapolis population has never been less than one-third African-American, and at the time of the Revolution the overwhelming majority were owned as slaves in large part by elite men like William Paca. Clearly, slavery and slave-ownership could not achieve legitimacy given the statements of liberty and equality embedded in the rhetoric of the Revolution, though Wolff (2005) reminds us that these qualities of personhood were never extended to the enslaved people who laboured throughout the colonies (see also Morgan 1975; Blumrosen and Blumrosen 2005). This problem became a factor of concern in the years between the Revolution and the American Civil War in 1860. However, the colonial landscapes built by slave owners included the place of the enslaved in the natural order: slaves were outside of society. As non-citizens they were, in the nature imagined and built by Paca, also non-persons (Frederickson 1987). To understand the gardens required the ability to see oneself as their appropriate subject; the viewer was the self, and simply to view them was an act of personhood (Leone 2005: 83). The enslaved were conceived in this order as the opposite of the garden viewer. They did not require techniques for self-discipline that might encourage understanding citizenship. Slaves were subjects produced by external legal discipline, and their violent suppression was defined in large part as a result of their attempt to assert personhood. The meaning of the Georgian landscapes is thus implicated in the exclusion of those unprepared for understanding how to use them. The brick wall around the Paca Garden is a solid form of this practice. Imagining the wall being built by Paca's slaves drives home the level of the assault on their personhood these constructions entailed (see Leone 2005: 199–236 for a discussion of alternative African-American landscapes in Annapolis).

BUILDING WITHIN HISTORY: 19TH CENTURY REINVENTION

During the 19th century, Annapolis languished as urban life in Maryland found a new focus at Baltimore. Annapolitans rationalised their marginality in the late 19th century by transforming their image of the city, drawing on its principal capital resource: its heritage (e.g. Riley 1995 [1887]). Like miners digging for coal or fishermen dredging oysters, the Annapolis elite extracted history from the landscape. It was a commodity available in vast quantities and to reap its benefits required, so it seemed, only the effort of digging it up and organising it for presentation to visitors. Income was not the only benefit: it was imagined that *being historic* could also be a viable approach to modernising the city. The effort to build a historic Annapolis for tourism, however, has had an enormous impact on the way lives can be led in the city.

As the first governor after the Revolution, William Paca presided over the building of a new state order, but the Annapolis he inherited was hardly prepared for growth in modern terms. By comparison with other towns, including the post-Revolutionary Virginian capital of Richmond, Annapolis lacked the merchant and labour population base required for growth in a market economy. Most working people in Annapolis were professionals, a group consisting predominantly of lawyers and bureaucrats (Matthews 2002: Table 1). The business of Annapolis was government, and the city was more a stage for political debate than a crucible for industrial production. Absent until the later 19th century were those interested in developing the infrastructure for a modern economy, which led Annapolis to fall rapidly behind other cities in terms of capital investment and value-added production.

Evidence of these effects is found in early-19th century Annapolis newspapers, which describe in detail the struggles surrounding productive growth in the state. The capital interests in Baltimore regularly appealed for state funding to support their commercial interests through initiatives in infrastructural investment. The most controversial event occurred in 1833 when the Chesapeake and Ohio (C&O) Canal Company sought to gain exclusive rights through a western

Maryland mountain passage known as Point of Rocks. Driven by the commercial efforts of Washington, DC, the C&O canal was in competition with the Baltimore-based Baltimore and Ohio Railroad. The state supported a shared passage that ultimately led to the canal's failure, as the railroad line was faster and considered more progressive. This decision also ensured the cancellation of a plan for a branch canal from Washington, DC to Annapolis that was envisioned to lead to commercial growth, by making Annapolis into an urban centre for southern Maryland. This failure was matched by a handful of similar episodes of underdevelopment that left Annapolis largely unchanged for as long as 50 years after the Revolution (Hurst 1981; Matthews 2002: 73–76).

Southern Marylanders in particular offered little support for development in Annapolis. Rural planters, most of whom were still slave owners, saw no advantage in developing a commercial hub in their region. They were by the 1830s already connected by steamboat to Baltimore and Norfolk, Virginia, so the creation of a new local commercial centre would simply be the introduction of a middleman into a system that was already working well. Furthermore, they were rightly concerned that the development of an urban centre based on modern commerce would challenge their paternalistic authority. Without a local alternative for work or for escape, enslaved labourers would remain dependent on their masters. As in medieval Europe, cities in general were mistrusted by the southern patriarchy as they encouraged freedoms that, as historian Barbara Fields has suggested, led 'slavery to be an attribute of individuals, not any longer of the system that organizes their labor' (Fields 1985: 52).

Annapolis was thus marginalised from both the progressive development of the north and the patriarchal conservatism of the south, and as a city with a premodern social structure and labour system, it thus had to find its own distinctive and independent way to exist in the modern world. This involved bringing the country into the city. This not only supported the city's position in the region, it also extended the inadvertent preservation of its landscape. Annapolis leaders did this in two ways. First, Annapolis emphasised itself as an alternative city form to the crowded, noisy, urban character of northern cities. Second, it found a way to make use of its landscape and its slim productive capacities to serve and profit from urban growth elsewhere.

Trying to keep pace with Baltimore, which had become known as the Monumental City because of its several monuments (including the first built to honour George Washington), Annapolis adopted the nickname 'the Ancient City' in the 1820s (Potter 1994; Matthews 2002: 63). Being 'ancient' in the sense of having a dignified association with the classical origins of republicanism and democracy, it was implied, brought a necessary corrective to unchecked urban growth. Its landscape represented a separate urban space that allowed for contemplation on the virtues and values of American democracy, but which required recognition from outside in order to flourish. These distinctive aspects of its urban form were emphasised in the pages of the *Maryland Gazette,* which printed a host of letters and articles describing Annapolis' historic landscape in glowing prose. One letter printed 2 May 1833 describes the campus of the city's St John's College as having '[o]ne venerable poplar tree standing, which has borne without hurt the blasts of centuries, and handsome rows of trees have been planted which will shortly spread their umbrageous branches over the grounds and give a delightful shade to citizens, strangers and students. No place can be more interesting or more healthy.' This appeal to the qualities of the historic landscape to be found in Annapolis took root and guided the development of the city as it straddled its marginal position between urban and rural Maryland.

In the realm of production, Annapolitan landowners turned to truck farming (market gardening), which was blossoming across the USA in the early 19th century as its growing urban centres became eager markets for fresh food. Several sources from Annapolis show truck farming's effects on the city and its landscape. Mid-19th century maps are often cited as evidence for the city's stagnant growth, but they also show the preservation of open land that supported truck farming around the commercial strip along Main Street. This activity kept valuable open space intact while also making use of the city's underemployed, unskilled labour force to nudge a way into the modern urban economic system. Figure 10.3, for example, shows an African-American labourer hoeing a small garden plot behind the William Paca house, which along with several adjacent properties remained large urban lots available for cultivation. From the 1830s newspapers started printing advertisements for seed and fertilisers as well as short promotional articles on the demand for products such as 'Annapolis Strawberries'.

The archaeological record supplies additional evidence of the 19th century rewriting of the Annapolis landscape. For example, fieldwork at the Bordley-Randall site, adjacent to State Circle, identified evidence of a radical landscape transformation associated with agricultural produce production. Excavations in the front yard of the late-18th century house revealed that during the immediate post-Revolutionary era a crushed-shell cart path ran across the site, connecting the stable to the kitchen wing. This path was buried in the 1840s when the property was acquired by Alexander Randall, a prominent local lawyer and politician. While burying the path served to eliminate a landscape feature related to productive household labour, the effort was part of a rational reworking of the landscape tied to truck farming. This particular example involved the elimination of productivity from the front of the house, which was reserved for the expression of elite refinement, and the development of the rear space for produce production. Excavation in the backyard identified a widely distributed deposit of pea-gravel that suggested a pattern of garden beds lined with walking paths (Matthews 2002: 87). In his extensive diaries after settling at the property, Randall recorded his annual produce production efforts, including what he sowed, the various gardeners he employed and his concerns about the weather and the market (Matthews 1998a: Appendix B). Randall later expanded this farming effort by purchasing three farms outside of the city and leading the effort to establish a cannery in Annapolis in the 1870s.

Such developments capitalised on what Annapolis had to offer without the requirement of additional investment. An unskilled enslaved labour force, large elite urban landholdings and a ready market for truck farm products in turn sustained the Annapolis elite landowning class in the city despite its awkward position in the region. Increasingly,

Figure 10.3 African-American gardener behind William Paca House, 1860. Photographer unknown. *Courtesy of the Maryland State Archives, Special Collections (Marrion E. Warren Collection), MSA SC 1890-2473-1.*

however, the economic gains that the Annapolis elite drew from its landscape came to be matched in importance by its symbolism, a process that intensified after the establishment of the United States Naval Academy on the site of a former fort on the banks of the Severn River in Annapolis in 1845. The Academy was a powerful federal presence that brought a great deal of attention and funding to Annapolis; and federal investment, building and provisioning for the Academy also brought the city substantial social capital. The Academy also offered unwanted competition to the local landed elite, spurring them to intensify the production of a distinctly Annapolitan identity tied to local heritage in an effort to stand apart. Parker Potter (1994, 1999) has remarked on the clear separation between the historic district and the Naval Academy in contemporary Annapolis. This distinction has its roots in the 1850s, immediately after the Academy was established.

The well-documented activities, and buildings, of Alexander Randall illustrate the changing attitudes to the historic landscape of Annapolis. During the 1850s Randall not only used his large plot for gardening, but, in order to increase the size of his manor house in part to make room for his 11 children, he also added a two-storey addition to the back of his house. The original house had been built by Thomas Bordley in the 1720s but the façade of Randall's addition copied a manor house that had been built in the city by the Acton family in the 1770s. In copying the Acton façade, re-creating an architectural feature that post-dated the standing structure it was attached to by 50 years, Randall was paying homage to Annapolis' Revolutionary roots, and was clearly citing not just a generic 18th century Annapolis past, but its particularly significant Golden Age.

At nearby St Anne's Cemetery Randall built another memorial, this time to his own family. After the untimely death of his wife and three of his children in the early 1850s, Randall purchased and built a new family burial plot. Besides sites for the children, however, Randall reburied his parents in the new plot in a crypt that he also planned for himself: a dynasty-creating act that connected the dead with the living in a particularly powerful way. It was not just architecture and landscape that made Annapolis history important, it was, especially for the sake of newcomers, the fact that the Randall family had a long-term residence in Annapolis, were 'old Annapolis' and would remain that way. Planning for his own burial of course only reinforced this carrying of the past into the future. What better way to establish the symbolic meaning of history in a place than to establish that their lives *and* deaths were tied to the very building of the landscape?

Through such acts of curation and reconstruction, therefore, an Annapolis with a new distinctive identity was created during the mid-19th century. The very particularly constructed historic Annapolis witnessed continued efforts to maintain the integrity of this urban landscape, efforts that involved continual acts of writing and rewriting the historic landscape. This textual metaphor extends easily into recent history and the movement to preserve the historic fabric of the city against overdevelopment tantamount to destruction. In the later 20th century, activities of the historic preservation movement continue to reveal this active process of place-making.

BUILDING TO RENEW HISTORY: 20TH CENTURY 'BLIGHT', URBAN RENEWAL AND HISTORIC PRESERVATION

During the 20th century, the desire for a historic city conflicted with the equally vigorous desire for Annapolis to be a progressive, modern city. A contributor to *Harpers Weekly* wrote of Annapolis:

> The objection to having a past is ... that it sometimes involves having no present. In ancient towns which it pleases susceptible writers to describe as fragrant with memories, the quaint inhabitants, sickening of picturesqueness, are generally to be found reviling their fate and longing for the vulgarity and newness of some nearby factory city. (Rhodes 1919: 641)

Some of the greatest entrepreneurs in Annapolis during the mid-20th century, many of them small business owners, had little regard for its 'memory infrastructure' (Mason 1999) of historic buildings. To the horror of preservationists – who by then had developed a particular landscape view in which the modern period had little place – merchants along the city's primary thoroughfares for automobile traffic applied a veneer-like surface of neon light to the Federal-period commercial buildings. Lighted signs projected over city streets, shone out from store windows, flashed like marquees or were fixed flat against building façades with no care or fondness for the ancient quality of the architecture, both in the Annapolis downtown area and along its major approaches from Baltimore and Washington. The vivid illumination of neon completely changed the impression of Annapolis streets, which were now encountered as a corridor of signs, at night a brilliant, chintzy white way. Brick structures that once themselves signed the ancientness of the 'Ancient City' were made into decorated sheds (Venturi et al. 1977), obscured by more clearly legible texts traced in light. Contributing in a more insidious way were the utility lines that powered this and other conveniences of mid-20th century America. Power and telephone lines formed a web-work above every single street and intruded into every vista, including every view of the State House and historic homes of 18th century patriots like William Paca, Charles Carroll and Samuel Chase. While neon was faddish, this infrastructure was more difficult to argue with.

Preservationists nonetheless struggled to undo these 20th century transformations, and this conflict came to a head during the 1960s, when the city began to receive funds to undertake urban renewal. Preservationists in particular saw an opportunity to use funds designated for urban renewal to bring about sweeping changes in the city and reverse the impact of the development of commercial street architecture. According to their plan, the downtown of the city would be administered as a historic district with stringent standards and protections. Overhead utility lines were considered especially damaging to the historic atmosphere of the city, and between the 1960s and 1990s vast resources – not to mention some 30 years of planning effort – were expended to relocate them into buried conduits beneath the streets and sidewalks. Much of this work was undertaken with support from the Department of Housing and Urban Development (HUD) under various projects made possible by federal housing legislation that sponsored urban renewal efforts in many American cities.

Urban renewal and progressive city planning in the United States during the 20th century entailed a dramatic engagement with place and the history of places. Historic preservation gained ground during this period with an approach that complemented progressive urban planning and existing rationales for urban renewal. Urban renewal saw dozens of capital projects in Annapolis, and money was frequently channelled from the city's Urban Renewal Authority (URA) into projects proposed by agents within the preservation movement. By seeking to demonstrate that historic preservation was an alternative to redevelopment that could also arrest urban decline and population flight, advocates for preservation gained considerable power in the city. Their efforts framed the historic landscape, including its reconstruction, as a means to revitalise the town's commercial economy at its centre.

In other cities, critics of 1970s and 1980s urban renewal in the United States have shown that an inordinate amount of decision-making power was left to municipalities where implementing renewal and redevelopment was concerned (Anderson 1964; Foard and Fefferman 1966; Saunders and Shackelford 1998; Schuyler 2002). Vast sums of money were invested for the purposes of renewal, yet aside from procedural guidelines all spending decisions were made locally. For many small cities, as for Annapolis, this was the first impetus to develop consistent city-wide zoning or master plans for the future development of their community. Such nationally organised and funded urban renewal programmes often provided opportunities for local historic preservation initiatives. This was certainly the case for Annapolis, where responses were influenced by the legacy of a perceived landscape on historical consciousness that guided the interest of the local elite.

In particular, the idea of urban blight was appropriated by leaders in the historic preservation movement, and was used to consolidate political power as well as public opinion. A woman named Ann St Clair Wright played a central role in this campaign and led the city's preservation activists during this period. Her public statements to the preservation community and to the press, as well as letters to the Mayor and City Council of Annapolis, agencies in the Maryland state government and members of the Maryland Senate, introduced the notion that the disfiguring utility lines amounted to a blight upon the city – a curious adaptation of a problematic word more often implying urban areas plagued by unemployment, crime, poverty and neglect (e.g. Wright 1963; original correspondence on file at the Historic Annapolis Foundation, Annapolis, MD, and archived in Ann St Clair Wright's papers in the Marylandia Collection, University of Maryland College Park). Preservationists declared a blighting influence not in areas of Annapolis that were historically neglected, but in the city's centre, which had received the most attention in terms of development and infrastructure in the project to modernise the city that lasted since the middle of the 19th century (Palus 2005).

The recent history of the movement to remove above-ground utilities from Annapolis streets illustrates this broader process of the creation and reworking of the historic landscape: the efforts to clear up what was called a 'wirescape' by Michael Dower of the English Civic Trust (Dower 1963), who consulted with Annapolis preservationists (Figure 10.4). The first attempt to put public utilities in Annapolis underground came in the early 1960s. In correspondence between the executive director of Historic Annapolis, Inc. (then the leading preservation advocacy group in the city, if not the state) and Dower at the Civic Trust in London, the possibility of redesigning declining urban centres such that they might compete with out-of-town shopping malls, one of which appeared on the margins of Annapolis around this time, was discussed. Preservationists sponsored multiple screenings of a film produced by the Civic Trust called *New Face for Britain*, which detailed the success of several such projects (Board of Directors 1962; Greenwood 1961; Kerr 1962). *The Evening Star*, a Washington, DC newspaper, announced that the 'British Renewal Plan Could Set Pattern Here' citing success at Norwich and Windsor in the UK (Anonymous 1962). Ann St Clair Wright, who presided over the Board of Directors at Historic Annapolis, Inc., introduced the film with lengthy comments about the condition and appearance of the major approaches to downtown Annapolis (Wright n.d.).

A picture was presented of visitors arriving in Annapolis by boat or by car, who encountered a forest of utility poles and their dense canopy of wires, alongside storefronts concealed by signs lit by neon coloured green, pink, red, or neon that didn't work, signs hanging over sidewalks, signs mounted to buildings, signs in windows advertising 'Pepsi Cola', or reading 'Air Conditioned for your Comfort'. The film screening and subsequent debates generated sufficient excitement for an initiative in 1961 to give the city's commercial streets a face-lift. Business owners along Main Street were to take relatively simple steps to improve the appearance of storefronts, the city was to replace street furniture such as benches and trashcans with something more suitable to the historic locale and utility companies were to relocate overhead wires underground. These steps were described in an unfulfilled plan prepared jointly by an association of residents called the Committee for Annapolis, the Chamber of Commerce and Historic Annapolis, Inc. (Annapolis Citizens et al. 1961).

Until the approach of the American Bicentennial in 1975 there was little progress on the kind of sweeping changes that preservationists envisioned (Apostol 1973). In the early 1970s the campaign to move public utilities underground was renewed in combination with related efforts at restoring street surfaces to more appropriate historic paving materials – another initiative to improve the face of things in order to prepare the city for what was anticipated as a nationwide celebration. At this point members of the state legislature allied with the preservation movement also involved themselves in the drive to place public utilities underground, and a law was drafted to address the issue in

Figure 10.4 Power lines along East Street in the historic district of Annapolis, with the Maryland State House in background. *Photograph by Matthew Palus.*

1974. This public law enabled some municipalities to establish zoning that required all *new* utilities to be placed underground. It soon became clear that the real expense of putting utilities underground would come not so much in the materials and necessary excavation but in the replacement of pavements and sidewalks. A budget prepared by the office of the Mayor of Annapolis in 1975 estimated a cost of more than $36 million over five years, from 1976 to 1980, to 'repair' roads and sidewalks with bituminous concrete or asphalt materials, while $60 million would be needed if the repairs used brick. The first phase of this project, undertaking work on only the choicest streets around the State House in Annapolis, was projected to cost around $2 million, plus $800,000 for State Circle and adjacent streets to be repaved in brick (Apostol 1975).

It was at this point that Historic Annapolis, Inc. also submitted a programme request for HUD funding to replace electric lights within the historic district with replica 18th century street lamps, which would themselves run on electricity but would support the ambiance preservationists sought to project downtown. The grant request submitted to HUD for partial funding described the project in stark terms: 'The scenic city ... is presently marred by unsightly utility poles and lines.... Placing the utility lines underground will enhance this scenic area and inspire citizens of all ages with the beauty of their community' (Wright 1975: 2).

Negotiations over how the expenses of moving public utilities underground would be met were intense, and involved a state agency responsible for adjusting rates charged by public utilities called the Maryland Public Services Commission. Utility companies entered into prolonged dispute over how much expense could be passed on to consumers, and how any rise in rates would be structured (Murphy 1975). Preservationists argued that because the changes would produce benefits across the state, owing to the historical importance of Annapolis for all of Maryland, rates should be adjusted throughout each utility company's customer base. Perhaps attempting to affect public support for the project, utilities pushed to adjust rates only for those living in areas that would be improved by the project. In the ensuing delay, HUD funding expired and evaporated. It was not until the mid-1980s that the underground cables project received funding from the State of Maryland that matched funds offered for the project through HUD's Community Block Development Program (City of Annapolis 1987). In time, city and state legislators came to employ the same terminology as preservationists: as one Maryland senator appealed to a state commissioner of public services, 'This area is blighted by the overhead lines and numerous visitors from Maryland, other States and overseas have remarked about the unsightly lines' (Winegrad 1985). In 1983, the projected total for all phases of work was $24 million for 'undergrounding' alone, with an additional $2.8 million to resurface streets and sidewalks in concrete and asphalt, or $19 million to repair streets and sidewalk in brick (Smith 1983). This work got underway in 1990 and continues to the present.

CONCLUSIONS

The continuing efforts to remove the 'wirescape' from the city of Annapolis is the most recent illustration of the complex negotiations involved in re-creating the urban landscape. The ongoing efforts by leading figures in Annapolis to control the city through defining the meaning of its historic landscape are part of a long line of conscious decisions since the 18th century, if not from the founding of Annapolis itself. Introducing their position as a debate over how to handle the 'blighting' of the historic landscape, preservationists proclaim their interests are the common interest of all Annapolitans who want to live in a prosperous, progressive community. William Paca and his contemporaries guided visitors through their city and its gardens to reveal the effects of their illusions that created the natural justification for their wealth and power; today preservationists walk us through Annapolis with the intention of revealing its hidden (but in their minds inherent) meaning and value. We suggest that this value is intentionally left ambiguous and underdetermined. It is at best what Potter (1994) describes as the inherent significance of history in the Annapolis landscape: that history is part of the nature of Annapolis. This effort mirrors that of the Revolutionary elite. Both then and now elite power lies in the claim not only that they have the best interests of Annapolis at heart, but that their public efforts and sacrifice, which on their own add up to a suite of credentials, establish them as representatives of what Annapolis truly is and should be.

We have considered how a landscape known today as 'historic Annapolis' was constructed. By tracing a series of particular efforts very closely bound to the local and regional political economy, we have aimed to show that the presentation of what is significant in Annapolis' history represents only a small part of the complex historical development of the urban landscape. In some ways, this is possible due to the city's stagnant economic growth over the last two centuries, which has led to the largely accidental preservation of its historic fabric. However, it is also tied to a persistent and successful effort by city leaders to maintain their authority by citing the significance of historic Annapolis to larger regional and national concerns, and by demonstrating their authority and power on the landscape. Today, this effort consistently establishes the preservation elite as a legitimate mediator between Annapolis and outsiders.

To criticise this stance we have shown that, like all modern urban landscapes, Annapolis has been constructed and re-constructed continuously since its founding. These efforts have been justified by a narrative that has argued for the preservation of the inherent qualities of the place. Of note is that these qualities, while presented as inherent, are clearly constructions of the moment as each generation cited their own version of its characteristics. The late-17th century founders reflected the precepts of the 'Glorious Revolution'; Revolutionary leaders built the perfect forms of the classical age; 19th century leaders built memorials to their own legacy; and preservationists rebuild the city to capture again its scenic, photographic charm, a meaning largely void of any historic content other than the aura of a significant past for it own sake. We maintain that Annapolis has experienced qualitatively the same issues surrounding modernisation as other American cities. The difference is merely one of scale. While cities are consistently rebuilt through demolition and reconstruction, much also persists, whether by default or through active preservation. In each instance described here a new Annapolis landscape was built for its residents to live in and for outsiders to visit: whether through the removal of more recent features such as above-ground cables, or in reconstructions. Through these constant negotiations, the ruins of Annapolis survive and emerge.

ACKNOWLEDGEMENTS

The Historic Annapolis Foundation in Annapolis, Maryland, allowed Matthew Palus to survey archived correspondence in 2001–2002 as part of a state-funded project to record oral histories of the historic preservation movement in Annapolis, which was funded by the Maryland Historical Trust. Portions of this essay could not have been written without access to those materials, and we are grateful to the Historic Annapolis Foundation and the Maryland Historical Trust for their support.

REFERENCES

Anderson, M. 1964. *The Federal Bulldozer*. Cambridge, MA: MIT Press.

Anonymous 1962. British Renewal Plan Could Set Pattern Here. *The Evening Star* 14 November 1962.

Apostol, J.C. 1973. John C. Apostol, Mayor of Annapolis, to the Legislative Council of the Maryland State Senate Economic Affairs Committee, 12 October 1973. Archived correspondence. Historic Annapolis Foundation, Annapolis, MD.

Apostol, J.C. 1975. John C. Apostol, Mayor of Annapolis, to George Lewis, Chairman, Council on the Capital City, Maryland Department of General Services, 1 December 1975. Archived correspondence. Historic Annapolis Foundation, Annapolis, MD.

Blumrosen, A.W. and R.G. Blumrosen 2005. *Slave Nation: How Slavery United the Colonies and Sparked the American Revolution*. New York: Sourcebooks.

Board of Directors, Historic Annapolis Foundation 1962. Memo RE: Proposed Commercial Street Revitalization Scheme, 3 December 1962. Archived correspondence. Historic Annapolis Foundation, Annapolis, MD.

City of Annapolis 1987. Comments of the City of Annapolis Presented to the Maryland Public Services Commission in Response to Order No. 67866 (In the matter of the application of Baltimore Gas and Electric Company for the determination of the method of computing and imposing charges for the net capital costs associated with the relocation underground of certain electric facilities in the City of Annapolis, Maryland). Archived correspondence. Historic Annapolis Foundation, Annapolis, MD.

Committee for Annapolis, Annapolis Chamber of Commerce, and Historic Annapolis, Inc. 1961. Pilot Proposal for Main Street Rejuvenation, Summer 1961. Archived correspondence. Historic Annapolis Foundation, Annapolis, MD.

Dower, M. 1963. Letter from Michael Dower, Civic Trust, to Robert J. Kerr, Historic Annapolis, Inc., 3 May 1963. Archived correspondence. Historic Annapolis Foundation, Annapolis, MD.

Fields, B.J. 1985. *Slavery and Freedom on the Middle Ground: Maryland during the Nineteenth Century*. New Haven, CT: Yale University Press.

Foard, A. A. and H. Fefferman 1966. Federal Urban Renewal Legislation. In J.Q. Wilson (ed) *Urban Renewal: The Record and the Controversy*. Cambridge, MA: MIT Press, pp. 71–125.

Frederickson, G.M. 1987. *The Black Image in the White Mind: The Debate on Afro-American Character and Destiny, 1817–1914*. Hanover, NH: Wesleyan University Press.

Greenwood, J. 1961. Letter from John Greenwood to Robert J. Kerr, Jr., 28 March 1961. Archived correspondence. Historic Annapolis Foundation, Annapolis, MD.

Hurst, H.W. 1981. The Northernmost Southern Town: A Sketch of Pre-Civil War Annapolis. *Maryland Historical Magazine* 76(3): 240–249.

Isaac, R. 1973. Religion and Authority: Problems in the Anglican Establishment in Virginia in the Era of the Great Awakening and the Parsons' Cause. *William and Mary Quarterly* 30: 3–36.

Jordan, D.W. 1979. Political Stability and the Emergence of a Native Elite in Maryland. In T.W. Tate and D L. Ammerman (eds) *The Chesapeake in the Seventeenth Century: Essays in Anglo-American Society*. Chapel Hill: University of North Carolina Press, pp. 243–273.

Kerr, R.J. 1962. Robert J. Kerr, Jr. to Michael Dower, 19 November 1962. Archived correspondence. Historic Annapolis Foundation, Annapolis, MD.

Kryder-Reid, E. 1998. The Archaeology of Vision in Eighteenth-Century Chesapeake Gardens. In P.A. Shackel, P.R. Mullins and M.S. Warner (eds) *Annapolis Pasts: Historical Archaeology in Annapolis, Maryland*. Knoxville: University of Tennessee Press, pp. 268–290.

Kulikoff, A. 1986. *Tobacco and Slaves: The Development of Southern Culture in the Chesapeake, 1680–1800*. Chapel Hill: University of North Carolina Press.

Leone, M.P. 1984. Interpreting Ideology in Historical Archaeology: Using Rules of Perspective in the William Paca Garden in Annapolis, Maryland. In D. Miller and C. Tilley (eds) *Ideology, Power and Prehistory*. Cambridge: Cambridge University Press, pp. 25–28.

Leone, M.P. 2005. *The Archaeology of Liberty in an American Capital: Excavations in Annapolis*. Berkeley: University of California Press.

Leone, M.P. and S.D. Hurry 1998. Seeing: The Power of Town Planning in the Chesapeake. *Historical Archaeology* 32(4): 34–62.

Leone, M.P. and P.A. Shackel 1990. Plane and Solid Geometry in Colonial Gardens in Annapolis, Maryland. In W. Kelso and R. Most (eds) *Earth Patterns*. Charlottesville: University of Virginia Press, pp. 153–167.

Leone, M.P., J. Stabler and A. Burlaga 1998. A Street Plan for Hierarchy in Annapolis: An Analysis of State Circle as a Geometric Form. In P.A. Shackel, P.R. Mullins and M.S. Warner (eds) *Annapolis Pasts: Historical Archaeology in Annapolis, Maryland*. Knoxville: University of Tennessee Press, pp. 291–306.

Mason, R.F. 1999. Memory Infrastructure: Preservation, 'Improvement' and Landscape in New York City, 1898–1925. Ph.D. dissertation, Columbia University.

Matthews, C.N. 1998a. Annapolis and the Making of the Modern Landscape: An Archaeology of History and Tradition. Ph.D. dissertation, Columbia University.

Matthews, C.N. 1998b. Part of a 'Polished Society': Style and Ideology in Annapolis's Georgian Architecture. In P.A. Shackel, P.R. Mullins and M.S. Warner (eds) *Annapolis Pasts: Historical Archaeology in Annapolis, Maryland*. Knoxville: University of Tennessee Press, pp. 244–267.

Matthews, C.N. 2002. *An Archaeology of History and Tradition: Moments of Danger in the Annapolis Landscape*. New York: Kluwer Academic/Plenum Publishers.

Miller, H. 1988. Baroque Cities in the Wilderness: Archaeology and Urban Development in the Colonial Chesapeake. *Historical Archaeology* 22(2): 57–73.

Morgan, E.S. 1975. *American Slavery American Freedom: The Ordeal of Colonial Virginia*. New York: Norton.

Morgan, E.S. and H.M. Morgan 1995. *The Stamp Act Crisis: Prologue to Revolution*. Chapel Hill: University of North Carolina Press.

Murphy, J.C. 1975. Letter from Assistant Attorney General John C. Murphy, State of Maryland, to Eugene M. Lerner, Annapolis City Attorney, dated 20 June 1975. Archived correspondence. Historic Annapolis Foundation, Annapolis, MD.

Paca-Steele, B. and A.S.C. Wright 1987. The Mathematics of an Eighteenth-Century Wilderness Garden. *Journal of Garden History* 6(4): 299–320.

Palus, M.M. 2005. Building an Architecture of Power: Electricity in Annapolis, Maryland in the 19th and 20th Centuries. In L.M. Meskell (ed) *Archaeologies of Materiality*. Oxford: Blackwell, pp. 162–189.

Placzek, A.K. 1965. Introduction. In A. Palladio *The Four Books of Architecture*. New York: Dover Publications, pp. v–vi.

Potter, P.B. 1994. *Public Archaeology in Annapolis: A Critical Approach to History in Maryland's Ancient City*. Washington, DC: Smithsonian Institution Press.

Potter, P.B. 1999. Historical Archaeology and Identity in Modern America. In M.P. Leone and P.B. Potter (eds) *Historical Archaeologies of Capitalism*. New York: Kluwer Academic/Plenum Publishers, pp. 51–79.

Reps, J. 1972. *Tidewater Towns*. Williamsburg, VA.: Colonial Williamsburg Foundation.

Rhodes, H. 1919. Annapolis and Annapolitans. *Harpers Monthly Magazine* 138(827): 641–654.

Riley, E. 1995 [1887]. *The Ancient City: A History of Annapolis, in Maryland 1649–1887*. Baltimore, MD: Clearfield Company.

Saunders, J.R. and R.N. Shackelford 1998. *Urban Renewal and the End of Black Culture in Charlottesville, Virginia*. Jefferson, NC: McFarland and Co.

Schuyler, D. 2002. *A City Transformed: Redevelopment, Race, and Suburbanization in Lancaster, Pennsylvania, 1940–1980*. University Park: Pennsylvania University Press.

Shellenhamer, J. 2004. *The Archaeology and Restoration of the William Paca Garden, Annapolis, Maryland, 1966–1990*. Report prepared for Archaeology in Annapolis. On file, Archaeology in Annapolis Laboratory, College Park. See also: www.bsos.umd.edu/anth/arch/PacaGarden/index.htm (last updated May 2004; consulted 1 July 2006).

Smith, D.L. 1983. Memorandum from David L. Smith to Richard Lazer Hillman, Mayor of Annapolis, 5 December 1983. Archived correspondence. Historic Annapolis Foundation, Annapolis, MD.

Venturi, R., D.S. Brown and S. Izenour 1977. *Learning from Las Vegas: The Forgotten Symbolism of Architectural Form*. Cambridge, MA: The MIT Press.

Winegrad, G.W. 1985. Letter from Maryland Senator Gerald W. Winegrad, to Frank Heintz, Chair, Maryland Public Services Commission, 10 September 1985. Archived correspondence. Historic Annapolis Foundation, Annapolis, MD.

Wittkower, R. 1971 [1962]. *Architectural Principles in the Age of Humanism*. New York: W.W. Norton.

Wolff, R. P. 2005. *Autobiography of an Ex-White Man: Learning a New Master Narrative for America.* Rochester, NY: University of Rochester Press.

Wright, A.S.C. 1963. Letter from St Clair Wright to L. Mercer Smith, Vice President of the Chesapeake and Potomac Telephone Company of Maryland, 15 March 1963. Archived correspondence. Historic Annapolis Foundation, Annapolis, MD.

Wright, A.S.C. 1975. Letter from St Clair Wright to Adm. (Ret.) James J. Stilwell, Annapolis Housing Committee Chairman, 22 December 1975. Archived correspondence. Historic Annapolis Foundation, Annapolis, MD.

Wright, A.S.C. n.d. Undated manuscript on file. Historic Annapolis Foundation, Annapolis, MD.

Wright, R. 1999. *A Report on the Preservation and Restoration of the William Paca House, 186 Prince George Street, Annapolis, Maryland.* Annapolis, MD: Historic Annapolis Foundation.

Yentsch, A.E. 1994. *A Chesapeake Family and their Slaves: A Study in Historical Archaeology.* Cambridge: Cambridge University Press.

CHAPTER 11

Colonialism and Landscape: Power, Materiality and Scales of Analysis in Caribbean Historical Archaeology

Mark W. Hauser and Dan Hicks

INTRODUCTION

This chapter considers the study of landscape and colonialism in the historical archaeology of the anglophone Caribbean. In doing so, it seeks to reveal both the potentials for, and challenges of, a 'postcolonial' archaeology of historical landscapes in the eastern Caribbean and Jamaica, as well as more generally in world archaeology. The idea of landscape was an important element in the development of post-medieval European colonialism, which imagined 'empty' landscapes especially through doctrines of *terra nullius* (unowned land): denying Indigenous property rights, creating new planned colonial landscapes and mapping and laying territorial claim to Indigenous land (Gosden 2004: 25–33). It is therefore remarkable that the study of colonial landscapes remains a relatively undeveloped area of archaeological research. Our central argument in this paper is that today landscape archaeologies and postcolonial archaeologies face a common challenge: the acknowledgement of material conditions and multiple geographical scales as well as purely ideational accounts of landscape or colonialism. Our uses of the terms 'colonial' and 'postcolonial' will overlap to some degree, as we follow Ann Stoler's observation that the postcolonial emphasis of contemporary situations and the studies of past colonial worlds necessarily collapse into each other (Stoler 2001: par. 2).

As the introduction to this volume argues, landscape archaeology is a complex field that 'refuse[s] to be disciplined' (Bender 2006: 304). In sociocultural anthropology as in prehistoric and historical archaeology, the Caribbean region represents an 'open frontier ... where boundaries are notoriously fuzzy' – a diverse and 'undisciplined' field that has rarely been able to contribute to metropolitan perspectives 'lessons learned on the frontier' (Trouillot 1992: 19–20, 35). After introducing the study of historical landscapes and colonialism in the island Caribbean, the chapter considers the parallel debates in socio-cultural anthropology, cultural historical geography and archaeology over how to weave together 'ideational' and 'material' approaches to both postcolonialism and the study of landscape. We then examine alternative approaches to power, domination and resistance in the archaeology of British colonial landscapes in the Caribbean, before introducing recent calls for a refocusing upon materiality and multiple scales of analysis in the study of colonialism and landscape, and discussing them in relation to alternative studies of 'landscapes' of mobility. Drawing examples from across the Caribbean (Figure 11.1), we hope to demonstrate that using landscape archaeology's ability to operate at multiple scales of analysis represents one important route through which Caribbean historical archaeologists might begin to develop a postcolonial archaeology – contributing to the efforts of an archaeology of colonialism to achieve 'a clearing of the ground in order to create new pasts to allow new futures' (Rowlands 1998: 332).

CARIBBEAN HISTORICAL LANDSCAPES AND COLONIALISM

The archaeological landscapes of European contact and colonialism in the Caribbean region are complex, ranging from urban, domestic, religious, commercial and maritime contexts to public buildings,

Figure 11.1 Map of the Caribbean showing the islands discussed in this chapter.

plantation landscapes, trading entrepôts, cemeteries and military sites (e.g. Barka 2001). They encompass Indigenous landscapes related to the Taino in the Greater Antilles and the Caribs in the Lesser Antilles, as well as the landscapes of African slaves and free black populations, European and Asian indentured servants, and European plantation owners, sailors, military personnel or, from the 18th century, European tourists. From the 1620s in the Leeward Islands and Barbados, and from the 1760s in the Windward Islands, the landscapes were increasingly cleared – of both vegetation and native populations – to make way for new plantation landscapes: a process that culminated in the extensive landscapes of sugar plantations (from the 1670s) and the quotidian violence of industrial and racial slavery (Hicks 2007). These plantations continue to dominate the contemporary landscapes, gradually covered over on many islands by the secondary regrowth of tropical vegetation, especially after decolonisation during the second half of the 20th century.

During this five-century sequence, Western ideas of landscape have developed in particular directions in the eastern Caribbean. Sociologist Mimi Sheller (2003) usefully discerns three broad historical phases in the European representation of Caribbean landscapes: a 17th century focus upon 'the "productions of nature" as living substances with particular kinds of utilitarian value' emerging from both establishment of early plantations and the collecting practices of early natural historians; an 18th century 'scenic economy', associated especially with the rise of sugar monoculture, in which 'tropical landscapes came to be viewed through a painterly aesthetic constructed around comparative evaluations of cultivated land versus wild vistas'; and a 19th- and 20th century 'romantic imperialism', emerging especially after slave emancipation, which returned to a stress on 'untamed tropical nature' which was now 'constructed around experiences of moving through Caribbean landscapes and of experiencing bodily what was already known imaginatively through literature and art' (Sheller 2003: 37–38).

FROM COLONIAL LANDSCAPES TO LANDSCAPES OF RESISTANCE IN CARIBBEAN HISTORICAL ARCHAEOLOGY

How might Caribbean archaeologists begin to work with Sheller's compelling sequence of the changing European ideas of landscape in the region? Such a task requires a broader consideration of the reception and development in archaeology of the diverse approaches to 'landscape' and 'postcolonialism' that have developed since the 1980s across the humanities and social sciences. One particular dominant model of postcolonialism has sought to examine 'colonial discourse' from the perspective of literary theory, influenced especially by the work of Edward Said (1978, 1993). The particular strengths of such literary approaches lie in their highlighting of the colonial dimensions of the production of knowledge: Said's notion of discourse sought to underline how European colonialism involved distinctive ideologies and technologies of knowledge construction as well as power and domination. Similarly, a focus upon 'reading' landscape developed from the influence of geographers such as Stephen Daniels and Denis Cosgrove (1988). In this vein, historical archaeologists have developed important studies of colonial discourse and ideas of landscape, especially from the perspective of interpretive archaeology. In South Africa, Margot Winer (1995) has demonstrated the role of picturesque pictorial and cartographic representations in the practice of 19th century colonialism in the Eastern Cape. Discussing Virginia and Ireland, Matthew Johnson has discussed how 'in colonial contexts ... the English settlers tended to perceive, if not actively construct, a *tabula rasa* whatever the reality' (Johnson 1996: 94).

In Caribbean historical archaeology, early attempts at examining the role of landscape in the processes of European colonialism were made by historical geographers, notably in David Watts' consideration of idealised sugar plantations (Watts 1987: 384–390) and Barry Higman's studies of plats (manuscript estate plans) of Jamaican plantations, dating from c. 1750 and 1880. In a series of papers (Higman 1986a, 1986b, 1987), and in his book *Jamaica Surveyed* (1988), Higman described how through surveying technologies and ideal models of designed plantation landscapes in the 'spatial economy' of sugar and coffee plantations, the location of the cane fields, the work areas, slave quarters or labourers' villages, slave gardens and the houses of overseers and planters were carefully arranged by plantation owners (cf. Higman 1999).

Watts and Higman highlighted how a new focus in the British Caribbean upon the management of plantation landscapes emerged during the first half of the 18th century, in tandem with a new literature on plantation management which offered advice on both the geographical positioning in the natural environment and the internal layout of an estate (Figure 11.2). Thus, Antiguan planter Samuel Martin's *Essay on Plantership* argued that a properly managed plantation 'ought to be considered as a well-constructed machine, compounded of various wheels, turning different ways, and yet all contributing to the great end proposed' (Martin 1765: 37), and included a focus upon detailed landscape design. Similarly, in his *The Jamaica Planter's Guide* (1823) Thomas Roughley not only recommended 'a well-contrived plan of the buildings, their relative, convenient and appropriate situations, one to another, should be digested, and laid out on a piece

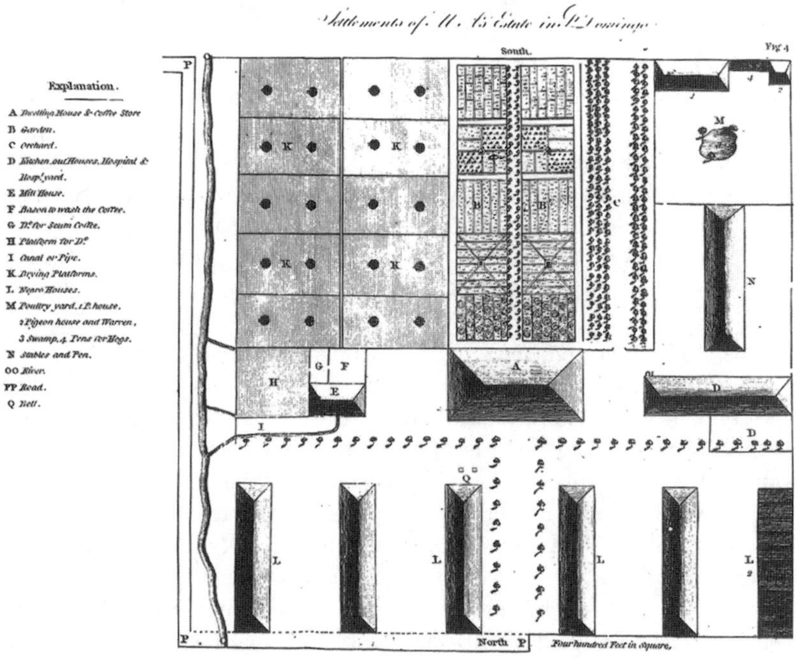

Figure 11.2 Ideal model of coffee plantation landscape. *From Laborie 1798.*

of paper, of a size sufficient to have the whole delineated upon it', but also:

> a situation, uniting within itself the blessings of a plenteous supply of wholesome water, on a piede of ground sufficiently large to admit building an extensive set of works, overseer's house, hospital or hot house, &c., with a large mill-yard, and being central among the surrounding cane cultivation, is a place most desirable. (Roughley 1823: 188)

Where Higman and Watts highlighted how the plantation map sometimes 'preceded the reality, enabling planter and surveyor to impose their ideal models of order on the landscape' (Higman 1988: 80), in historical archaeology and sociocultural anthropology such perspectives have since the 1980s been increasingly complemented by others who have emphasised alternative 'readings' of colonial landscapes. Indeed, the potential for colonial landscapes to be read in alternative ways was at the heart of Ian Hodder's critique of Mark Leone's classic archaeological study of the William Paca Garden in Annapolis, Maryland (see review by Hicks 2005). Thus, Barbara Bender has not only drawn attention to the partiality of the 'Western gaze', but has called for archaeologists to subvert it by mapping alternative, local perspectives upon landscape (Bender 1999). Discussing colonial history, 'the construction of place' and 'the politics of landscape', anthropologist Howard Morphy (1993) has demonstrated how alternative, Aboriginal constructions of time and space existed in white colonial landscapes. In historical archaeology, a more explicit response to 'ideational' approaches is Martin Hall's study of the order and form of the colonial built environment in South Africa and the Chesapeake (Hall 2000), which drew upon James Scott's distinction between 'public' and 'hidden transcripts' (J.C. Scott 1990) to accommodate alternative readings of landscape.

It is clear, then, that geography, anthropology and archaeology can highlight not only the role of landscape in the colonial process, but also moments at which colonial ideas are contested. Archaeological studies of Caribbean plantation landscapes have often directly addressed these issues of power, domination and resistance. In his study of the landscape of coffee plantations in the Yallahs region of Jamaica between 1790 and 1865, James Delle drew upon the work of Lefebvre (1991) and, to a lesser extent, Foucault (1979) to examine how on the plantation space represented 'a class of material culture ... used to manipulate human behaviors' and especially to establish and maintain conditions of inequality (Delle 1998: 37). Importantly, however, Delle combined this analysis with the suggestion that 'the schemes proposed by the planters never fully materialized' (Delle 1998: 217) and a focus upon how the 'spatialities' of the plantation were contested and resisted by slaves and workers (Delle 1998: 200). However, Delle's exploration of these themes was held back by his study's limited engagement with

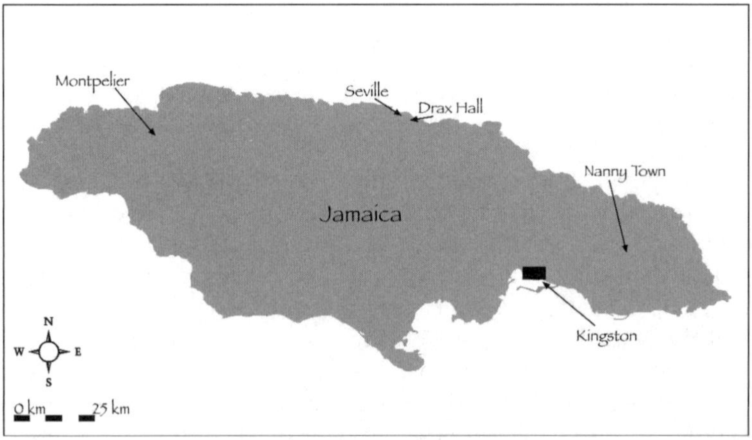

Figure 11.3 Map of Jamaica showing the locations of the Jamaican sites discussed in this chapter.

archaeological material, which was grounded in the study of historical maps of Jamaica pioneered by Higman rather than the use of the results of archaeological fieldwork.

Delle's perspectives were extended by Teresa Singleton's study of 'slavery and spatial dialectics' at the El Padre coffee plantation in Cuba during the first half of the 19th century (Singleton 2001). Singleton examines how the construction during the 1840s of masonry structures (*barracones*) in which plantation slaves' living areas were 'separated into prison-like cells' (Singleton 2001: 102) attempted to deter slave resistance by constructing 'spatial order', reinforced through surveillance, but how such initiatives were not wholly effective in the face of a range of 'everyday and cultural resistance' to slavery (Singleton 2001: 110).

A number of other studies have sought to identify 'African' dimensions of plantation landscapes by examining slave quarters or provision grounds, or sometimes in clear situations of active resistance such as Maroon settlements (Agorsah 1994). Some of the earliest work in this field was Douglas Armstrong's excavations at the 18th and 19th century slave village at Drax Hall in Jamaica between 1980 and 1983 (Armstrong 1990: 87). Building upon Sidney Mintz's (1974) account of the transformations of African culture in the colonial situations of the Caribbean, Armstrong's principal focus was upon material culture recovered from house yard contexts, which he used to discuss the 'lifeways' and foodways of slaves (Armstrong 1990: 133, 210). However, in subsequent work with Kenneth Kelly at the Seville plantation, a few miles from Drax Hall, Armstrong focused directly upon landscape by tracing shifts in household and village arrangements between the 1670s and the late 19th century, showing how house placement and architecture were 'governed by the choices and actions of African Jamaican residents', rather than solely by the plantation owner (Armstrong and Kelly 2000; cf. Armstrong and Hauser 2004). The relationships between archaeologies of the African Diaspora and archaeologies of landscape have been explored by the contributors to Jay Haviser's collection of studies, *African Sites Archaeology in the Caribbean* (Haviser 1999). The most successful work in this field presents comparative studies, such as Kofi Agorsah's use of archaeological and ethnographic material from Efutu in Ghana to inform the study of continuities and discontinuities in African traditions in the archaeology of the Maroon community at Nanny Town in the Blue Mountains in eastern Jamaica (Agorsah 1993). However, such approaches have often failed to engage with the complexity of African identities in the New World. Meanwhile, in Montserrat, cultural geographer Lydia Pulsipher has since the 1980s directed a major field project with Conrad M. Goodwin at Galways Plantation, the results of which have inspired studies of 'the human landscape surrounding Galways' (Pulsipher 1994: 203). Combining oral history with archaeological survey, they have explored the 'ideational role' of slave gardens as places in which slaves not only escaped the surveillance of the planter but also 'construct[ed] a decent life for themselves within a hostile system' (Pulsipher 1994: 217; cf. Pulsipher and Goodwin 1999: 15; Pulsipher and Goodwin 2001).

DISCUSSION: POWER AND MATERIALITY

The shift sketched above in Caribbean archaeology, in studies of both landscape and colonialism, from studies of domination to an acknowledgement of resistance and alternative voices, was influenced in part by a similar shift during the 1980s in both North American critical archaeology and British post-processualism, where analysis of 'Ideology' and 'Power' (Miller and Tilley 1984) was reoriented onto *Domination and Resistance* (Miller et al. 1987; Shackel et al. 1998). Since then, a number of scholars have criticised the focus upon 'power' that characterised all these studies. One critique has come from British archaeologist Sarah Tarlow, who in her review of Delle's (1998) study of Jamaican coffee plantations and some of the results of the Archaeology in Annapolis project (Shackel et al. 1998) complained that:

> little attention is given to the more subtle kinds of power, or different kinds of identity for which 'power' may not always be the most appropriate way of thinking.... [There are] particular problems in the consideration of intentionality.... [I]t is frequently implied or stated that the use of material culture, spatial ordering or architecture was part of a conscious strategy on the part of power elites to augment or maintain their social positions, or the tactics of non-dominant groups to resist élite ideologies.... [T]he language of tactics, strategy and intention is everywhere. This leaves us in a position where the only way to characterise the powerful capitalists is as pantomime villains, gleefully rubbing their hands and cackling as they dispossess the peasantry or mystify a few relationships of inequality. Similarly, all we see of the landless poor is their acceptance, rejection or redefinition of an ideology created by their 'betters'. This is something of a caricature ... but the serious point remains that we see people whose actions are always directed either above or below themselves and motivated by power interests. (Tarlow 1999: 468)

While the call for more subtle approaches to human agency is well taken (Johnson 2000: 230), it is hard to reconcile Tarlow's perspectives with the practice of archaeology in the colonial landscapes of the Caribbean and the Chesapeake. In these situations, the acknowledgements of the quotidian violence of plantation life, physical punishment, sexual violence against enslaved women and the role of both urban and rural landscapes in the horror of forced migrations and the emergence of modern racism represent crucial aspects of archaeological work (Epperson 1990; Wilkie 1994; Farnsworth 2000). As Julian Thomas has observed, a humanistic archaeology that seeks to disengage from materiality and to focus solely upon 'the human body, individuality, the emotions, and the cognitive operations of the mind' or human agency in the abstract risks obscuring the unevenness of power relations and material conditions (J. Thomas 2000: 155). Laurie Wilkie and Kevin Bartoy (2000) sought to present a more balanced approach than Tarlow's above, calling for the 'complex combination' of Marxist and more 'humanistic' approaches, while from the Left Matthew Palus and other archaeologists have importantly pointed out how models of domination and resistance can presume 'a balance that may not actually' have existed (Palus et al. 2006: 90).

In this context Lu Ann DeCunzo's recent consideration of the archaeology of institutions is of particular significance to archaeological studies of plantation slavery. Discussing archaeological studies of prisons, workhouses, schools, hospitals and asylums, including her own work at the Magdalen Society Asylum in Philadelphia, DeCunzo (2006) presents a compelling critique of the Foucauldian emphasis upon social control in such studies, drawing upon Lynn Meskell's work (1996) to call for new studies of experience and embodiment in institutional situations informed by feminist and disability studies. She encourages us to refocus upon human engagements with materiality in such built environments. Her discussion with Julie Ernstein of Delle's study of Jamaica is illuminating:

> Through maps and plans, Delle analysed coffee planters' *conceptions* of space, but his discussions of plantation landscapes remain purely descriptive. Studies such as Delle's seek to illustrate wider social and historical processes, but provide little sense of how archaeology can explore the particular materialities that these processes constituted in specific situations. (DeCunzo and Ernstein 2006: 260)

This call for a focus upon embodiment and materiality in landscape archaeology, in contrast with conventional Foucauldian understandings of the influence of material culture and the built environment upon social life as purely constraining (cf. Hicks and Horning 2006), chimes with recent calls from cultural historical geographers, sociocultural anthropologists and archaeologists in relation to both colonial discourse *and* landscape (although the issues have rarely been considered together) to complement purely 'textual' models by acknowledging materiality. Among the most significant contributions in relation to colonialism are Nicholas Thomas' critique of Said and other postcolonial theorists from the perspective of anthropological material culture studies (N. Thomas 1994) and Chris Gosden's archaeological study of how 'colonialism is about material culture' (Gosden 2004: 1; cf. Gosden 2001). As Peter Pels has suggested, the study of colonial constructions of landscape forms part of an increasing stress on 'the nonverbal, tactile dimensions of social practice' which also include 'the exchange of objects, the arrangement and disposition of bodies, clothes, buildings, and tools in agricultural practices, medical and religious performances, regimes of domesticity and kinship, [or] physical discipline' (Pels 1997: 169).

Recent work on landscape and colonialism in cultural historical geography highlights the potential here. Andrew Sluyter has highlighted the close relationships between ideas of landscape and European colonialism (1999, 2001), but despite calling for a focus upon both 'material' and 'conceptual' dimensions of colonial landscapes, and upon Indigenous, precolonial landscapes, he offers little sense of the detailed contexts of impact or interaction – just a broad sense of how 'materially, Europeans introduced disease; natives died and desettled the landscape; Europeans resettled and built fences' (Sluyter 2001: 419). More successful is the study of the history of cadastral mapping, land rights and the Aboriginal experience of racial segregation in New South Wales by Denis Byrne (2003). Byrne observes how 'an indigenous minority's presence in and movement through a landscape of settler colonialism is spatially controlled or constrained by the colonizers', and how 'the minority group subverted that system of spatial control, transgressing its numerous finely drawn boundaries' (Byrne 2003: 170). He suggests that the continued mobility of Aboriginal populations, despite administrative attempts at containment, contributed to colonial landscapes characterised by white 'nervousness' (Byrne 2003: 188) and Aboriginal resistance – whether 'fence-jumping', orchard raiding, or 'sheer obdurate persistence' (Byrne 2003: 183). More radical, and still more successful, is Jane Jacobs' vision of British and Australian cities in colonial perspective. Jacobs seeks to move 'towards a cultural politics of place as opposed to a reading of a textualised landscape', 'to move beyond the spatial rhetoric of colonial and postcolonial theory and return it to "real" geographies' by engaging with 'the materiality of place' (Jacobs 1996: 9, 159, x). The result is the exposure of an 'unruly imperialism' (Jacobs 1996: 159), in which colonial initiatives were refracted according to local situations.

Such work reminds us that the alternative Western narratives of Caribbean landscapes as wild or domesticated, productive or hostile, developed not simply through the representation of the Caribbean in writing or painting but in a complex range of material practices: from Hans Sloane's collecting of natural specimens (Sloane 1707) to Georgian map-making (Higman 1986a, 1988), or contemporary archaeological practices – a history of land clearance, land grants, land holdings, land prices, land as a resource, designed estate landscapes, colonial governmentality and post-emancipation planned villages, and more recently abandoned landscapes (Watts 1987). The success and reception of these strategies varied. And the contemporary practice of archaeology, like that of sociocultural anthropology, is not disconnected from these legacies, and demands an awareness of both the intimate relationships that exist between the history of the discipline and the history of European colonialism since the 17th century (van Dommelen 1997; Rowlands 1998; Gosden 2004), and also the intimate relationships between representation and the production of knowledge on the one hand, and power and inequality on the other, in both past and present (Lawrence and Shepherd 2006; compare Trouillot 2003).

It also problematises two-dimensional models of dominating/coloniser and dominated/colonised by emphasising material and geographical complexity and contingency. In Caribbean studies, the region's lack of fit with conventional models of premodern and modern, or Old and New World, situations has been acknowledged since the 1960s, especially in accounts of its precocious 'modernity' (Mintz 1985). '[N]either "primitive" nor "civilised", neither "non-Western" on the conventional criteria nor unambiguously "Western" (in short, neither fish nor fowl), the Caribbean has never quite fit[ted] securely within any anthropological agenda' (D. Scott 2004a: 192). Indeed, David Scott has suggested that the Puerto Rico Project, with which Sidney Mintz's part of Caribbean anthropology began, represented an important part of the shift away from a Boasian model of 'salvage ethnography'. Instead, it embraced the study of 'modern' situations and of the global connectedness of places like the Caribbean through familiar, everyday substances such as sugar (D. Scott 2004a: 206), and promoted an awareness of how the heterogeneity of the Antilles problematises conventional models of 'Culture Area' in American historical anthropology (Mintz 1966).

This has two implications for Caribbean archaeology. Firstly, it requires the breakdown of unhelpful boundaries between prehistoric and historical situations that several Caribbean scholars have called for (Sued-Badillo 1992; Wilson 1997: 206), recognising how, as Charles Cobb (2006) has observed, such a division reinforces postcolonial accounts that envisage change, movement and complexity as distinctively modern or Western, retaining in Indigenous prehistories what Trouillot has called anthropology's 'Savage Slot' (Trouillot 1991). As Stephen Silliman has suggested, the study of 'the social and physical landscapes of contact' is a promising place from which to challenge rigid distinctions between 'European' and 'native' archaeologies (Lightfoot 1995), or between 'contact' and 'colonialism' (Silliman 2005): 'The notion of "landscape" is a useful metaphor to organise the issues because it combines the physical and the social, local and global, setting and outcome, and spatiality and materiality' (Silliman 2004: 280).

Secondly, it leads away from conventional archaeological studies of creolisation (see review by Gundaker 2000) to a sense of colonial identities (those of both colonised and colonisers) as emergent through the geographies of colonial relations. Such a position has been explored recently in archaeologies of Roman Gaul and Britain (Jones 1997; Woolf 1997; Gosden 2001: 247), and holds similarities to Homi Bhabha's suggestion that the 'location' of any 'culture' is not fixed, but 'interstitial' in both time and space (Bhabha 1994: 235). In a Caribbean context, it builds upon sociologist Paul Gilroy's study of *The Black Atlantic* (1993) which sought to 'decent[er] African-*American* narratives, bringing the Caribbean, Britain, and Europe into the picture' (Clifford 1994: 316), by exploring alternatives to anticolonial nationalist or pan-African perspectives on the one hand, and antinationalist postcolonial accounts of broad global contexts on the other. An awareness of diaspora can expose in particular landscapes the persistent 'copresence of "here" and "there"', focusing our attention upon 'the routing of diaspora in specific maps/histories' (Clifford 1994: 318–319). However, the multiple, contrasting landscapes of Caribbean history also expose what Jonathan Friedman sees as the dangers of Clifford's (1997) focus on 'routes' and 'roots' – that 'the *metaphor* of fixity or dwelling vs movement' serves to obscure the parallel sense of 'the verb, to rout' and 'the real social issue of displacement' (Friedman 2002: 22). The challenge for Caribbean historical archaeologists is not simply to move beyond nationalist histories to accounts of the complexity of colonial entanglements and ethnogenesis, but to find ways of examining simultaneously local and transnational contexts in a manner that has not characterised previous archaeological accounts of Atlantic diasporas (Agorsah 1996; Orser 1998). In the next section, we suggest that using landscape archaeology is one way of doing this.

CARIBBEAN LANDSCAPES AND SCALES OF ANALYSIS

So, how might historical archaeologists start to work with the complex materialities of colonialism, exploring the emergent, explicit and objectified dimensions of Caribbean history rather than producing static 'totalising models of implicit cultures' (N. Thomas 1992: 227) and their movement? There is no single, or simple, answer to the question, but in this section we want to point to a few recent studies that point to the potential of harnessing archaeological approaches to landscape, situation and multiple scales of analysis.

We take as our point of departure the history of the development of Caribbean prehistory. The study of human mobility has been a central trope in this field since the 1950s, mainly due to the influence of Yale anthropologist Irving Rouse. His research into the origins of the Indigenous populations encountered by Europeans in the Columbus era set the agenda for a generation of Caribbean prehistorians. South American prehistory had been an important region for the development of culture historical perspectives in New World archaeology, especially through Julian Steward's *Handbook of South American Indians* (1946–1959) which drew upon the 'direct historical approach' (Steward 1942) to map pre-contact cultural groups through ethnohistory. The *Handbook*'s interpretive framework combined evolutionism with diffusionism (Neves 1999: 217), and its fourth volume presented what Irving Rouse termed a 'circum-Caribbean theory' of a single, highland centre of a South American pre-Columbian Indigenous 'culture sphere' (Rouse 1953). For the Antilles, this implied that the region witnessed the spread of a single 'Formative' culture from Venezuela that correlated with the Saladoid horizon, as the second of three prehispanic 'waves of cultural influence' (Steward 1948: 23).

Testing this theory against archaeological data, Rouse (1953: 198) demonstrated how archaeological material from the Greater Antilles indicated a much more complex range of mobilities. Rouse's approach was still grounded in a normative and diffusionist model of culture history. It bore similarities to Steward's contemporary development of the notion of 'multilinear evolution' in cultural change (Steward 1955a), especially in his use of 'modal analysis' to produce a chronology of ceramic traditions from which cultural affiliations were inferred (Rouse 1992: 58), presenting highly variable material as homogeneously as possible (Rouse 1939, 1964, 1986: 7). However, Rouse (1948) provided the first sustained overview of the 'tribes and cultures of the Antilles', which discerned three 'linguistic-cultural groups' – the Ciboney, Arawak and Carib. Ideas of regional settlement patterns and 'cultural ecology' that were associated with Steward (1937, 1955b) and were developed by the later processual archaeology (Rouse 1956) were used. The physical geography of the Lesser and Greater Antilles was considered as 'a series of stepping stones between the northeastern part of South America and the peninsulas of Florida and the Yucatan, projecting from North and Middle America respectively' (Rouse 1964: 499).

Accounts developed of how South American populations (with subsistence practices based on tropical forest environments) adapted to the maritime environments of the island Caribbean, combining Steward's notion of 'cultural ecology' with a Binfordian emphasis upon adaptation. Later studies examined 'the cultural and biological processes that produced the observed patterns of expansion' (Keegan 1995: 400), including the overexploitation of marine resources (Wing 1989; Wing and Wing 1995) and the dynamic relationship between humans and their terrestrial environment, especially in relation to aboriculture and emergent agriculture (Newsom 1993; Newsom and Deagan 1994). Rather than exploring 'origins', Keegan (1996a) considered the reasons for, and manners of, migrations into the island Caribbean, sought to develop statistical models to account for them

(Keegan 1995) and has played a crucial role in drawing together a rigorous account of the complex sequence of Caribbean prehistoric and early Indigenous archaeology (Keegan 1994, 1996b, 2000).

While the descriptive focus and normative and evolutionary conceptions of culture inherent in Rouse's work are unsatisfactory, it inspired Caribbean prehistorians to focus upon the relationships between human mobility, material culture and 'landscape'. In contrast, in historical archaeology these issues, and the maritime dimensions of the Antillean string of islands – the principal landfall for Old World vessels travelling to the New World – have been forgotten. However, a number of recent studies have started to redress this imbalance, in relation to Indigenous, African-Caribbean and European landscapes. In what remains of this section, we want to review some of these.

In studies of early colonial contact and encounter, the influence of Indigenous populations upon post-Columbian situations has led scholars away from models of replacement, adaptation and assimilation. Samuel Wilson's account of a 'cultural mosaic' in the Caribbean region in the 15th century AD (Wilson 1993) is more satisfactory, although as William Keegan has observed this pre-Columbian diversity was not 'the product of discrete entities' but resulted from 'matrices of connections linking communities with one another' (Keegan 1996: 277). An awareness of human geography in these studies necessarily includes an appreciation of the huge transformations in demography that took place immediately following Spanish conquest, especially through enslavement and disease (Newson 1995: 41), a process which Alfred Crosby (1986) famously described as one of *Ecological Imperialism*. These environmental and landscape contexts have been fruitfully explored in relation to the interconnectedness of the island landscapes of the Antilles by Antonio Curet, whose study of the site of Tibes in western Puerto Rico explores how the distribution of Ostinoid-type material culture relates to 'constant and intensive interaction' worked out through local interactions with specific environments (Curet 2005: 225, 143). Lennox Honychurch's study of the site of Indian River, on the northwestern coast of Dominica, demonstrates the potential of the archaeological study of sites at which such interactions took place. Honychurch explored its role as a 'maritime crossroad' at which Indigenous Carib populations traded with Europeans between the 16th and 19th centuries (Honychurch 1997). He suggested that Caribs 'at certain key locations around refreshment stops such as the Indian River, manned tobacco markets so to speak' (Honychurch 1997: 299). Complementing conventional models that define the history and archaeology of Caribs in terms of resistance, Honychurch seeks to 'accord the Caribs some element of rational decisiveness in their exchange mechanisms with Europeans' (Honychurch 1997: 301).

In their work on the historical archaeology of Atlantic trade in the Bahamas, Laurie Wilkie and Paul Farnsworth (1999) have pointed to the complexity of African identities in the New World, and have called for 'multiscalar explorations' of the construction of New World Creole identities. They have used their study of the Clifton Plantation on New Providence in the Bahamas to build upon the potential of African diaspora archaeologies to operate 'at multiple spatial scales, ranging from the household to the quarters to the plantation' and beyond (Wilkie and Farnsworth 2005: 3). Mark Hauser's study of locally produced utilitarian pottery in Jamaica during the 18th century shows how the domestic economies of slaves living on multiple plantations can be examined (2001). A system of Sunday markets and itinerant traders, known in Jamaica as Higglers, linked archaeological contexts of domestic servants and free labourers in the cities of Port Royal and Spanishtown with rural estates located in the central corridor and northern coast of the island. By examining the local materials traded through the system of street markets and Higglers, archaeologists might start to map the production and movements of objects onto human landscapes. Wilkie and Farnsworth's multiscalar approach can thus help landscape archaeology to examine the role of undocumented (etc).

European colonial landscapes can also be studied at a range of scales of analysis. The clearest example of the potential here is the study by Susanne Seymour, Stephen Daniels and Charles Watkins of the close relationships between the development of Sir George Cornewall's Moccas estate in Herefordshire and his La Taste estate in Grenada (1771–1819). By 'examining landed estates and their owners in an imperial context', Seymour et al. (1998: 341) demonstrate a series of 'overlapping concerns ... especially in terms of the management of land, labour and finance' at the two sites. Similarly, Dan Hicks has demonstrated the potential of studying landscapes in southwest Britain as part of broader Atlantic landscapes that encompassed the Caribbean and West Africa (Hicks 2003). Like the examples above, these were places at which the global was enacted – an observation that leads us away from opposing local and global contexts, to explore how multiple scales of analysis are imagined and worked out in landscapes in the past and the disciplinary present (Hicks 2005), and to consider the potential for postcolonial archaeologies of Europe.

We are suggesting that the dual ability of landscape archaeology to operate at a series of geographical scales and also to retain a sense of position or situatedness makes it a powerful tool in the development of a postcolonial archaeology. From this perspective, the practice of landscape archaeology (Figure 11.4) may lead away from hierarchical conceptions of archaeological scale to 'flatter' ontologies in which both large and small scales can be explored through unified archaeological methodologies (Hicks forthcoming). Archaeology can examine what Barbara Bender (2001) has called 'landscapes on-the-move'. The examples cited above, drawn from a contrasting range of Indigenous, African and European situations in the colonial Caribbean, point to the potential of landscape approaches to provide ways of putting into practice the crucial task highlighted by Mary Beaudry of:

> Accept[ing] the challenge to push our interpretations beyond bland generalizations about 'colonial' or 'imperial' material culture and colonial interactions, to ply back and forth between our study of imperial power and settler colony to ferret out how in each instance material culture in all its manifestations ... figured in colonial discourses as part of local knowledge in multifarious contexts. (Beaudry 2003: 294)

Figure 11.4 Avery Trim (St Lucia National Trust) and Rebecca Craig (Museum of London Archaeological Services) undertaking a drawn landscape survey at Balenbouche Estate, St Lucia, in January 2001. *Photograph by Mick Worthington.*

CONCLUDING THOUGHTS

Little has changed in Caribbean archaeology in the 15 years since Jalil Sued-Badillo, addressing the Society for American Archaeology in 1992, suggested that 'archaeological research in the Caribbean, still dominated by a North American orientation and North American capital, has simply refused to address the subject of colonialism and underdevelopment in the region' (Sued-Badillo 1992: 603). In exploring the contemporary intersections between the study of landscape and colonialism in the anglophone Caribbean, we have suggested that archaeological approaches to landscape represent one way of achieving the acknowledgement of situation and position that a post-colonial archaeology demands. Just as 'because one cannot simply demarcate a past colonialism from struggles in the present … the anthropology of colonialism systematically interrogates contemporary anthropology as well as the colonial circumstances from which it emerged' (Pels 1997: 164), so a postcolonial archaeology must be reflexive and situated, rather than ahistorical or conducted *in vacuo*. Landscape archaeology's abilities to develop situational studies grounded in materiality, rather than only borrowing the ideational perspectives or textual metaphors of other disciplines, offer one way of achieving this: an observation that suggests that perhaps British interpretive archaeologists passed on too quickly from the 'highly empirical school of "landscape archaeology", dedicated to the surveying and mapping of upstanding cultural features (boundaries, field systems, deserted villages) [which had] emerged as a complement to excavation' during the 1970s (*pace* J. Thomas 1993: 19–20).

Landscape archaeology in the postcolonial Caribbean can explore the plantation, the slave garden or the urban landscape as what David Scott has termed a 'problem-space': a contemporary site of historical contingency, argument and intervention (D. Scott 2004b: 3–4). In doing so, previous models of power, domination and resistance are replaced by an account of colonialism as both destructive and creative (Gosden 2004: 25; Silliman 2005: 67), in which humans and objects were bound up together in transatlantic choreographies. Such a postcolonial archaeology aims to acknowledge the complexities of apparently 'shared landscapes', and the challenges in realising historical archaeology's potential to generate 'shared history' (Harrison 2004), while recognising the extreme power relations, violence and horror of plantation slavery, and the continued hardships and inequalities in many parts of the contemporary, postcolonial Caribbean region.

We suggest that archaeologists of colonialism need to find ways to place the study of *materiality* – both the permeabilities between humans, objects and places, and also the contemporary sense of what 'matters' – at the heart of their studies, rather than using objects and landscape to illustrate historical accounts, elite ideas or political observations already developed elsewhere. The methods of landscape archaeology hold the potential to be used in the crucial task of mapping the permeabilities between human and object movements, and the role of landscapes (plantations, harbours, cities) in imagining and bringing about a diversity of Atlantic worlds (Hicks in prep.). This involves moving beyond Nicholas Thomas' (1991) account of objects as 'entangled' in known historical social relations, in which the artefact or the landscape 'is eviscerated in the all-powerful context of history or culture', towards more complex and messy materialities (Pinney 2005: 259). The contemporary political dimensions of this refocusing upon materiality include a new awareness of the processes of inequality, especially when such work is conducted by working 'from the margins' (Tsing 1994: 279). As Jane Jacobs puts it, 'it is not solely that discursively constituted notions of identity have material effects, but that the sheer uneven materiality of the lives of people affected by imperialism must inform the moral and ethical function of critical postcolonial studies' (Jacobs 1996: 29). Discussing the social contexts of archaeological practice in Puerto Rico, Jaime Pagán-Jiménez contrasts 'centric' perspectives produced by scholars in the West with alternative, 'eccentric' theoretical positions developed from Latin American situations (Pagán-Jiménez 2004). We hope that the particular vision of a postcolonial landscape archaeology presented here will contribute to the *de*-centring of mainstream impulses in 'world historical archaeology', and to the continued development of distinctively archaeological and materially focused contributions to historical anthropology.

ACKNOWLEDGEMENTS

We are indebted to our colleagues Mary Beaudry, Antonio Curet, Ken Kelly, Laura McAtackney, Francois Richard, Jim Symonds and Joshua Torres for their comments on an earlier draft of this chapter, the limitations of which remain our own responsibility.

REFERENCES

Agorsah, E.K. 1993. Archaeology and Resistance History in the Caribbean. *African Archaeological Review* 11: 175–195.

Agorsah, E.K. 1994. Archaeology of Maroon Settlements in Jamaica. In K. Agorsah (ed) *Maroon Heritage: Archaeological, Ethnographic and Historical Perspectives*. Kingston, Jamaica: Canoe Press, pp. 163–187.

Agorsah, E.K. 1996. The Archaeology of the African Diaspora. *African Archaeological Review* 13(4): 221–224.

Armstrong, D.V. 1990. *Old Village and the Great House*. Urbana: University of Illinois Press.

Armstrong, D.V. and M.W. Hauser 2004. An East Indian Laborer's Household in 19th Century Jamaica: A Case for Spatial, Chronological, and Material Analysis in Determining Ethnicity. *Historical Archaeology* 38(2): 9–21.

Armstrong D.V. and K.G. Kelly 2000 Settlement Patterns and the Origins of African Jamaican Society: Seville Plantation, St Ann's Bay, Jamaica. *Ethnohistory* 47: 369–397.

Barka, N.F. 2001. Time Lines: Changing Settlement Patterns on St Eustatius. In P. Farnsworth (ed) *Island Lives: Historical Archaeologies of the Caribbean*. Tuscaloosa: University of Alabama Press, pp. 103–141.

Beaudry, M.C. 2003. Concluding Comments: Disruptive narratives? Multidimensional perspectives on 'Britishness'. In S. Lawrence (ed) *Archaeologies of the British. Explorations of Identity in Great Britain and its Colonies, 1600–1945*. London: Routledge (One World Archaeology 46), pp. 291–295.

Bender, B. 1999. Subverting the Western Gaze: Mapping Alternative Worlds. In P. Ucko and R. Layton (eds) *The Archaeology and Anthropology of Landscape*. London: Routledge, pp. 31–45.

Bender, B. 2001. Landscapes on-the-move. *Journal of Social Archaeology* 1(1): 75–89.

Bender, B. 2006. Place and Landscape. In C. Tilley, W. Keane, S. Küchler, M. Rowlands and P. Spyer (eds) *Handbook of Material Culture*. London: Sage, pp. 303–314.

Bhabha, H. 1994. *The Location of Culture*. London: Routledge.

Byrne, D.R. 2003. Nervous Landscapes: Race and Space in Australia. *Journal of Social Archaeology* 3(2): 169–193.

Clifford, J. 1994. Diasporas. *Cultural Anthropology* 9(3): 302–338.

Clifford, J. 1997. *Routes*. Cambridge, MA: Harvard University Press.

Cobb, C. 2006. Archaeology and the 'Savage Slot': Displacement and Emplacement in the Premodern World. *American Anthropologist* 107(4): 563–574.

Crosby, A.W. 1986. *Ecological Imperialism: The Biological Expansion of Europe, 900–1900*. Cambridge: Cambridge University Press.

Curet, L.A. 2005. *Caribbean Paleodemography: Population, Culture History, and Sociopolitical Processes in Ancient Puerto Rico*. Tuscaloosa: The University of Alabama Press.

Daniels, S. and D. Cosgrove 1988. *The Iconography of Landscape: Essays on the Symbolic Representation, Design and Use of Past Environments*. Cambridge: Cambridge University Press.

DeCunzo, L.A. 2006. Exploring the Institution: Reform, Confinement, Social Change. In M. Hall and S. Silliman (eds) *Historical Archaeology*. Oxford: Blackwell, pp. 167–189.

DeCunzo, L. and J.H. Ernstein 2006. Landscapes, Ideology and Experience in Historical Archaeology. In D. Hicks and M.C. Beaudry (eds) *The Cambridge Companion to Historical Archaeology*. Cambridge: Cambridge University Press, pp. 255–270.

Deagan, K.A. and J.M. Cruxent 2002a. *Columbus's Outpost among the Taìnos: Spain and America at La Isabela, 1493–1498*. New Haven: Yale University Press.

Deagan, K.A. and J.M. Cruxent 2002b. *Archaeology at La Isabela: America's first European Town*. New Haven: Yale University Press.

Delle, J. 1998. *An Archaeology of Social Space: Analyzing Coffee Plantations in Jamaica's Blue Mountains*. New York: Plenum (Contributions to Global Historical Archaeology).

Epperson, T.W. 1990. Race and the Disciplines of the Plantation. *Historical Archaeology* 24(4): 29–36.

Farnsworth, P. 2000. Brutality or Benevolence in Plantation Archaeology. *International Journal of Historical Archaeology* 4(2): 145–158.

Ferguson, L. 1992. *Uncommon Ground: Archaeology and early African America*. Washington, DC: Smithsonian Institution Press.

Foucault, M. 1979. *Discipline and Punish: The Birth of the Prison* (trans. A. Sheridan). London: Methuen.

Friedman, J. 2002. From Roots to Routes: Tropes for Trippers. *Anthropological Theory* 2(1): 21–36.

Gilroy, P. 1993. *The Black Atlantic: Modernity and Double-Consciousness*. Cambridge, MA: Harvard University Press.

Gosden, C. 2001. Postcolonial Archaeology: Issues of Culture, Identity and Knowledge. In I. Hodder (ed) *Archaeological Theory Today*. Cambridge: Polity Press, pp. 241–261.

Gosden, C. 2004. *Archaeology and Colonialism: Culture Contact from 5000 BC to the Present*. Cambridge: Cambridge University Press.

Gundaker, G. 2000. Discussion: Creolization, Complexity and Time. *Historical Archaeology* 34(3): 124–133.

Hall, M. 2000. *Archaeology and the Modern World: Colonial Transcipts in South Africa and the Chesapeake*. London: Routledge.

Harrison, R. 2004. *Shared Landscapes: Archaeologies of Attachment and the Pastoral Industry in New South Wales*. Sydney: University of New South Wales Press (Studies in the Cultural Construction of Open Space 3).

Hauser, M.W. 2001. *Peddling Pots: Determining the Extent of Market Exchange in Eighteenth Century Jamaica Through the Analysis of Local Coarse Earthenware*. Unpublished Ph.D. Dissertation, Syracuse University, Syracuse, NY.

Haviser, J. (ed) 1999. *African Sites Archaeology in the Caribbean*. Princeton: Markus Wiener.

Hicks, D. 2003. Archaeology Unfolding: Diversity and the Loss of Isolation. *Oxford Journal of Archaeology* 22(3): 215–229.

Hicks, D. 2005. 'Places for Thinking' from Annapolis to Bristol: Situations and Symmetries in 'World Historical Archaeologies'. *World Archaeology* 37(3): 373–391.

Hicks, D. 2007. *'The Garden of the World': An Historical Archaeology of Sugar Landscapes in the Eastern Caribbean*. Oxford: Archaeopress (Studies in Contemporary and Historical Archaeology 3, British Archaeological Reports International Series 1632).

Hicks, D. forthcoming. Commentary: The Multiscalar Debate in Historical Archaeology. *International Journal of Historical Archaeology*.

Hicks, D. in prep. *A Material Atlantic: Historical Archaeology and the British Atlantic World, AD 1600–1800*. Cambridge: Cambridge University Press. Manuscript in preparation.

Hicks, D. and A. Horning 2006. Historical Archaeology and Buildings. In D. Hicks and M.C. Beaudry (eds) *The Cambridge Companion to Historical Archaeology*. Cambridge: Cambridge University Press, pp. 273–292.

Higman, B.W. 1986a. Plantation Maps as Sources for the Study of West Indian Ethnohistory. In D. Wiedman (ed) *Ethnohistory: A Researcher's Guide*. Williamsburg, VA: Department of Anthropology, College of William and Mary (Studies in Third World Societies Publication Number 35), pp. 107–136.

Higman, B.W. 1986b. Jamaican Coffee Plantations 1780–1860: A Cartographic Analysis. *Caribbean Geography* 2: 73–91.

Higman, B.W. 1987. The Spatial Economy of Jamaican Sugar Plantations: Cartographic Evidence from the 18th and 19th Centuries. *Journal of Historical Geography* 13(1): 17–19.

Higman, B.W. 1988. *Jamaica Surveyed: Plantation Maps and Plans of the 18th and 19th Centuries*. Kingston: Institute of Jamaican Publishers.

Higman, B.W. 1999. *Montpelier, Jamaica: A Plantation Community in Slavery and Freedom, 1739–1912*. Kingston: University of the West Indies Press.

Honychurch, L. 1997. Crossroads in the Caribbean: A Site of Encounter and Exchange on Dominica. *World Archaeology* 28(3): 291–304.

Jacobs, J.M. 1996. *Edge of Empire: Postcolonialism and the City*. London: Routledge.

Johnson, M.H. 1996. *An Archaeology of Capitalism*. Oxford: Blackwell.

Johnson, M.H. 2000. Self-Made Men and the Staging of Agency. In M. Dobres and J. Robb (eds) *Agency in Archaeology*. London: Routledge, pp. 213–231.

Jones, S. 1997. *The Archaeology of Ethnicity*. London: Routledge.

Keegan, W.F. 1994. West Indian Archaeology 1. Overview and Foragers. *Journal of Archaeological Research* 2(3): 255–284.

Keegan, W.F. 1995. Modeling Dispersal in the Prehistoric West Indies. *World Archaeology* 26(3): 400–420.

Keegan, W.F. 1996a. West Indian Archaeology 2. After Columbus. *Journal of Archaeological Research* 4(4): 265–294.

Keegan, W.F. 1996b. Columbus Was a Cannibal: Myth and the First Encounters. In R.L. Paquette and S.L. Engerman (eds) *The Lesser Antilles in the Age of European Expansion*. Gainesville: University Press of Florida, pp. 18–32.

Keegan, W.F. 2000. West Indian Archaeology 3. Ceramic Age. *Journal of Archaeological Research* 8(2): 135–167.

Laborie, P.J. 1798. *The Coffee Planter of Saint Domingo*. London: T. Cadell and W. Davies.

Lawrence, S. and N. Shepherd 2006. Historical Archaeology and Colonialism. In D. Hicks and M.C. Beaudry (eds) *The Cambridge Companion to Historical Archaeology*. Cambridge: Cambridge University Press, pp. 69–86.

Lefebvre, H. 1991. *The Production of Space*. Oxford: Oxford University Press.

Lightfoot, K.G. 1995. Culture Contact Studies: Redefining the Relationship between Prehistoric and Historical Archaeology. *American Antiquity* 60: 199–217.

Martin, S. 1765. *An Essay on Plantership, humbly inscribed to his Excellency George Thomas, Esq*. (4th ed). Antigua: Samuel Clapham.

Meskell, L. 1996. The Somatization of Archaeology: Institutions, Discourses, Corporeality. *Norwegian Archaeological Review* 29(1): 1–16.

Miller, D. and C. Tilley (eds) 1984. *Ideology, Power and Prehistory*. Cambridge: Cambridge University Press.

Miller, D., M. Rowlands and C. Tilley (eds) 1987. *Domination and Resistance*. London: Routledge (One World Archaeology 3).

Mintz, S.W. 1966. The Caribbean as a Socio-Cultural Area. *Cahiers d'Histoire Mondiale* 9: 912–937.

Mintz, S.W. 1974. *Caribbean Transformations*. Baltimore: Johns Hopkins University Press.

Mintz, S.W. 1985. *Sweetness and Power: The Place of Sugar in Modern History*. London: Penguin.

Morphy, H. 1993. Colonisation, History and the Construction of Place. In B. Bender (ed) *Landscape: Politics and Perspectives*. Oxford: Berg, pp. 205–243.

Neves, E.G. 1999. Changing Perspectives in Amazonian Archaeology. In G.G. Politis and B. Alberti (eds) *Archaeology in Latin America*. London: Routledge, pp. 216–243.

Newsom, L. 1993. Native West Indian Plant Use. Unpublished Ph.D. dissertation, Department of Anthropology, University of Florida Gainesville.

Newsom, L. and K. Deagan 1994. *Zea mays* in the West Indies: The Archaeological and Early Historic Record. In S. Johannessen and C.A. Hastorf (eds) *Corn and Culture in the Prehistoric New World*. Minneapolis: University of Minnesota Press, pp. 203–218.

Newson, L. 1995 Indian Population Patterns in Colonial Spanish America. *Latin American Research Review* 20(3): 41–74.

Orser, C.E. 1998. The Archaeology of the African Diaspora. *Annual Review of Anthropology* 27: 63–82.

Pagán-Jiménez, J.R. 2004. Is All Archaeology at Present a Postcolonial One? Constructive Answers from an Eccentric Point of View. *Journal of Social Archaeology* 4(2): 200–213.

Palus, M.M., M.P. Leone and M.D. Cochran 2006. Critical Archaeology: Politics Past and Present. In M. Hall and S. Silliman (eds) *Historical Archaeology*. Oxford: Blackwell, pp. 84–104.

Pels, P. 1997. The Anthropology of Colonialism: Culture, History and the Emergence of Western governmentality. *Annual Review of Anthropology* 26: 163–183.

Pinney, C. 2005. Things Happen: Or, from Which Moment Does that Object Come? In D. Miller (ed) *Materiality*. Durham, NC: Duke University Press, pp. 256–272.

Pulsipher, L.M. 1991. Galways Plantation, Montserrat. In H.J. Viola and C. Margolis (eds) *Seeds of Change*. Washington, DC: Smithsonian Institution Press, pp. 139–159.

Pulsipher, L.M. 1994. The Landscapes and Ideational Roles of Caribbean Slave Gardens. In N.F. Miller and K.L. Gleason (eds) *The Archaeology of Garden and Field*. Philadelphia: University of Pennsylvania Press, pp. 202–222.

Pulsipher, L.M. and C.M. Goodwin 1999. 'Here Where the Old Time People Be': Reconstructing the Landscapes of the Slavery and Post-Slavery Era on Montserrat, West Indies. In J.B. Haviser (ed) *African Sites Archaeology in the Caribbean*. Kingston, Jamaica: Ian Randle, pp. 9–37.

Pulsipher, L.M. and C.M. Goodwin 2001. 'Getting the Essence of It': Galways Plantation, Montserrat, West Indies. In P. Farnsworth (ed) *Island Lives: Historical Archaeologies of the Caribbean*. Tuscaloosa: University of Alabama Press, pp. 165–203.

Roughley, T. 1823. *The Jamaica Planter's Guide*. London: Longman, Hurst, Rees, Orme and Brown.

Rouse, I. 1939. *Prehistory in Haiti: A Study in Method*. New Haven: Yale University Publications in Anthropology, No. 21.

Rouse, I. 1948. The Ciboney. In J.H. Steward (ed) *Handbook of South American Indians. Volume 4: The Circum-Caribbean Tribes*. Washington, DC: United States Government Printing Office (Bulletin of the Smithsonian Institution, Bureau of American Ethnology 143), pp. 497–503.

Rouse, I. 1953. The Circum-Caribbean Theory: An Archaeological Test. *American Anthropologist* 55(2): 188–200.

Rouse, I. 1956. Settlement Patterns in the Caribbean Area. In G.R. Willey (ed) *Prehistoric Settlement Patterns in the New World*. New York: Wenner Gren Foundation for Anthropological Research (Viking Fund Publications in Anthropology 23), pp. 165–172.

Rouse, I. 1964. Prehistory of the West Indies. *Science* 144(3618): 499–513.

Rouse, I. 1986. *Migrations in Prehistory: Inferring Population Movement from Cultural Remains*. New Haven: Yale University Press.

Rouse, I. 1992. *The Tainos: Rise and Decline of the People Who Greeted Columbus*. New Haven: Yale University Press.

Rowlands, M. 1998. The Archaeology of Colonialism. In K. Kristiansen and M. Rowlands (eds) *Social Transformations in Archaeology: Global and Local Perspectives*. London: Routledge, pp. 327–333.

Said, E. 1978. *Orientalism*. New York: Vintage Books.

Said, E. 1993. *Culture and Imperialism*. London: Chatto and Windus.

Scott, D. 2004a. Modernity that Predated the Modern. *History Workshop Journal* 58(1): 191–210.

Scott, D. 2004b. *Conscripts of Modernity: The Tragedy of Colonial Enlightenment*. Durham, NC: Duke University Press.

Scott, J.C. 1990. *Domination and the Arts of Resistance. Hidden Transcripts*. New Haven: Yale University Press.

Seymour, S., S. Daniels and C. Watkins 1998. Estate and Empire: Sir George Cornewall's management of Moccas, Herefordshire and La Taste, Grenada, 1771–1819. *Journal of Historical Geography* 24(3): 313–351.

Shackel, P.A., P.R. Mullins and M.S. Warner (eds) 1998. *Annapolis Pasts: Historical Archaeology in Annapolis, Maryland*. Knoxville: University of Tennessee Press.

Sheller, M. 2003. *Consuming the Caribbean: From Arawaks to Zombies*. London: Routledge.

Silliman, S.W. 2004. Social and Physical Landscapes of Contact. In T.R. Pauketat and D. DiPaolo Loren (eds) *North American Archaeology*. Oxford: Blackwell, pp. 273–296.

Silliman, S.W. 2005. Culture Contact or Colonialism? Challenges in the Archaeology of Native North America. *American Antiquity* 70(1): 55–74.

Singleton, T.A. 2001. Slavery and Spatial Dialectics on Cuban Coffee Plantations. *World Archaeology* 33(1): 98–114.

Sloane, H. 1707. *A Voyage to the Islands of Madeira, Barbados, Nieves, St Christopher and Jamaica, with the natural history of the herbs and trees, four-footed beasts, fishes, birds, insects, reptiles, etc.* London.

Sluyter, A. 1999. The Making of the Myth in Post-Colonial Development: Material-Conceptual Landscape Transformation in Sixteenth-Century Veracruz. *Annals of the Association of American Geographers* 89: 377–401.

Sluyter, A. 2001. Colonialism and Landscape in the Americas: Material/Conceptual Transformations and Continuing Consequences. *Annals of the Association of American Geographers* 91(2): 410–428.

Steward, J.H. 1937. Ecological Aspects of Southwestern Society. *Anthropos* 32: 87–104.

Steward, J.H. 1942. The Direct Historical Approach to Archaeology. *American Antiquity* 7(4): 337–343.

Steward, J.H. (ed) 1946–1959. *Handbook of South American Indians*. Washington, DC: United States Government Printing Office (Bulletin of the Smithsonian Institution, Bureau of American Ethnology 143) (seven volumes).

Steward, J.H. 1948. The Circum-Caribbean Tribes: An Introduction. In J.H. Steward (ed) *Handbook of South American Indians. Volume 4: The Circum-Caribbean Tribes.* Washington, DC: United States Government Printing Office (Bulletin of the Smithsonian Institution, Bureau of American Ethnology 143), pp. 1–41.

Steward, J.H. 1955a. *Theory of Culture Change: The Methodology of Multilinear Evolution.* Urbana: University of Illinois Press.

Steward, J.H. 1955b. Region – An Heuristic Concept. *Rural Sociology* 20: 297–298.

Stoler, A.L. 2001. Tense and Tender Ties: The Politics of Comparison in North American History and (Post)Colonial Studies. *Journal of American History* 88(3): 94 pars. http://www.historycooperative.org/journals/jah/88.3/stoler.html (Consulted 10 June 2006).

Sued-Badillo, J. 1992. Facing up to Caribbean History. *American Antiquity* 57(4): 599–607.

Tarlow, S. 1999. Capitalism and Critique. *Antiquity* 73: 467–470.

Thomas, J. 1993. The Politics of Vision and the Archaeologies of Landscape. In B. Bender (ed) *Landscape: Politics and Perspectives*. Oxford: Berg, pp. 1–17.

Thomas, J. 2000. Reconfiguring the Social, Reconfiguring the Material. In M.B. Schiffer (ed) *Social Theory in Archaeology*. Salt Lake City: University of Utah Press (Foundations of Archaeological Enquiry).

Thomas, N. 1991. *Entangled Objects: Exchange, Material Culture and Colonialism in the Pacific*. Cambridge, MA: Harvard University Press.

Thomas, N. 1992. The Inversion of Tradition. *American Ethnologist* 19(2): 213–232.

Thomas, N. 1994. *Colonialism's Culture: Anthropology, Travel and Government*. London: Polity Press.

Trouillot, M-R. 1991. Anthropology and the Savage Slot: The Poetics and Politics of Otherness. In R. G. Fox (ed) *Recapturing Anthropology*. Santa Fe: School of American Research, pp. 17–44.

Trouillot, M-R. 1992. The Caribbean Region: An Open Frontier in Anthropological Theory. *Annual Review of Anthropology* 21: 19–42.

Trouillot, M-R. 1995. *Silencing the Past: Power and the Production of History*. Boston: Beacon Press.

Trouillot, M-R. 2003. *Global Transformations: Anthropology and the Modern World*. Basingstoke: Palgrave Macmillan.

Tsing, A.L. 1994. From the Margins. *Cultural Anthropology* 9(3): 279–297.

Van Dommelen, P. 1997. Colonial Constructs: Colonialism and Archaeology in the Mediterranean. *World Archaeology* 28(3): 31–49.

Watts, D. 1987. *The West Indies: Patterns of Development, Culture and Environmental Change since 1492*. Cambridge: Cambridge University Press.

Wilkie, L.A. 1994. 'Never Leave Me Alone': An Archaeological Study of African American Ethnicity, Race Relations and Community at Oakley Plantation. Unpublished Ph.D. dissertation, Archaeology Program, University of California, Los Angeles.

Wilkie, L.A. and K.M. Bartoy 2000. A Critical Archaeology Revisited. *Current Anthropology* 41(5): 747–777.

Wilkie, L.A. and P. Farnsworth 1999. Trade and the Construction of Bahamian Identity: A Multiscalar Approach. *International Journal of Historical Archaeology* 3(4): 283–320.

Wilkie, L.A. and P. Farnsworth 2005. *Sampling Many Pots: An Archaeology of Memory and Tradition at a Bahamian Plantation*. Boca Raton: University Press of Florida.

Wilson, S.M. 1993. The Cultural Mosaic of the Prehistoric Caribbean. In W. Bray (ed) *The Meeting of Two Worlds: Europe and the Americas 1492–650*. Oxford: Oxford University Press (Proceedings of the British Academy 81), pp. 37–66.

Wilson, S.M. 1997. The Legacy of the Indigenous People of the Caribbean. In S.M. Wilson (ed) *The Indigenous People of the Caribbean*. Gainesville: University Press of Florida, pp. 206–213.

Winer, M. 1995. The Painted, Poetic Landscape: Reading Power in Nineteenth Century Textual and Visual Representations of the Eastern Cape Frontier. In M.E. D'Agostino, E. Prine, E. Casella and M. Winer (eds) *The Written and the Wrought: Complementary Sources in Historical Anthropology*. Berkeley: Department of Anthropology, University of California, Berkeley (Kroeber Anthropological Society Papers 79), pp. 74–109.

Wing, E. 1989. Human Exploitation of Animal Resources in the Caribbean. In C.A. Woods (ed) *Biogeography of the West Indies*. Gainesville, FL: Sandhill Crane Press, pp. 137–152.

Wing, E.S. and S.R. Wing 1995. Prehistoric Ceramic Age Adaptation to Varying Diversity of Animal Resources along the West Indian Archipelago. *Journal of Ethnobiology* 15: 119–148.

Woolf, G. 1997. Beyond Romans and Natives. *World Archaeology* 28(3): 339–350.

INDEX

A New Future for Long Kesh/ Maze 50
abolition (of slave trade) 186
Abomey *see* Bénin
Adams, Dominic 42, 47–48
Adams, Gerry 37, 40, 42, 46
Aelian (Claudius Aelianus) 214
Africa *see* Bénin, Botswana, Comoros, Ghana, Guinea, Kenya, Madagascar, Mozambique, Namibia, Nigeria, Senegal, Somalia, Swahili Coast, Tanzania, Togo, Sierra Leone, South Africa, Zimbabwe
African Burial Ground *see* New York City
African Diaspora 176, 189, 262
African Diaspora archaeology 172, 188–189, 257, 262, 265. *See also* Posnansky, Merrick
African Sites Archaeology in the Caribbean 257
African-American archaeology 80, 188, 235–236
African-American history 80, 94, 235, 238, **239**
  decentring 262
Agorsah, Kofi 257, 262
agricultural landscapes *see* rural landscapes
air photograph interpretation *see* landscape archaeology, techniques used within
Alcock, Susan 194, 200, 201, 212
Aldred, Oscar 129, 131
Allada (polity) *see* Bénin
Allen, James de Vere 58
Alpers, Edward A. 64
American archaeology *see* landscape archaeology, regional traditions of
Amstutz, Mark 34
An Archaeological Perspective 198
Anderson, Martin 242
Andrews, Malcolm 196, 206
Annales school of historical writing 128, 201, 205, 220n3. *See also* Braudel, Ferdinand; landscape archaeology, themes within
Annapolis (Maryland) 18, 226–247 *passim*
  Anglican Church 227–228
  Ann Arundel Town 227
  as 'the Ancient City' 237
  baroque landscape of, 227–230
  Bordley-Randall site 238–239
  Committee for Annapolis 244
  Department of Housing and Urban Development (HUD) 242
  Georgian landscape of 230–236
  Historic Annapolis Foundation 231, 243–245
  Historic District, as symbol of American history 18
  Historic District, creation of 228, 240, 242, 246–247
  'Historic District', street paving of 244–245
  Maryland Public Services Commission 245
  Naval Academy 240
  'Nicholson Plan' for (1696) 227
  19th century landscape of 236–241
  St Anne's Cemetery 240

St John's College 238
State Circle 228–230
State House 227–230, 241, **244**, 245
Stoddert Survey of (1718) 228, 229, 230
storefronts, appearance of 243–244
street furniture, design of 245
'truck farming' 238
20th century landscape of 241–246
Urban Renewal Authority 242
utility wires 242–244
see also Archaeology in Annapolis, William Paca House and Garden
Anne Arundel Town (Maryland) *see* Annapolis
Anschuetz, Kurt F. 14, 172, 177
anthropology, sociocultural 14, 21, 25, 251, 255, 256, 260, 261, 267
Antrop, Marc 137
Apostol, John C. (Mayor of Annapolis) 244–245
Arawak 263
archäeologische Landesaufnahme see landscape archaeology, techniques used within
archaeology *see* archaeology, themes within; critical archaeology; ethnoarchaeology; historical archaeology; household archaeology; Indigenous archaeology; interpretive archaeology; landscape archaeology; medieval archaeology; Marxist archaeology; post-processual archaeology; processual archaeology; spatial archaeology; world archaeology, idea of
archaeology, themes within *see* colonialism, archaeology of; ethnicity, archaeology of; identity, archaeology of; Islam, archaeology of; natural places, archaeology of; prisons, archaeology of; slavery, archaeology of
Archaeology in Annapolis 37, 258. *See also* Leone, Mark P.
Archaeology of Mediterranean Landscapes 195
Archaeology of Time 203
Archeologiczne Zdjęçie Polski (Archaeological Record of Poland) 152
architectural design, neo-classical 230
architecture, impermanent 60–61, 68–69
Argolid Exploration Project (AEP) 17, 199, 207–217 *passim*
arm rings, sapropelite (Iron Age) 157, 158, 160
Armstrong, Douglas 257
and Kenneth G. Kelly 188
Asbury, Herbert 82
Ashmore, Wendy 21, 172, 174, 175, 176, 177
Ashworth, Gregory J. 38, 52
Aston, Mick 21, 128
Athanassopoulos, Effie-Fontie 194, 195
Atlantic Africa 172–189. *See also* West Africa, South Africa
Atlantic Ocean 188, 262, 265, 268. *See also* slave trade, African diaspora
Australia 14, 15, 176, 260

Bahamas 265
Bailey, Geoff 195, 200, 201
notion of 'time perspectivism' 202

Baltimore (Maryland) 236
as 'the Monumental City' 237
Baltimore and Ohio Railroad 237
Banda Chieftaincy *see* Ghana
Bannister, Nicola 131
Bantu *see* Hambukushu
baraza (stone bench) 65–66, **68**
Barbot, Jean 182
Barford, Paul 152, 153
Barka, Norman 253
Barker, Graeme 197, 201
and David Mattingly 195
Barrett, John 17, 22
Bartoy, Kevin 259
Beaudry, Mary C. 25, 266
Bednarik, Robert G. 102
Begho (archaeological site) *see* Ghana
Belfast (Good Friday) Agreement 34, 39, 51
Belfast *see* United Kingdom
Belfast Telegraph 51
Belgium 136–137
Bell, Michael M. 18–19
Bender, Barbara 14, 16, 18, 23, 70–71, 172, 177, 251, 255–256, 266
and Margot Winer 16
Bénin 20, 177, 181
Abomey **173**, 184
Abomey Plateau 185–186
Allada (polity) 181
Bight of Bénin 181
Cana 181
Dahomey (polity) 20, 184, 185–186
Hueda (polity) 20, 181–185
Ouidah 20, **173**, 177, 181–185
Savi 20, **173**, 177, 180, 181–185, **183**
Benjamin, Walter 70
Bennett, Norman R. 62, 64
Benoît, Catherine 188
Beresford, David 37
Beresford, Maurice 127
Bergson, Henri 196
Bhabha, Homi 262
Bialor, Perry A. 208, 219
Bight of Bénin *see* Bénin
bin Jair, Abdalla 65
Binford, Lewis 197–198, 200, 201
influence upon Caribbean archaeology 263
Bintliff, John 198, 201, 220n3
biodiversity 124
biographical approaches *see* landscape archaeology, themes within
Black Atlantic 262
Bladensburg (Maryland) 233
Blakey, Michael L. 80, 94
Blanton, Richard E. 199
Bloody Sunday 42
Blumrosen, Alfred W. and Ruth G. Blumrosen 235
Bohemia (Czech Republic) 19, 149, 151, 152, 153, 156, 157–166, **158**
Bylany (archaeological site) 153
České Budějovice 161
Hluboká nad Vltavou (archaeological site) 161, **162**, **163**, **164**
Loděnice (archaeological site) 19, 155, 157, **158**, **159**, 160
Mšecké Žehrovice (archaeological site) 157
Schwarzenberg Deer Parks (archaeological site) 161, **162**, **163**, **164**
Böhm, Jaroslav 147
Botswana 101, 110. *See also* Tsodilo (World Heritage Site)
government of 106, 107, 110, 115, 116
Maun 102
Monuments and Relics Act 106
National Conservation Strategy 110
Trans-Kalahari Highway 103

Botswana National Museum 104, 105, 108, 109, 113, 114, 115
Bowden, Mark 127
Bowker, Geoffery C. 198, 200, 218
Bradley, Andrea 137
Bradley, Richard 14, 16, 200
Braudel, Ferdinand 198, 201, 216, 220n3
Bredwa-Mensah, Yaw 177, 179
Brillouin, Leon 221n5. *See also* negentropy
Brink, Yvonne 187
British army 45–46
British-Czech Ancient Landscape Reconstruction (ALRB) Project 151
Broadening the Battlefield: The H-Blocks and the Rise of Sinn Féin 37
Bronze Age (Europe) 19, 146, 157, 160–166, 194
Brook, George A. 104
Brown, Jessica 124
Brown, Ras Michael 188
Bunce Island (Sierra Leone) 177
Bunce, Robert 131
burial mounds 19, 146, 157, 160–166
Burma-Thailand Railway 13
Burton, Sir Richard 57
Bylany (archaeological site) *see* Bohemia
Byrne, Denis 14, 260

Cadw 128
Calvert, Cecil (first proprietor of Maryland colony) 227
Campbell, Alec C. 104, 105, 106, 108
Campbell, Jane 64
Cantwell, Anne-Marie 75, 80
Caribbean 13, 251–274 *passim*, **252**
  archaeology, North American domination of 267
  as a study region 251, 261
  prehistory of 262–264
  see also Arawak; Caribs; Ciboney; Cuba; Jamaica; Leeward Islands; Puerto Rico; Windward Islands; landscape archaeology, regional traditions of
Caribs 263–264
Caroll, Charles of Carrollton 233, 234
Carver, Martin 102
Cavanagh, William G. 194, 195, 202
central Europe *see* Czech Republic; Germany; Poland
ceramic scatters 194, 213–214
Černý, Ervín 149
České Budějovice *see* Bohemia (Czech Republic)
Chake Chake *see* Pemba Island (Zanzibar)
Chami, Felix 57, 58, 71
Chang, Kwang-chih 13, 148
characterisation *see* landscape archaeology, techniques used within
Chase, Samuel 233, 234
Cheek, Charles D. 80
Cherry, John 194, 195, 197, 198, 199, 200, 201, 220n2, 220n4
Chesapeake 231, 232, 233, 234, 256, 258. *See also* Maryland; Virginia
Chesapeake and Ohio (C&O) Canal Company 236–237
China 13, 95, 149
Chittick, Neville 57
Chwaka (archaeological site) *see* Pemba Island (Zanzibar)
Ciboney 263
cities *see* urban landscapes
Civic Trust, The 243
Clare, Thomas 131

Clark, Jo 120, 134, 138
Clarke, David L. 13, 148, 220n4
Clarke, Liam 37
Clifford, James 262
Clifton Plantation (New Providence) 265
cloves 55–56
  production/processing of 66–68
  trade in 62
Clwyd-Powys Archaeological Trust (CPAT) 136
Cobb, Charles 261
Collect Pond (Lower Manhattan) 17, 75–97 *passim*, **76**, **78**, **79**, **81**
  as 'cultural frontier' 77
  infilling of 80, **81**, 83, 84–85, **86**, 89
  long-term development of 90–93
collecting (in natural history) 260
colonial landscapes 20–21, 187–188, 235, 251–268 *passim*
colonialism, archaeology of 61, 253–258, 265–266, 267–268
colonialism, European 20–21, 61, 62, 75, 77, 176, 251–268 *passim*. *See also* decolonisation
colonialism, Middle Eastern 57
communism 156
community area theory 19, 146, 147, 154–155, 167. *See also* Kuna, Martin; Neustupný, Esven
community involvement *see* heritage, community involvement in
Comoros 55, **56**
Conkey, Meg 172
contested landscapes *see* heritage, politics of
Coogan, Tim Pat 37
Coones, Paul 153
Cooper, Frederick 62, 63, 64, 66
Corded Ware 146
  cemeteries 151
Cornwall Archaeological Unit 129
Cosgrove, Denis 22, 123, 254
COST A27 (funding programme) 120
Council of Europe 120, 121, 122
Countryside Agency 128
Countryside Commission 128, 129
Cozzens, Issachar 83, 85
craft production 114–116
Crawford, O.G.S. 127
creolisation 261–262
critical archaeology 258
Crosby, Alfred W. 264
Croucher, Sarah 20, 64, 65, 66, 69
Cuba 257
'cultural ecology' 263
cultural geography *see* geographical studies
culture history 147, 151, 262–263
Curet, Antonio 264
Czech Republic 146, 161
  archaeological thought in 148
  Czechoslovakia 147
  Prague **158**
  South Moravian Hills 146
  see also Bohemia (Czech Republic)
Czechoslovakia *see* Czech Republic

da Gama, Vasco 62
Dahomey (polity) *see* Bénin
Danckaerts, Jaspar 77
Daniels, Stephen 22, 123, 254, 265
Danish Gold Coast *see* Ghana
Darling, Patrick 180
Darlington, John 131
Darvill, Tim 128, 129
Davis, Jack L. 195
De Cunzo Lu Ann 259

and Julie Ernstein 15, 259
De Maret, Pierre 105, 108
Deagan, Kathleen A. 264
débitage 151
decolonisation 57–58, 253
DeCorse, Christopher R. 175, 177, 178, 179
Deeben, Jos 156
Deetz, James F. 172, 175, 187
Delle, James 68, 69, 172, 180, 188, 256, 257, 258, 259
dendrochronology *see* landscape archaeology, techniques used within
Denmark 136, 156, 165
Description of Greece 212
desk-based archaeological research *see* landscape archaeology, techniques used within
Devon Oral History approach 139
Dickens, Charles 17, 82, 95
Digitalt Atlas over Kulturmiljøer i Danmark (DAKD) 136
direct historical approach *see* landscape archaeology, techniques used within
disability studies 259
Divuyu (Botswana, archaeological site) 106
Dixon, Philip 135
documentary archaeology *see* landscape archaeology, techniques used within
Domesday Book and Beyond: Three Essays in the Early History of England 127
domestic space *see* household archaeology
Domination and Resistance 258
Dominica *see* Leeward Islands
Donley-Reid, Linda W. 65, 66
Doukellis, Panagiotis 135
drawn landscape survey *see* landscape archaeology, techniques used within
Drax Hall *see* Jamaica
Dreslerová, Dagmar 19, 154
Drewal, Henry John 102
Dunga (archaeological site) *see* Unguja Island (Zanzibar)
Dutch East India Company 187
dwelling perspective *see* Ingold, Tim
Dyson-Bruce, Lynn 131

earthworks 151
earthworks survey *see* landscape archaeology, techniques used within
East Africa *see* Kenya; Mozambique, Somalia; Swahili Coast; Tanzania. *See also* landscape archaeology, regional traditions of
Ecological Imperialism 264
ecological studies 123
Ecosurv 106
Edgeworth, Matt 22
Edmonds, Mark 23
Edroma, Eric L. 101
Efutu *see* Ghana
Eggers, Hans J. 147
El Padre coffee plantation (Cuba) 257
Elmina Castle *see* Ghana
embodiment 15, 17, 259
empiricism *see* landscape archaeology, empiricism in
England *see* United Kingdom
English Heritage 128, 129, 138
entropy 200, 202, 203, 210–211, 218, 221n5. *See also* negentropy (negative entropy)
environmental psychology 123
EPCL (European landscape network) 131

Epperson, Terrence P. 258
Ermischer, Gerhard 135
Ernstein, Julie 15, 259
Escobar, Arturo 25
ethnicity, archaeology of 57, 147, 176, 188
ethnoarchaeology 178
ethnography *see* landscape archaeology, techniques used within
ethnohistorical research *see* landscape archaeology, techniques used within
Eurocentrism 23
Europae Archaeologiae Consilium 120
Europe *see* Belgium, Czech Republic, Denmark, Germany, Greece, Italy, Poland, United Kingdom. *See also* landscape archaeology, regional traditions of
European Association of Archaeologists (EAA) 120
European Landscape Convention (ELC) 18, 120, 121–122, 124, 129, 134–137
European Pathways to the Cultural Landscape (funding programme) 120
European Union 120
Evening Star (Washington DC newspaper) 243
everyday life *see* landscape archaeology, themes within
excavation *see* landscape archaeology, techniques used within
Expédition Scientifique de Morée (1829–1831)198

Fairclough, Graham 15, 18, 120, 122, 126, 128, 129, 131, 135, 137, 138, 139
and Andrew Wigley 131
Farnsworth, Paul 258, 265
Faulkner, Neil 117
Fefferman, Hilbert 242
feminist theory 15, 23–24, 259
field boundaries 149
Fields, Barbara 237
fieldwalking *see* landscape archaeology, techniques used within
fieldwork, archaeological 16, 22–23. *See also* landscape archaeology, techniques used within
Filip, Jan 147
Finberg, Herbert P.R. 128
Five Points neighbourhood *see* New York City
Flannery, Kent V. 198
Fleisher, Jeff B. 57, 58, 59, 60, 61, 63–64, 65
Fleming, Andrew 22
Fletcher, Roland 16
Florida **252**, 263
Foard, Ashley A. 242
Fontein, Joost 102
Foucault, Michel 15, 31, 204, 256
emphasis upon social control 259
Foundation for the Advancement of Mesoamerican Studies 13
Fourmont, Étienne (Abbé Fourmont) 211
Frederickson, George M. 235
Fresh Water *see* Collect Pond (Lower Manhattan)
Friedman, Jonathan 262
Fry, Gary 124
Futa Jallon *see* Guinea

Galaty, Michael 19
Galways Plantation (Montserrat) 257–258
Gangs of New York (movie) 17, 82

gardens *see* landscapes, historic
Gaul 262
Gede (archaeological site) *see* Kenya
Geismar, Joan 81, 83
Gell, Sir William 211–212
geographical information systems (GIS) *see* landscape archaeology, techniques used within
geographical studies 14, 21, 38, 68, 96, 117n1, 123, 127, 147, 174, 252, 254, 257, 259, 260, 263. *See also* New Geography
geomorphological studies *see* landscape archaeology, techniques used within
Germany 134–135, 146, 152, 156, 165
  Aldenhovener Platte 153, 154
  Hessen 154
  Niederlausitz 153
  Rhineland 150, 153, 156
Ghaidan, Usam 66
Ghana 180
  Banda Chieftaincy **173**, 180
  Begho (archaeological site) 178
  Danish plantations in 179–180
  Efutu (archaeological site) 257
  Elmina Castle (slave fort) **173**, 177–180, **179**
ghosts (of place) 18–19
Gilchrist, Roberta 24
Gilroy, Paul 262
GIS *see* landscape archaeology, techniques used within
Glassman, Jonathon 62, 66, 70
Goodwin, Conrad M. 189, 257–258
Gorée Island *see* Senegal
Gorman, Alice 14
Gosden, Chris 20, 251, 259, 261, 262, 267
  and Lesley Head 14
Graham, Brian 38, 52
Gramsch, Alexander 150
Grau-Møller, Per 136
Great Awakening 234
Great Zimbabwe (Zimbabwe, archaeological site) **173**, 175
Greater Antilles *see* Caribbean
Greece 194. *See also* Keos; Melos; southern Argolid
Greene, Sandra 175
Grenada *see* Windward Islands
Guinea 181
  Futa Jallon **173**, 186
  Rio Pongo **173**, 181, 186
Gundaker, Grey 261

Hall, Martin 122, 175, 187, 188, 256
Hallewas, Daan P. 156
Hambukushu **104**, 107–116
Handbook of South American Indians 263
Harding, Sandra 23
Harper's Weekly 241
Harraway, Donna 23, 24
  'partial perspective', idea of 24–25
Harris, Edward 85
  Harris Matrix **86**
Harrison, Rodney 267
  idea of 'shared landscapes' 19
Hartsock, Nancy 23
Harvey, David C. 139
Hatchery West, Illinois 197
Hauser, Mark 188, 265
  and Dan Hicks 20, 23
Haviser, Jay 257
Hayden, Dolores 76
Hayles, N. Katherine 204
Head, Lesley 14
Hensel, Witold 153
heritage 15–16, 18, 21, 25, 31, 50, 52, 101–102, 106, 109,

110–117, 120, 123, 125, 129, 136, 138, 200, 219, 236, 240
community involvement in 15–16, 116–117, 139
definitions of 102
management of 18, 120–139 *passim*, 152, 156
ownership of 111–116, 121
politics of 15–16, 30–52, 226–250
see also heritage landscapes; heritage tourism; planning, spatial; significance (of heritage); value (of heritage)
heritage landscapes 13, 15–16, 30–54, 101–119, 226–250
as 'common culture' 122, 139
integrity of 129
management of 125, 156–157
setting (of sites or monuments) 128
see also heritage; landscapes, historic
heritage tourism 16, 108, 109–116, 139
and local crafts 114–116
Herring, Peter 120, 129, 131, 132, 133
Hessen *see* Germany
Hicks, Dan 14, 19, 20, 21, 23, 25, 188, 253, 255, 265, 266, 268
and Mary C. Beaudry 25
and Audrey Horning 259
Higman, Barry 23, 254, 255, 257, 260
idea of 'spatial economy' 254
Hingley, Richard 135
Historic Annapolis *see* Annapolis (Maryland)
historic landscape characterisation (HLC) *see* landscape archaeology, techniques used within
historic land-use assessment 135
historic preservation movement 242–246
historical archaeology 13, 20–21, 36–37, 137, 147, 172, 175, 176, 188, 251–268 *passim*
in Africa 176
industrial archaeology 147
relationship with prehistoric archaeology 176, 261
see also African American archaeology; African Diaspora archaeology
historical geography *see* geographical studies
Hluboká nad Vltavou (archaeological site) *see* Bohemia (Czech Republic)
HMP Maze *see* Long Kesh/Maze Prison (Northern Ireland)
Hodder, Ian 22, 24, 197, 201, 220n3, 255
and Clive Orton 172
Hoffmann, Tobias 95
Holland 134, 156
Holtorf, Cornelius 16
Honychurch, Lennox 264
Hope Simpson, Richard 220n2
Horning, Audrey 259
Horton, Mark C. 20, 59–60
and John Middleton 58, 69
Hoskins, W.G. 127
household archaeology 65–66
Hueda (polity) *see* Bénin
Hurry, Silas D. 227–229, 231
Hurst, Harold W. 237
Hurst, John 127

Ideas of Landscape 21
identity, archaeology of 70–71
Ilare District *see* Nigeria
indentured servants 253

Indian Ocean 55, **56**, 57, 188. *See also* Comoros, Madagascar, Swahili Coast
Indian River (Dominica) 264
Indigenous archaeology 15, 20, 57–63, 176, 186, 253, 260–264, 266
Indigenous people 77
Indigenous property rights 251
industrial archaeology *see* historical archaeology
Ingle, Marjorie 80
Ingold, Tim 14, 16–17, 201
  taskscape, idea of 16
interpretive archaeology 17, 21–24, 174, 177, 254, 267. *See also* post-processual archaeology
Irish News 34
Irish Republican Army (IRA) 33–34, 39
Iron Age (Africa) 106
Iron Age (Europe) 19, 146–147, 157–160
Isaac, Rhys 234
Islam, archaeology of 57, 58, 59–60, 66, 70. *See also umma* (Islam)
Isle of Man **126**
Italy 95, 194
IUNC (World Conservation Union) 124
ivory, trade in 62

Jacobs, Jane 260, 268
  notion of 'unruly imperialism' 260
Jacobsen, Thomas W. 207, 209, 210
Jakob, Karl H. 152
Jamaica 69, 251, 254, **255**, 256–259, 265
  Drax Hall plantation 257
  'Higglers' (itinerant traders) 265
  Nanny Town 257
  Port Royal 265
  Seville plantation 257
  Spanishtown 265
  Yallah's region 256
  see also Maroon settlements
Jamaica Planter's Guide 254–255
Jamaica Surveyed 254
Jameson, Michael H. 199, 201, 207–209, 211–214, 216, 217, 219. *See also* Argolid Exploration Project
Jameson, Virginia 207
Jankuhn, Herbert 147, 148, 151, 152
Jarman, Neil 39, 43
John Milner Associates, Inc. 81, 82
Johnson, Matthew 21–22, 127, 254, 258
Johnson, Nick 133
Jones, Siân 262
Jongman, Rob 123
Jordan, David W. 233
Journal of the Medieval Settlement Research Group 21

Keegan, William F. 263–264
Keiller, Alexander 127
Keitumetse, Susan O. 15, 102, 106, 107, 108, 110–111, 114–116
Kelly, Kenneth G. 20, 173, 175, 177, 181, 182, 184, 185, 188
Kenya **56**, 57
  Gede (archaeological site) 57
  Manda (Lamu Archipelago) 57
  Mombasa 62
  Shanga (Lamu Archipelago) 20, 58, 59
Keos (Greece) 197, 199
  Laconia 202
Kersten, Karl 152
Khoekhoen pastoralists (Namibia) 187
Kiekebusch, Albert 152

Kilwa region *see* Tanzania
Kinahan, Jill 187
Kirkman, James 57
Knapp, A. Bernard 201, 220n3
  and Wendy Ashmore 172
knowledge, situated 14, 24
Kossina, Gustav 147
Kostrzewski, Józef 147
Krotz, Esteban 25
Kryder-Reid, Elizabeth 184, 230
Kulikoff, Allan 233
Kuna, Martin 19, 146, 151, 152, 153, 154, 155, 156, 161, 163
!Kung San **104**, 107–115
Kusimba, Chapurukha M. 58, 71

La Taste Estate (Grenada) 265
La Tène *see* Iron Age (Europe)
Laconia Rural Sites Project 202. *See also* Cavanagh, William G.
Lambrick, George 127
land clearance 260
Landesaufnahme, archäeologische see landscape archaeology, techniques used within
Landmarks project (Europe) 120
Landmarks Preservation Commission (New York) 80
landscape, Western ideas of
  and colonialism 20–21, 23
  as ideational 21, 251, 256, 258, 267
  as *tabula rasa* 254
  as *terra nullius* (unowned land) 251
  as visual ('Western gaze') 20, 23, 61, 71, 122, 128, 196, 253, 255–256
  brought to Africa 187–189
  'god trick' 23
  'jeweller's eye' view 23
  wilderness 94, 97, 253–254
  see also colonial landscapes; landscapes
landscape architecture 123, 128, 139
Landscape History 21
landscape painting **78**, 96–97
landscapes
  and aesthetics 122
  and emotion 15
  and memory 77–80, 93–94
  as materially complex 195, 260–261
  as non-verbal 259–260
  as palimpsest **126**, 127, 207, 218
  as preserving earlier landscape forms 235
  as standpoints 14, 21–25
  as understood in many different ways 69, 173
  aura of 18–19
  boundaries of 30
  definitions of 121–122
  impermanent landscapes 60–61, 68–69
  'quality' of 123
  mobility in 252, 260, 262, 264, 266
  perception of 122, 124
  'reading' landscapes 22
  ubiquity of 122
  see also landscape, Western ideas of; landscapes, historic; soundscapes; spacescapes; taskscape, idea of; wirescapes
landscapes, historic
  active creation of 236–247 *passim*
  as a commodity 236
  as 'setting' for monuments 128
  baroque designed landscapes 184, 227–229
  gardens 36–37
  neo-classical ('Georgian') designed landscapes 70, 184, 187, 230, 234–235

parterre gardens 231
see also colonial landscapes; heritage landscapes; landscapes; maritime landscapes; plantations; rural landscapes; urban landscapes
landscape archaeology *see* landscape archaeology, regional traditions of; landscape archaeology, techniques used within; landscape archaeology, themes within; community area theory; *Landschaftsarchäologie*; *Siedlungsarchäologie*; settlement patterns
landscape archaeology, regional traditions of
Caribbean region 253–258, 262–266
continental Europe 19, 134–137, 146–157 *passim*, 166–167
East Africa 20, 55–61, 71
Mediterranean region 194–195, 197–200
Middle East 14
North America 148, 172, 174
southern Africa 174–177, 187–188
United Kingdom 14, 21–22, 148, 174
West Africa 20, 172–181
landscape archaeology, techniques used within 14, 124–125
air photograph interpretation 156
archäeologische Landesaufnahme, (archaeological area scan) 19, 146, 150, 152–153, 157
dendrochronology 152
desk-based research 152
direct historical approach 263
diversity of 122–124
documentary archaeology 17, 31, 36, 39, 58, 82–85, 102
drawn landscape/earthworks survey 64–65, 127–128, 131, 194, **266**
ethnography 110–116
ethnohistorical research 64, 176, 178
excavation 17, 65–68, 76–77, 80–93 *passim*, 97, 153–154, 158, 200
fieldwalking 19, 157–160, 198, **199**, 212–214
geomorphological studies 77, 80–81, 85–89
GIS, analytical use of 19, 132, 136, **138**, 156, 161–166, **165**
historic landscape characterisation (HLC) 18, 120–139 *passim*
linguistic history 58
oral history/oral testimony 31, 36, 39, 47–49, 58, 139, 176, 178
predictive modelling 157–158
probabilistic survey 197–198, 208
radiocarbon dating 85–89
spatial analysis 146, 172, 174
tree surveys 131
landscape archaeology, themes within
alternatives to site- or monument-based approaches 122, 133, 149, 157
designed landscapes 227–228
diversity of 14, 23–25, 134
empiricism 14, 21–22, 24–25
everyday life 16, 70–71
interdisciplinary landscape studies 121
landscape formation 202
'life histories' or 'biographies' of landscapes 16, 22, 134, 207, 217, 218
long-term change 16, 17, 76, 82–93 *passim*, 149, 177, 194, 195, 201, 219

resistance (against social control) 21, 35, 70, 189
scales of analysis in 14, 24, 121–122, 125–127, 149–150, 262–266
social control 31, 39, 42, 46–47, 52, 68
'social landscapes' 60, 70
time in 94–221 *passim*
unity of 14
see also archaeology, themes within
Landschaftsarchäologie 150, 167
Lane, Paul 176
LaRoche, Cheryl J. 94
Late Stone Age (Africa) 106
Late Woodland Period (North America) 86, 88, 90
Latin American archaeology 268
Latour, Bruno 204, 205, 214, 217. *See also* Serres, Michel
LaViolette, Adria 58, 59, 60, 65
Law, Robin 175, 182, 184
Lawrence, Arnold W. 180
Lawrence, Susan 261
Layton, Robert H. 14, 135, 172
LE:NOTRE Thematic Network Project in Landscape Architecture 123
Leake, William M. 198, 210–211
Leeward Islands 253
Dominica 264
Montserrat 257–258
St Lucia **266**
Lefebvre, Henri 256
Leicester School of Local History 127
Leone, Mark P. 37, 175, 184, 227–231, 233–236
and Silas D. Hurry 177
Lesser Antilles *see* Leeward Islands; Windward Islands
Leverhulme Trust 178
Lewis-Williams, J. David 104, 112
'life histories' *see* landscape archaeology, themes within
Lightfoot, Kent 261
Lin, Maya 96
linguistic history 58
Lipe, William D. 102
Loděnice (archaeological site) *see* Bohemia (Czech Republic)
Long Kesh, RAF base at 31, 35
Long Kesh/Maze Prison (Northern Ireland) 15, 30–52 *passim*
academic studies of 37–39
artefacts made within (by prisoners) 40–42
army base at 31, 45–46
buildings archaeology of 35–36
depicted in murals 43–44
'dirty protests' at 33
future of 49–51
Gardiner Report on (1975) 32
H-Blocks of 32, **33**, 34, 35, 37, 42, **44**, 45, 47–48
Hennessey Report on (1984) 35
Hunger Strikes at 18, 33–34, 37, 39, 41–44, 49
Masterplan Consortium Panel 50–51
name of 34
nissen huts at 31, 32, 35, 44–45
prison officers of 39, 45, 47–48, 49
prisoners in 30, 33, 35, 39, 47–49
Loyalist banners (Ulster) **41**
Loyalist paramilitaries (Ulster) 34, 44, 49
Lucas, Gavin 187–188, 197, 200, 201, 203, 204, 205
Lüning, Jens 150, 154

McAtackney, Laura 15, 49, 50
McCormick, Jonathan 43
McDonald, William A. 198
and George R. Rapp 194, 198
MacEachern, Scott 178

McGlade, James 204–205, 220n1
  and Sander van der Leeuw 204–205
MacInnes, Lesley 120, 121, 131
McIntosh, Roderick J. 172, 174, 175
McIntosh, Susan 172, 175
McKeown, Laurence 37
McNab, Andrew 127
Madagascar 55
Magdalen Society Asylum (Philadelphia) 259
Mahonda (archaeological site) *see* Unguja Island (Zanzibar)
Maitland, Frederic W. 127–128
Mangole (Botswana, archaeological site) 109
Manda (Lamu Archipelago) *see* Kenya
maps and plans (historical) 251, 254, 255, 259, 260, 262
maritime landscapes 122, 216–217, 252
Markell, Ann 187
Maroon settlements 257
Martin, Samuel 254
Marxist archaeology 259
Maryland 233
  'native elite' of (18th century) 233
  plantation owners in (19th century) 237
  see also Annapolis; Baltimore; Bladensburg; Chesapeake; St Mary's City
Maryland Gazette 238
Masao, Fidelis T. 112
Mason, Randall F. 241
materiality 21, 22, 38, 52, 124, 252, 258–262, 267–268
  double meaning of 25–26
Matlapeng, Geoffrey 15
Matless, David 123
Matopo (Zimbabwe, archaeological site) 104
Matsikure, Joymore 102
Mattingly, David 195
Matthews, Christopher 18, 230, 236–237, 239
Maze Consultation Panel (MCP) 50
Maze Prison *see* Long Kesh/ Maze Prison (Northern Ireland)
medieval archaeology 127, 136, 147, 149, 151, 154, 158
Mediterranean region 134, 194–220 *passim*. *See also* landscape archaeology, regional traditions of
Meehan, Martin 47–48
Meier, Thomas 135
Melos (Greece) 198, **199**, 200–201, 219
  Phylakopi (archaeological site) 200
memory 13, 16, 31, 51–52, 77–80, 93–94
  commemoration 38
  'distributed memory' 207, 218
  remembrance 17
Mendoni, Lila 135
Merriman, Nick 117
Meskell, Lynn 177, 259
Mesolithic (Europe) 209
Mesolithic-Neolithic transition 209–210
Metheny, Karen B. 36–37, 76, 184
Mexico 13
Mgoli (archaeological site) *see* Pemba Island (Zanzibar)
Middle East 14, 57. *See also* landscape archaeology, regional traditions of
Middle Stone Age (Africa) 106
Middleton, John 58, 69
  and Jane Campbell 64
migration 262–264
Miller, Daniel 258
Miller, Henry 227–228

Milne, Claudia 77, 80
Minnesota Messenia Expedition 198
Mintz, Sidney W. 257, 261
missionaries 186
mobility, landscapes of *see* landscapes, types of
Mogadishu (Somalia) 55
Monamo, Leseka 15, 108, 109, 112
Monroe, J. Cameron 175, 177, 181, 185
Montserrat *see* Leeward Islands
monuments, biographies of 16. *See also* landscape archaeology, themes within
Moore, Henrietta L. 70
Morgan, Edmund S. 234, 235
Morgan, Helen M. 234
Morphy, Howard 256
Morris, Ian 128, 194
mosques 57, 70
Mowlam, Mo 34
Mozambique **56**, 64
  Sofala 55
Mšecké Žehrovice *see* Bohemia (Czech Republic)
Mtambile (archaeological site) *see* Pemba Island (Zanzibar)
Muir, Richard 123
murals 43–44
Museum of London Archaeological Services **266**
Muscat *see* Oman
Museums 152, 205, 217, 231. *See also* Botswana National Museum; Tsodilo Hills World Heritage Site
Myers, Garth 68–69

Namibia 105, 187
Native American archaeology 90. *See also* Indigenous Archaeology
natural places, archaeology of 14
Nazism 147
Ndoro, Webber 102
negentropy (negative entropy) 203, 207, 210–211, 217, 218
  defined in the work of Leon Brouillon and Michel Serres 221n5
  see also entropy
Neolithic (Europe) 14, 146, 150, 152, 153, 209. *See also* Mesolithic-Neolithic transition
Netherlands *see* Holland
Neustupný, Esven 19, 146, 148, 149, 154–157, 160
  idea of *sféra jinosti* 155
Neves, Eduardo G. 263
Neville, Christopher P. 77, 80, 94, 96
New Amsterdam *see* New York City
New archaeology *see* processual archaeology
New Face for Britain (film) 243
New Geography 148, 153. *See also* geographical studies
New South Wales 14, 260
New York City 17, 75–97 *passim*, **91**, **92**
  African Burial Ground 17, 75, 78, 80, 93, 94, 95, 96, 97
  African Zion Church 95
  Battery Park 97
  Bowery Theatre 77
  Bull's Head Tavern 77
  Castello Plan of (1660) 91
  Catimut's Hill 90, **91**, **92**, 93
  City Hall 85, **92**
  Dutch settlers in 75–77, 81–84, 85, 88, 91, **92**, 94
  East River 75, 83
  English settlers in **76**
  Five Points neighbourhood 17, 80, 82, **91**, **92**, 93–94, 95, 97

Foley Square 77, 80, 81, **92**
geology of 82–83
Hudson River 75, 82, 83, 90
Jewish Burial Ground 78
Lispenard's Meadows **81**, 91
Lyne Plan of (1728) 83, 91
Maerschalck plan of (1754) **79**, 80, 83, 84
Manhattan Company (waterworks) 83
Metropolitan Corrections Center (MCC) Tunnel 77, 81–82, 85, **86**, **87**, 88–90, 96, 97
New Amsterdam 75
prehistory of 85–88, 90–91, **92**
Ratzer Plan of (1767) 84
tanning industry in 77, 84, 85, 88, 89, 93, 95
Taylor and Roberts' *New and Accurate Plan* of 84
Vingboom map of (1639) 82
World Trade Center 97
see also Collect Pond (Lower Manhattan)
New York Times 96
Newsom, Lee A. 263
and Kathleen A. Deagan 263
Newson, Linda A. 264
Nicholson, Francis (Governor of Maryland, 1694–1699) 227
Nigeria 180
early urban centres in 180
Ilare District **173**, 180–181
Niger Delta 175
noiseuse see Serres, Michel
Norfolk, Virginia 237
Norman, Neil 20, 185
and Kenneth G. Kelly 173, 175, 177, 181, 185, 188
North America 15, 176. *See also* landscape archaeology, regional traditions of; United States
Northern Ireland Prison Service (NIPS) 35
Northern Ireland *see* United Kingdom
Notsé *see* Togo
Nqoma (Botswana, archaeological site) 106
Nugent, Kieran 33
Nurse, Derek 58
and Thomas Spear 58
Nyame Akuma 174

Oaxaca (Mexico) 13
Office of First Minister and Deputy First Minister (Northern Ireland) 51
Ogundiran, Akinwuni O. 175, 180, 181
Olivier, Laurent 201, 218
Olsen, Bjørnar 195, 196, 197, 205
Olwig, Karen F. 123
Oman 62
Muscat 62
Omani presence in Zanzibar 62, 64
On the Nature of Animals 214
One World Archaeology (book series) 25
oral history/oral testimony *see* landscape archaeology, techniques used within
Ordnance Survey 127, 131
Orejas, lmudena 135
Orser, Charles E. 262
Orton, Clive 172
Ostinoid-type material culture (Puerto Rico) 264
Otis, James 234
Ottoman period (in Mediterranean archaeology) 194
Ouidah *see* Bénin
Out of Time: Irish Republican Prisoners Long Kesh 1972–2000 37

Paca, William (Governor of Maryland) 233, 234–235, 236. *See also* William Paca House and Garden (Annapolis)
Paca-Steele, Barbara 231
Pagán-Jiménez, Jaime 268
Palang, Hannes 124
palimpsest, landscape as *see* landscapes
Palladio, Andrea (Renaissance architect) 230
Palus, Matthew 18, 243, 259
pan-Africanism 262
Paris 70
Patterson, Thomas C. 102
Pausanius 211–212
Pels, Peter 259–260, 267
Pemba Island (Zanzibar) 20, 55, **56**, 59, 60, 62, **63**, 64–67
  Bweni 66, 67, **68**, 69
  Chake Chake **63**
  Chwaka (archaeological site) 59
  clove production on 62–64
  Mgoli 64–66, **67**, 69
  Mtambile (archaeological site) **63**, 64–68
  Piki (archaeological site) **63**, 64–65
  Pujini (archaeological site) 59
performance art 38
Perlès, Catherine 209
Perry, Warren R. 80
Peru 172
  Viru Valley 172
Pettegrew, David 200
phenomenology 14, 22, 37
Piki (archaeological site) *see* Pemba Island (Zanzibar)
Pinney, Christopher 268
Placzek, Adolf K. 230
planning, spatial 120, 124, 125, 138
plantation archaeology *see* plantations. *See also* landscape archaeology, themes within
plantations 20, 21, 55, 62–71 *passim*, 177, 179–180, 187, 253, 254–258
plough damage (to buried archaeological remains) 151
Poland 146, 147, 152, 153, 156
politics *see* heritage, politics of
Posnansky, Merrick 178
postcolonial theory 251–254, 262. *See also* colonialism, archaeology of; decolonisation
Postgate, Nicholas 14
post-processual archaeology 13, 17, 21–22, 148, 174, 258. *See also* interpretive archaeology
Potter, Parker B. 237, 240, 246
Poznañ University 147
PPG 15 (Planning Policy Guidance: Planning and the Historic Environment) 129
predictive modelling 157–158
Preucel, Robert W. 177
Price, T. Douglas 209
Prigogine, Ilya 203, 204, 205
prisoners of war 13
prisons, archaeology of 15, 31, 259. *See also* Long Kesh/ Maze Prison (Northern Ireland)
Pritchett, W. Kendrick 198
probabilistic survey *see* landscape archaeology, techniques used within
processual archaeology 22, 24, 61, 174, 175, 263
protected areas 124
Puerto Rico 261, 264, 268
  Tibes (archaeological site) 264
Puerto Rico Project 264
Pujini (archaeological site) *see* Pemba Island (Zanzibar)
Pulsipher, Lydia 189, 257, 258
  and Conrad M. Goodwin 189, 257–258

Pungetti, Gloria 123
Purbrick, Louise 39

Quarcoopome, Nii O. 175

RAC (Royal African Company) 182
racial segregation 14
Rackham, Oliver 128
radiocarbon dating 85–89
Randall, Alexander 238, 240
Rankin-Hill, Lesley 80
Rapp, George R. 194, 198
Redman, Charles M. 197
reflexivity 20, 23, 24, 172, 267
Reformation, Protestant 227
Regan, Madeleine 13
Reid, Andrew 176
Reid, Bryonie 38
remembrance *see* memory
Renfrew, Colin 200, 220n4
  and Malcolm Wagstaff 198
Reps, John 228
resistance (against social control) *see* landscape archaeology, themes within
Restrepo, Eduardo 25
Revolution, American 226, 235
Rhineland *see* Germany
Rhodes, Harrison 241
Ribeiro, Gustavo L. 25
Richard, François G. 175, 177
Richmond (Virginia) 236
Riley, Elihu S. 236
Riley, Mark 139
Rio Pongo *see* Guinea
Rippon, Stephen 120, 126, 128, 130, 135
Robbins, Lawrence H. 104, 105, 106
Robertshaw, Peter 176
Robertson, Alexander **78**, 96–97
Robinson, Nicholas A. 101
rock art 13, 103–104. *See also* Tsodilo Hills World Heritage Site (Botswana)
Romanticism 22
Roughley, Thomas 254–255
Rouse, Irving 262–264
  'circum-Caribbean' theory 263
Rowlands, Michael 252, 261
Royal Air Force (RAF) 31, 35
Royal Commission on the Ancient and Historical Monuments of England 127
Rubertone, Patricia 15
Ruiz del Árbol, María 135
Runnels, Curtis N. 207
rural landscapes 20, 55, 58, 61, 71, 94, 122, 127, 132, 187, 202, 234, 237, 238, 258, 265. *See also* plantations

Sádlo, Jiří 146
Said, Edward 254, 259
St Lucia National Trust **266**
St Lucia *see* Leeward Islands
St Mary's City (Maryland) 227, 233
  18th century history of 233
  17th century planned landscape of 227
  State House 227
Salmond, Anne 22
San *see* !Kung San
Sands, Bobby 33, 39, 42, 43
  death of 33–34
Sarlöv-Herlin, Ingrid 122
Saunders, James R. and Renae N. Shackelford 242
Savi *see* Bénin
scales of analysis 121–122, 125–127, 135–136, 149–150, 262–266
Scazzosi, Lionella 124, 135
Schaafsma, Curtis F. 117
Schama, Simon 94, 123

Scheick, Cherie 172, 177
Schiffer, Michael 25, 200
Schleidt-Peñalva, Maria L. 81
Schmidt, Peter R. 23, 102
Schou, Annette 136
Schrire, Carmel 187
Schuldenrein, Joseph 17, 77, 81, 88, 90
Schuyler, David 242
Schwarzenberg Deer Parks *see* Bohemia (Czech Republic)
science studies 22
Scorsese, Martin 82
Scotland *see* United Kingdom
Scott, David 261, 267
Scott, James 256
Scott, Patricia 187
Selman, Paul 122
Senegal 177
  Gorée Island (slave fort) **173**, 180, 182
  Senegal Valley 175
Senses of Place: Senses of Time 38
Serres, Michel 203, 204, 205, 210, 219
  noiseuse, concept of 216
  with Bruno Latour 17, 195, 196, 203, 205, 206
setting (of sites or monuments) *see* heritage landscapes
settlement patterns 13, 148, 172, 174, 179. *See also* landscape archaeology
Seville plantation *see* Jamaica
Seymour, Suzanne 265
Shackel, Paul 230, 231, 258
Shanga (Lamu Archipelago) *see* Kenya
Shanks, Michael 194, 203, 208, 214, 217, 218, 220n1
  and Christopher Tilley 195, 197
shared landscapes *see* Harrison, Rodney
shell middens 88
Shellenhamer, Jason 231
Sheller, Mimi 253
Shepherd, Nick 25, 261
Sheriff, Abdul 62, 63, 64, 68, 70
Sherratt, Andrew G. 22, 146
Siedlungsarchäologie (settlement archaeology) 147–148, 151. *See also* Kossina, Gustav; Jankuhn, Herbert
Siedlungskammer (micro-regions) 19, 146, 153–154
Sierra Leone 177
significance (of heritage) 15, 26, 38, 39–42, 101, 107, 113, 114, 128, 129, 135, 246–247. *See also* heritage; value (of heritage)
Silliman, Stephen 261, 267
Singleton, Theresa 257
Sinn Féin 37, 39, 40
Sites and Monument Records (SMRs) (United Kingdom) 132–133
slave gardens *see* landscapes, types of
slave trade, Atlantic 178, 180, 181–186. *See also* African Diaspora
slave trade, Indian Ocean 62, 64
slavery, archaeology of 62–71, 77, 94, 187–188, 235, 237, 253–258
  intangibility of 69
  slave accommodation 257
  slave burials 70, 75
  slave gardens 258
  slave house yards 257
  urban slavery 94
Sloane, Hans 260
Slovenia 13
Sluyter, Andrew 260
Smith, Claire 13, 15

Snodgrass, Anthony 198
social landscapes *see* landscape archaeology, themes within
Society for American Archaeology 267
Society of Africanist Archaeologists 174
Somalia 55, **56**
Soudský, Bohumil 153
Soumonni, Elisée 175
soundscapes 13
South Africa 105, 175, 176, 187, 254, 256
  Drankensburg (archaeological site) 104
  Dwars Valley 187–188
  Vergelegen (archaeological site) 187
South America, prehistoric archaeology in 262–263
South Carolina 13
southern Africa *see* Namibia; South Africa; Zimbabwe
southern Argolid (Greece) 17, 207–217 *passim*
  Bay of Lorenzo **207**, 216
  Bisti of Ermioni (archaeological site) **207**, 210–211, 216, 217
  Ermioni Magoula (archaeological site) **207**, 216
  Franchthi Cave (archaeological site) **207**, 209–210
  Halieis (archaeological site) **207**, 216
  Iliokastro (archaeological site) **207**, 212
  Koiladha Bay **207**, 216
  Loutro, well at **207**, **217**, 218
  Monastery of Ayioi Anayiroi **207**, 212
  Panayitsa (archaeological site) **207**, 216
  Philanoreia (archaeological site) **207**, 217
  Pron of Ermioni (archaeological site) **207**, 210, 214
  Soros (archaeological site) **207**, 212
spacescapes 14
spatial analysis 146, 172, 174
  multivariate spatial analysis 159
  see also landscape archaeology, techniques used within
spatial archaeology 13, 22, 148. *See also* landscape archaeology; spatial analysis
spatial planning *see* planning, spatial
Spear, Thomas 57, 58
Stahl, Anne 175, 176, 177, 180
Stamp Act (1765) 234
standpoint epistemology 14, 21–25
Star, Susan L. 198
Stengers, Isabelle 203, 204, 205
Steward, Julian 263
Stoler, Ann L. 251
stone tools 151
stonetowns (Swahili coast) 56–60, 71
stonetowns *see* Swahili coast
Stoumann, Ingrid 136
Strathern, Marilyn 24
stratigraphy 85–89, 97
Sued-Badillo, Jalil 267
sustainability in heritage management 101–102, 112, 117
Sutton, Susan B. 207
Swahili coast 20, 55–71 *passim*, 175. *See also* stonetowns (Swahili coast)
Swahili identity 55, 57
Swahili origins, archaeological study of 57–59

Taino 253
Tanganyika *see* Tanzania
Tanzania **56**, 60, 64

Kilwa Kisiwani 60
Kilwa region 57, 60–61
Tanganyika 57
see also Pemba Island (Zanzibar); Unguja Island (Zanzibar); Zanzibar
Tarlow, Sarah 258–259
taskscape, idea of *see* Ingold, Tim
Tawana Land Board (Botswana) 108
temporality 16–19. *See also* time
Ten Men Dead 37
terra nullius (unowned land) *see* landscape, Western ideas of
The Billy-Boy: The Life and Death of the LVF Leader Billy Wright 37
The Mediterranean and the Mediterranean World in the Age of Philip II 198
Thiaw, Ibrahima 177
Thomas, Julian S. 14, 21, 22, 61, 71, 177, 258–259, 267
Thomas, Nicholas 262
critique of Edward Said 259
notion of 'entangled objects' 268
Tilley, Christopher 14, 37, 122, 195, 197
time 16–19, 121, 194–225 *passim*
as a quality of landscapes 17
as non-linear 196, 204, 206–210 *passim*, 217–218
clock time 195, 203
'dynamic time' 204
'entropic time' 195, 200–203, 217
'ethnographic time' 201
future, the 122
modernist or Newtonian conception of 203
past and present, relationship between 121, 195, 220
'percolation' of 17, 203–206, 210, 217–220
speed of 204
time depth 137
'time perspectivism' 202
see also Braudel, Ferdinand; *Annales* school of historical writing
Tode, Alred 152
Togo 178
Notsé 173, 178
Togola, Tereba 175
tourism *see* heritage tourism
towns *see* urban landscapes
Townsend, Andrew 13
Tranter, Margery 127
tree surveys *see* landscape archaeology, techniques used within
Troubles (Northern Ireland) 30, 34, 36, 37, 40, 50
Trouillot, Michel-Rolph 251, 261
on anthropology's 'savage slot' 261
Tsing, Anna 268
Tsodilo Hills Management Plan: Scheme for Implementation 106
Tsodilo Hills World Heritage Site (Botswana) 15, 102, **103**, **104**, **105**, 106–117
criteria for inscription on World Heritage List 106–107
development of heritage landscape 108
Divuyu (archaeological site) 106
Mangole (archaeological site) 109
museum at 108, 109, 113–116
Nqoma (archaeological site) 106
rock art at 103–104, **105**, 107, 110 111, 112, 113, 115
Tsodilo, Mountain of the Gods 106
Turkey 194, 197

Rough Silicia (landscape survey) 202
Turner, Sam 15, 18, 120, 126, 131, 132

Ucko, Peter 14, 135, 172
umma (Islam) 59
UNESCO (United Nations Educational, Scientific and Cultural Organisation) 15, 101, 106–108, 110, 113, 115
  'Agenda 21' guidelines 101
  World Heritage Committee 106
  World Heritage list, criteria for 15, 101, 106–107, 110
Unguja Island (Zanzibar) 20, 55, **56**, 62, **63**, 64–65
  clove production on 62–64
  Dunga (archaeological site) 64
  Mahonda (archaeological site) 64
  Zanzibar Town 62, **63**
United Kingdom 138, 262
  Belfast 38, 39, 43, 46, 51
  Cornwall 129, 131
  Devon 127, **130**, **133**, **138**, 139
  Gwent Levels (south Wales) 130
  Midlands 127
  Moccas Estate (Hertfordshire) 265
  Ulster 15, 34, 38, 44
  Scotland 135–136
  Somerset 127
  Wales 134, 136
  see also Belfast (Northern Ireland); landscape archaeology, regional traditions of; Long Kesh/Maze prison
United States *see* Chesapeake, Maryland, New York City, South Carolina, Virginia, Washington DC
Upper Palaeolithic (Europe) 146, 151, 209
Upton, Dell 175
urban landscapes 17–18, 20, 56–60, 71, 75–97 *passim*, 122, 175, 178, 180, 226–247, 252, 260
uwezo (status, Swahili) 68–69

value (of heritage) 15, 39–42, 101–102, 128, 135, 137, 139
  as a commodity 18
  local values 107–116
  sense of place 75
  'universal' values 106–107
  see also heritage; significance (of heritage)
van Andel, Tjeerd 207
van der Leeuw, Sander 204–205
van Dommelen, Peter 261
van Dyke, Ruth M. 201
Venclová, Natalie 155, 157, 160
Venetian period (in Mediterranean archaeology) 194
Venturi, Robert 241
Verger, Pierre 184
vibanda (impermanent houses, Swahili) 68–69
Virginia 254. *See also* Chesapeake; Richmond (Virginia)
Visweswaran, Kamala 24

Wagstaff, Malcolm 198
Wahle, Ernst 147
Wales *see* United Kingdom
wall murals (Ulster) 42–44
Wall, Diana di Zerega 75, 80
Wallerstein, Immanuel M. 25
Wandsnider, LuAnn 194, 195, 200–202, 204–205
Washington DC 237
Watkins, Charles 265
Watts, David 254, 255, 260
WCED (World Commission on Environment and Development) 117
Webmoor, Timothy 197

West Africa 172–189 *passim*, 265. *See also* Bénin; Ghana; Guinea; landscape archaeology, regional traditions of; Nigeria; Senegal; Sierra Leone; Togo
West Africa Trade Project 178
Western Gaze *see* landscape, Western ideas of
whaling 187
Whatmore, Sarah 21
Wickham-Jones, Caroline R. 121
Wigley, Andrew 131
wilderness *see* landscape, Western ideas of
Wilkie, Laurie 258
  and Kevin Bartoy 259
  and Paul Farnsworth 265
Willey, Gordon 148, 172
William Paca House and Garden (Annapolis) 231, **232**, 233–236, 238, **239**, 241, 246, 255
Williams, Howard 16
Williamson, Tom 21
Willroth, Karl-Heinz 165
Wills, Patrick 131
Wilshusen, Richard 14, 172, 177
Wilson, Patrick 47–49
Wilson, Samuel M. 261
  'cultural mosaic', contact-period Caribbean as 264
Windward Islands 253
  Grenada 265
  St Lucia **266**
Winer, Margot 16, 254
wirescapes 18, 243, **244**, 245–247
Witmore, Christopher L. 17, 195, 197, 198, 203, 204, 205, 206, 208, 213, 214, 216, 218
Wittkower, Rudolf 230
Wolff, Robert P. 235
Woolf, Greg 262
World Archaeological Congress 13
world archaeology, idea of 19, 25
World Conservation Union *see* IUNC
World Heritage list *see* UNESCO
world heritage, idea of 102, 110–111, 116–117. *See also* significance (of heritage); value (of heritage)
World Trade Center 97
World War I 44
Wright, Ann St Clair 231, 243, 245
Wright, Henry T. 58, 59
Wylie, Alison 24–25
  'mitigated objectivism', idea of 24
Wylie, Donovan 37
Wynne-Jones, Stephanie 59, 60–61

Yamin, Rebecca 17, 78, 80, 81, 82, 85, 95
  and Karen B. Metheny 36–37, 76, 184
Yarrow, Duncan 22
Yentsch, Anne 230
Yovogan (Dahomean Governor) 184. *See also* Bénin

Zangher, Eberhart 195
Zanzibar (Tanzania) 55, **56**, 57, 62–71 *passim*. *See also* Unguja Island (Zanzibar), Pemba Island (Zanzibar)
Zanzibar Clove Plantation Survey (ZCPS) 64–68
Zanzibar Town *see* Unguja Island (Zanzibar)
Zimbabwe 105, **173**, 175
Zimmerman, Andreas 150, 154
Zvelebil, Marek 149

About the Contributors

SARAH CROUCHER is assistant professor at Wesleyan College;

DAGMAR DRESLEROVÁ is head of the Department of Spatial Archaeology in the Institute of Archaeology of the Academy of Sciences of the Czech Republic, Prague.

GRAHAM FAIRCLOUGH is head of Characterisation in English Heritage's Strategy Department.

MARK HAUSER is a visiting associate professor of anthropology at the University of Notre Dame, Indiana.

DAN HICKS is Lecturer-Curator in Archaeology of the Modern Period, School of Archaeology and Pitt Rivers Museum, University of Oxford. He is a Fellow of St Cross College, Oxford and a Research Fellow in Archaeology at Boston University.

SUSAN KEITUMETSE is a research fellow at the Harry Oppenheimer Okvango Research Centre, University of Botswana.

KENNETH G. KELLY is associate professor of anthropology at the University of South Carolina.

MARTIN KUNA is assistant director of the Institute of Archaeology of the Academy of Sciences of the Czech Republic, Prague.

LAURA MCATACKNEY is a doctoral student in archaeology at the University of Bristol.

S. GEOFFREY MATLAPENG is the site manager of the Tsodilo World Heritage Site for Botswana National Museum.

CHRISTOPHER MATTHEWS is an associate professor of anthropology at Hofstra University.

LESEKA MONAMO is a postgraduate student at the University of Botswana.

NEIL L. NORMAN is a doctoral student, Department of Anthropology, University of Virginia and Doctoral Fellow, Center for Historical Research, Ohio State University.

MATTHEW PALUS is a doctoral student in anthropology at Columbia University.

JOSEPH SCHULDENREIN is president of Geoarchaeology Research Associates, Inc.

SAM TURNER is a lecturer in archaeology at Newcastle University.

CHRIS WITMORE is a postdoctoral research associate at the Joukowsky Institute for Archaeology and the Ancient World at Brown University.

REBECCA YAMIN is a principal archaeologist and anthropologist with John Milner Associates, Inc.